New Paradigms in Ergonomics

The systems in which we work continue to evolve, creating emergent problems and often strengthening intractable issues. In order to remain relevant and impactful, the discipline of ergonomics needs its paradigms to evolve too. The aim of this book is to provide researchers and practitioners with new paradigms in the form of ideas, concepts, theories, methods, practices and values. The chapters take the reader on a journey through underlying theories, new ways to apply those theories and emerging domains in which ergonomics is expected to play a greater role. Readers of this book will be inspired by these new paradigms in ergonomics and seek to push the boundaries even further. The lifeblood of the science depends on continual evolvement and developments to take on the challenges we face in complex sociotechnical systems design and evaluation. Perhaps the most significant take-home message from this book is the demonstration of how theory maps onto practice. As such, the only remaining paradigm shift is for these ideas, concepts, methods and practices to be taken up more widely and the discipline advanced, until the next paradigm shift occurs.

The chapters were originally published as a special issue in the journal *Ergonomics*.

Neville A. Stanton is a Chartered Psychologist, Chartered Ergonomist and Chartered Engineer. He holds the Chair in Human Factors Engineering in the Faculty of Engineering and the Environment at the University of Southampton, UK. His research interests include the development and validation of Ergonomics and Human Factors methods.

Paul M. Salmon is Professor of Human Factors and Director of the Centre for Human Factors and Sociotechnical Systems at the University of the Sunshine Coast, Australia. He has a PhD in Human Factors from Brunel University London, UK, and currently holds a prestigious Australian Research Council Future Fellowship.

Guy H. Walker is Associate Professor within the Institute for Infrastructure and Environment at Heriot-Watt University, Edinburgh, UK. He has a PhD in Human Factors from Brunel University London, UK, is a Fellow of the Higher Education Academy and is a member of the Royal Society of Edinburgh's Young Academy of Scotland.

New Paradigms in Ergonomics

Edited by
Neville A. Stanton, Paul M. Salmon and Guy H. Walker

Routledge
Taylor & Francis Group

LONDON AND NEW YORK

First published 2019
by Routledge
2 Park Square, Milton Park, Abingdon, Oxon, OX14 4RN, UK

and by Routledge
52 Vanderbilt Avenue, New York, NY 10017

First issued in paperback 2020

Routledge is an imprint of the Taylor & Francis Group, an informa business

Chapters 1–9, 11–12 © 2019 Taylor & Francis
Chapter 10 © 2016 Sarah Sharples and Robert J. Houghton. Originally published as Open Access.

British Library Cataloguing-in-Publication Data
A catalogue record for this book is available from the British Library

ISBN 13: 978-0-367-57095-8 (pbk)
ISBN 13: 978-0-8153-8218-8 (hbk)

Typeset in Myriad Pro
by codeMantra

Publisher's Note
The publisher accepts responsibility for any inconsistencies that may have arisen during the conversion of this book from journal articles to book chapters, namely the possible inclusion of journal terminology.

Disclaimer
Every effort has been made to contact copyright holders for their permission to reprint material in this book. The publishers would be grateful to hear from any copyright holder who is not here acknowledged and will undertake to rectify any errors or omissions in future editions of this book.

Contents

THEME 3
New paradigms in domains and values

Citation Information

The chapters in this book were originally published in *Ergonomics*, volume 60, issue 2 (February 2017). When citing this material, please use the original page numbering for each article, as follows:

Chapter 1

New paradigms in ergonomics
Neville A. Stanton, Paul M. Salmon and Guy H. Walker
Ergonomics, volume 60, issue 2 (February 2017) pp. 151–156

Chapter 2

Quantum ergonomics: shifting the paradigm of the systems agenda
Guy H. Walker, Paul M. Salmon, Melissa Bedinger and Neville A. Stanton
Ergonomics, volume 60, issue 2 (February 2017) pp. 157–166

Chapter 3

Nonlinear dynamical systems for theory and research in ergonomics
Stephen J. Guastello
Ergonomics, volume 60, issue 2 (February 2017) pp. 167–193

Chapter 4

Fitting methods to paradigms: are ergonomics methods fit for systems thinking?
Paul M. Salmon, Guy H. Walker, Gemma J. M. Read, Natassia Goode and Neville A. Stanton
Ergonomics, volume 60, issue 2 (February 2017) pp. 194–205

Chapter 5

Quantitative modelling in cognitive ergonomics: predicting signals passed at danger
Neville Moray, John Groeger and Neville Stanton
Ergonomics, volume 60, issue 2 (February 2017) pp. 206–220

Chapter 6

Beyond human error taxonomies in assessment of risk in sociotechnical systems: a new paradigm with the EAST 'broken-links' approach
Neville A. Stanton and Catherine Harvey
Ergonomics, volume 60, issue 2 (February 2017) pp. 221–233

For any permission-related enquiries please visit:
http://www.tandfonline.com/page/help/permissions

Notes on Contributors

Melissa Bedinger is Research Associate in the Institute for Infrastructure and Environment at Heriot-Watt University, Edinburgh, UK.

Maarten De Vos is Associate Professor of Engineering and Lecturer in Engineering at Brasenose College, University of Oxford, UK.

Sean Gallagher is Associate Professor in the Industrial and Systems Engineering Department at Auburn University, USA.

Ivan Gligorijević is based in the Faculty of Engineering at the University of Kragujevac, Serbia.

David Golightly is Senior Research Fellow in the Faculty of Engineering at the University of Nottingham, UK.

Natassia Goode is Senior Research Fellow in Human Factors and Sociotechnical Systems at the University of the Sunshine Coast, Australia.

John Groeger is Professor of Psychology at Nottingham Trent University, UK.

Stephen J. Guastello is based in the Department of Psychology at Marquette University, Milwaukee, USA.

P. A. Hancock is Distinguished Research Professor in the Department of Psychology at the University of Central Florida, Orlando, USA.

Catherine Harvey is Research Fellow in the Human Factors Research Group at the University of Nottingham, UK.

Robert J. Houghton is Assistant Professor in the Faculty of Engineering at the University of Nottingham, UK.

Branislav Jeremić is based in the Faculty of Engineering at the University of Kragujevac, Serbia.

Waldemar Karwowski is Pegasus Professor and Chair of Industrial Engineering at the University of Central Florida, Orlando, USA.

Vanja Ković is Assistant Professor in the Department of Psychology at the University of Belgrade, Serbia.

Ryan Lumber is Early Career Research Fellow in Psychology at De Montfort University, Leicester, UK.

Ivan Mačužić is Assistant Professor in the Department for Production Engineering at the University of Kragujevac, Serbia.

Marta Maspero is based in the Department of Life Sciences at the University of Derby, UK.

Pavle Mijović is based in the Department for Production Engineering at the University of Kragujevac, Serbia.

Neville Moray was Professor in the Department of Psychology at the University of Surrey, Guildford, UK.

Gemma J. M. Read is Senior Research Fellow in Human Factors and Sociotechnical Systems at the University of the Sunshine Coast, Australia.

NOTES ON CONTRIBUTORS

Miles Richardson is Head of Psychology at the University of Derby, UK.

Paul M. Salmon is Professor of Human Factors and Director of the Centre for Human Factors and Sociotechnical Systems at the University of the Sunshine Coast, Australia.

Ben D. Sawyer is based in the Applied Experimental and Human Factors Psychology program at the University of Central Florida, Orlando, USA.

Mark C. Schall Jr. is Assistant Professor of Industrial and Systems Engineering at the University of Auburn, USA.

Sarah Sharples is Professor of Human Factors in the Faculty of Engineering at the University of Nottingham, UK.

David Sheffield is Associate Head of the Centre for Psychological Research at the University of Derby, UK.

Neville A. Stanton is a Chartered Psychologist, Chartered Ergonomist and Chartered Engineer. He holds the Chair in Human Factors Engineering in the Faculty of Engineering and the Environment at the University of Southampton, UK.

Vicki Staples is Programme Leader for the MSc in Health Psychology at the University of Derby, UK.

Petar Todorović is based in the Faculty of Engineering at the University of Kragujevac, Serbia.

Guy H. Walker is Associate Professor within the Institute for Infrastructure and Environment at Heriot-Watt University, Edinburgh, UK.

Petros Xanthopoulos is Assistant Professor of Decision and Information Sciences at Stetson University, DeLand, USA.

New paradigms in ergonomics

Neville A. Stanton, Paul M. Salmon and Guy H. Walker

New paradigms

A paradigm is a world view underlying the theories and methodology of a particular scientific discipline; in our case, ergonomics. The ability to engage in new thinking and new paradigms is a critical aspect for the lifeblood of any discipline. Thankfully, ergonomics has a long history of innovations in theory, methodology, science and application. Once regarded as an offshoot of experimental psychology, physiology, biomechanics, engineering and computer science, ergonomics now draws from a much broader range of insights to become a discipline in its own right. This is reflected in the development of models and methods that are unique to the discipline. Accident causation models, for example, first emerged in the early 1900s (Heinrich 1931) but have since evolved to consider entire systems and emergent properties therein (e.g. Leveson 2004; Perrow [1984] 1999; Rasmussen 1997; Reason 1990). Methodologies have moved from focusing on human tasks (Taylor 1911) and work (Gilbreth 1912) to entire systems and the constraints shaping behaviour in the world (e.g. Vicente 1999). Human performance itself has shifted from a world view in which it was mathematically characterised in terms of, for example, response times, sensitivity or decision bias (e.g. Green and Swets 1966) to more complex forms of distributed and situated cognition (e.g. Hutchins 1995).

The systems in which ergonomists work continue to evolve, creating emergent problems and often strengthening intractable issues. In order to remain relevant and impactful, the discipline needs its paradigms to evolve too. The aim of this special issue is to provide researchers and practitioners with an opportunity to present and discuss contemporary, forecasted and required paradigm shifts in ergonomics.

Contributions to this *New Paradigms in Ergonomics* special issue have been grouped into three themes: new paradigms in theories and methods; new paradigms in practice; and new paradigms in domains and values. These themes take the reader on a journey through underlying theories, news ways to apply those theories, emerging domains in which ergonomics is expected to play a greater role and to what ultimate ends. In the new paradigms, in theories and methods theme, are papers on 'Quantum ergonomics' (Walker et al. this issue), 'Nonlinear dynamical systems for theory and research in ergonomics' (Guastello), 'Fitting methods to paradigms' (Salmon et al. this issue), 'Quantitative modelling in cognitive ergonomics' (Moray, Groeger, and Stanton, this issue) and 'Beyond human error taxonomies in assessment of risk in sociotechnical systems' (Stanton and Harvey, this issue). In the new paradigms, in practice theme, are papers on 'Detection of error-related negativity in complex visual stimuli' (Sawyer et al. this issue), 'Towards continuous and real-time attention monitoring at work' (Mijovic et al. this issue), 'Musculoskeletal disorders as a fatigue failure process' (Gallagher and Schall, this issue). In the new paradigms, in domains and values theme, are papers on 'The field becomes the laboratory?' (Sharples and Houghton), 'Imposing limits on autonomous systems' (Hancock, this issue) and 'Nature' (Richardson et al. this issue). Summaries of each of the contributions together with conclusions for the special issue are in the following sections of this editorial.

New paradigms in ergonomics theory and methods

Walker, Salmon, Bedinger and Stanton (Quantum ergonomics: Shifting the paradigm of the systems agenda) draw inspiration from the world of quantum physics to confront some difficult truths about ergonomics paradigms. Walker et al. present three case studies to demonstrate 'quantum effects' in ergonomics science. In the first case study, they show how the nanoscale, individual behaviour of drivers cannot reveal the macro-scale behaviour of the entire road traffic network. The case study reveals the unexpected reverse finding; that more route guidance (in effect, better situational awareness) on the part of individual drivers does not necessarily result in lower CO_2 emissions for the whole road network. Sometimes it is worse. 'Catastrophically counter-intuitive' findings like these show the difficulty in predicting system-wide emergent properties (often the thing we really want to know) simply from studying the behaviour of individuals and magnifying the resultant findings. In the second case study of a submarine control room, Walker et al. present insights into the behaviour of a command team system that have been hitherto unforthcoming from studying individual operators. The system dynamics have enabled the development of a submarine command room simulator that is currently being used to generate new understanding of how these teams work together in challenging environments (Stanton 2014). Walker et al. argue that we need systems approaches if we are to remain an effective discipline in the new increasingly complex and networked world. They argue that we need to look at new approaches, such as phase spaces (see Stanton, Walker, and Sorensen 2012; Guastello, this issue), as a means of offering some rigour in naturalistic studies (see Sharples and Houghton this issue). In the final case study, Walker et al. present findings from a large-scale operation field trial (see Stanton et al. 2009). Using network analysis methods, they were able to show how the organisational dynamics changed over time. Walker et al. comment that ergonomic methods need to be fit-for-purpose (see Salmon et al. this issue) and offer the maximum of insight for the minimum of effort. This is in contrast to a 'one size fits all' approach to method (and theory) selection. The discipline of ergonomics is good at responding to problems in the world – which are in a constant state of flux – so the methods need to adapt and develop to meet this evolving demand. The overall message is one of hope: methods currently exist to enable these challenging systems problems to be tackled in practical real-world circumstances, provided the necessary paradigm shift can be made to occur.

Guastello (Nonlinear dynamical systems for theory and research in ergonomics) also argues that the complexity of sociotechnical systems requires a paradigm shift in ergonomics theory (see Walker et al. 2010 this issue) and practice (see Salmon et al. this issue). He proposes the non-linear dynamical systems approach as an alternative paradigm that can cope with this complexity, as it addresses:

- structures and patterns of variability;
- underlying dynamics and system changes;
- both internal and external dynamics;
- effects of state-dependent changes; and
- both top-down and bottom-up emergent properties in systems.

Guastello explains how systems do not just have one stereotypical response, rather there are a multiplicity of behaviours with different patterns and outcomes. The effects can be both large and small, depending upon the state of the system. Guastello brings many different systems concepts and methods together under the non-linear dynamical systems framework with the view that these will help ergonomics explain complex systems behaviour. One example is the use of phase spaces to visualise the dynamical processes in systems (Walker et al. this issue; Stanton, Walker, and Sorensen 2012). He provides numerous examples of the practical application of non-linear dynamical systems methods, including (but not limited to): accidents, biomechanics, performance variability, resilience and team coordination and workload. The non-linear dynamical systems approach has much to offer systems ergonomics, both as a theoretical construct and practical methods.

Salmon, Walker, Read and Stanton (Fitting methods to paradigms: are ergonomics methods fit for systems thinking?) question whether current ergonomics methods really are fit for the new systems paradigm which, from the above, is clearly in the ascendancy. In many ways, ergonomics methods have always been about system interactions, from individuals to teams to organisations (Stanton et al. 2013). Yet Salmon et al. argue that the increasing complexity of the modern world (Walker et al. 2010), and challenges that ergonomics faces, may have left the methods wanting. They show that

despite the dramatic reductions in accidents over the decades the statistics are plateauing. It is suggested that we may have reached the limits of deterministic methods, so new approaches are required. Salmon et al. explore how well our current methodological toolkit can cope with modern day problems by focusing on five key areas within the ergonomics paradigm of systems thinking: normal performance as a cause of accidents, accident prediction, system migration, systems concepts and ergonomics in design. The ergonomics methods available for pursuing each line of inquiry are explored, along with their ability to respond to key requirements. Salmon et al. come to the conclusion that, although our current suite of ergonomics methods is highly useful, there is work to be done. For example, they conclude that, although providing rich outputs, some of our existing accident analysis methods do not describe accident causation in a manner that is congruent with contemporary models (e.g. Rasmussen 1997). With regard to accident prediction, it is concluded that we currently do not have appropriate methodologies for predicting systemic accidents (although see Stanton and Harvey, this issue, for new developments). Likewise, Salmon et al. argue that assessing the migration of performance towards and away from safety boundaries, a key systems thinking concept, is outside of the capabilities of our current methodological toolkit. They also suggest that further ergonomics problems and constructs may be suited to systems level analysis, and that few ergonomics methods are actually being used directly in system design processes. Salmon et al. close the article in an upbeat manner, highlighting that many seemingly appropriate methods already exist, both in ergonomics and other disciplines, and that research is underway to develop some of the methods required (see Salmon 2016a,b). In closing, they map out the prerequisites for methods development in systems ergonomics. If our discipline is to maintain currency and rise to the contemporary and future design challenges, we need to develop methods that have the entire sociotechnical system as the unit of analysis.

Moray, Groeger and Stanton (Quantitative modelling in cognitive ergonomics: predicting signals past at danger) state that the discipline of ergonomics is sufficiently mature to enable quantitative modelling of performance. A case study of the activities of the Thames Trains driver in the Ladbroke Grove rail accident is presented. All three authors were expert witnesses in this case, so they combined their knowledge for this paper. Moray et al. present accounts of the accident background, context, infrastructure and timeline together with 'black box' data from the train. The accident raises questions about why the driver behaved in a particular way on that fateful day, as well as the performance of signals and warnings inside and outside the train cab. Eye movement data were used as the basis for the development of a cognitive model of driver visual attention. The model accounts for attention both inside and outside the train cab. It revealed the difficulty in acquiring the signal in a relatively short amount of time, due to the number of signals, their placement and the speed of the train as well as initial masking of the signals.

The ambiguity of the automatic warning system horn further compounded the problem coupled with a strong expectation of the signal being in a non-red aspect and glare from the sun. The quantitative model shows how prediction of visual attention can be used to determine risks associated with signal sighting, which can in turn be used to support guidelines and in-cab display design.

Stanton and Harvey (Beyond human error taxonomies in assessment of risk in sociotechnical systems: a new paradigm with the EAST 'broken-links' approach) introduce a new paradigm for risk assessment based on the Event Analysis of Systemic Teamwork (EAST) method. They argue that the approach offers a fundamentally different way of thinking about risk in systems, and is dramatically different from existing human error taxonomies. Rather than treat accidents as the result of 'human error' the EAST broken-links approach treats them as information communication failures. Stanton and Harvey provide examples of information communication failures in the Herald of Free Enterprise capsize at Zeebrugge, the British Midland crash at Kegworth and the Thames Train collision at Ladbroke Grove. All of these accidents have the common feature that key information was not communicated to an appropriate agent by the system in an effective manner at the right time. Stanton and Harvey present a case study of an EAST analysis of a RAF Hawk and RN Frigate system (see Salmon et al. this issue, for a call on understanding normative systems). The EAST model comprised normative task, information and social networks together with the composite model. Stanton and Harvey demonstrate that by systematically breaking the task and social links in the networks, risks are revealed by information not being communicated between tasks and social agents. One hundred and thirty-seven risks were revealed by breaking only 12 task and 19 social links. Stanton and Harvey show how the emergent information communication failures transpose risk around the network of agents, actors and artefacts. Furthermore, they show how reducing risk in one part of the system increases risk in another part. For future work, Stanton and Harvey plan to break and/or add multiple links simultaneously to show how compounded information communication failures and short circuits could affect system performance.

New ergonomics paradigms in practice

Sawyer, Karwowski, Xanthopoulos and Hancock (Detection of error-related negativity in complex visual stimuli: a new neuroergonomic arrow in the practitioner's quiver) have established a link between the error-related negative evoked response potential (ERN-ERP) and the complex task of identifying a motorcycle in a busy visual scene. The innovation in this work is to take visual search tasks beyond simple letter arrays and icon images in binary forced-choice tasks to a more naturalistic task. Motorcycle detection is notoriously difficult and undoubtedly the cause of accidents. Sawyer et al. transferred this task into the laboratory alongside a more traditional letter flanker task (where the target letter was flanked on either side by non-target letters). They demonstrated that both tasks produce similar error rates and workload ratings. Sawyer et al. found that the ERN-ERP waveforms were very similar for the letter flanking and motorcycle identification tasks. In both tasks, the wave was flat for the correct response and had a pronounced negative deflection for the incorrect response. Sawyer et al. propose that this finding shows the ERN-ERP has potential for practical application and they plan to use the procedure for dynamic environments in the future. Potential applications include use in training and naturalistic human–computer interfaces.

Mijovic, Kovic, De Vos, Macuzic, Todorovic, Jeremic and Gligorjevic (Towards continuous and real-time attention monitoring at work: reaction time versus brain response) state that wearable electroencephalogram (EEG) has made objective, online, continuous, monitoring of attentional state a practical possibility. They have applied the use of EEG to examine the P300 (an ERP wave component associated with the process of decision-making) whilst people performed simulated simple and repetitive assembly tasks. The purpose of the research was to test the relationship between attention level (as measured by the P300 amplitude) and the time it takes to complete a task cycle on the assembly line. The simulated assembly task was designed to replicate monotonous work, comprising: picking up two items, connecting them together, placing them in a machine, pressing a pedal to crimp the parts together, removing them from the machine and placing them in a bin. The entire task cycle took less than 10 s and a shift would involve over 2500 cycles. The EEG data were collected simultaneously. The study revealed that the P300 was correlated with attention on the task. High P300 values were found with shorter cycle times and vice versa. Mijovic et al. are keen to point out that there are inter-individual differences which led to some inconsistencies in the data, but the findings hold true for the collective group level. This study has also shown that continuous monitoring of attention is possible and that neuroergonomic approaches have the potential to be used in the workplace (see also Sawyer et al. this issue).

Gallagher and Schall (Musculoskeletal disorders as a fatigue failure process: evidence, implications and research needs) state that musculoskeletal disorders are a major cause of workplace injury and, although much is known about the causes, little theoretical work has been presented. By way of contrast, fatigue failure in materials is well established and well understood. Fatigue failure in materials is a function of differential loading characteristics and the frequency of cycles. This is analogous to the cumulative trauma disorder in human tissue, which has led Gallagher and Schall to the fatigue failure hypothesis. Some evidence from recent studies has lent support to this hypothesis, that musculoskeletal tissue also shows signs of fatigue failure. In particular, a large epidemiological study has shown that forceful repetition leads to carpal tunnel syndrome. Gallagher and Schall offer a unifying framework for musculoskeletal disorder risk factors together with validated methods for assessing the risk of cumulative damage. They argue that the fatigue failure process offers the possibility to assess risk and develop effective interventions. Ultimately, musculoskeletal disorders could be predicted as our understanding of the fatigue failure process develops. This requires a much larger data-set of fatigue failure on musculoskeletal tissues (much like the data held on material fatigue). It also requires a better understanding of the dynamical properties of the musculoskeletal system (see Guastello, this issue). Gallagher and Schall are working towards a fatigue failure theory of musculoskeletal disorders and a consequent reduction in workplace injury.

New ergonomics paradigms in domains and values

Sharples and Houghton (The field becomes the laboratory? The impact of the contextual digital footprint on the discipline of E/HF) argue that the ubiquity of data collection in the digital world has had a dramatic effect on the volume of information available for ergonomics research and development. They go so far as to suggest it could even change the nature of how the discipline goes about its practice in terms of new methods and insights (see Walker et al. this issue; Guastello, this issue; Salmon et al. this issue). The digital exhaust (called 'contextual user data' by Sharples and Houghton) that is emitted by people in their work, social and home lives could provide a richer understanding about the behaviour of individuals and collectives, but with that comes a range of ethical, moral and privacy issues. Sharples and Houghton present three scenarios, a hospital as a digital workplace, distributed crowd-sourced activity, and managing journeys. In each of these scenarios, Sharples and Houghton show the potential for the technology to be both beneficial (improving human well-being) and detrimental (making work harder, less enriching, disengaging and potentially harmful). They argue that, used appropriately, ergonomics methods could make use of the contextual user data in a manner to improve the design of work, social and home lives consistent with the values of the discipline. Big data analytics are likely to play an important role in the discipline, and its methods, going forward.

Hancock (Imposing limits on autonomous systems) warns of the impending march from automated rule following technologies that require human supervision to autonomous systems, those intelligent enough to not necessarily require human supervision). He states that it is necessary for the ergonomics community to engage with the debate, and to rigorously research and design these systems if we are to ward off the (many) potential pitfalls of the technology. There are some foreseeable problems, such as the so-called 'mode error', where human supervisors confuse the mode the automated (or autonomous) system is in (e.g., Stanton, Dunoyer, and Leatherland 2011). More worryingly are the unforeseeable problems, those that emerge through complex system interactions (Walker et al. this issue; Guastello, this issue). Hancock is quick to point out that each potential benefit afforded by technology is accompanied by drawbacks (see Sharples and Houghton, this issue). He goes much further to suggest that

automation could have deeper and more profound societal effects. One of these is to deepen the societal divide between rich and poor. History has shown that technology can have dramatic effects on work, social and leisure lives. Hancock warns that technological Darwinism may not always be to the benefit of society at large. By way of mitigation, he argues for an ethical approach to automation, which the ergonomics community should be leading. Autonomy brings moral philosophy into sharp focus, with Hancock's depiction of the ultimate delegation of authority to automation that could lead to very inhuman acts. He argues that, as ergonomics is about human betterment especially with technological development, we should not shirk our responsibilities., Thus the paradigm is not merely theoretical or methodological, but strikes to the heart of the discipline's underlying value base.

The final paper of this special issue is provided by Richardson, Maspero, Golightly, Sheffield, Staples and Lumber (Nature: a new paradigm for well-being and ergonomics) who propose 'nature' as a new paradigm for ergonomics that is required to promote physical and mental well-being as a core principle of the discipline. They argue that the progressive urbanisation and removal of our experience with the natural world (a global trend) has a negative effect on health and well-being. From a review of the corpus of evidence, Richardson et al. are able to demonstrate the benefits that arise from the 'nature experience'. This evidence shows that people who live in urban environments generally have poorer health than those in rural settings. Green spaces in urban environments have offset some of those negative effects for city dwellers, leading to improvements in mood, mental health, well-being, anxiety, vitality, job and life satisfaction. Similar benefits have been found for physical health as well, with reports on improvements in heart rate, blood pressure, muscle fatigue, sickness and stress. Even virtual scenes of nature seem to have short-term restorative effects, presumably because of the associations they evoke. Richardson et al. argue that the degree of connectedness people feel with nature has an effect on their psychological well-being in the workplace. They offer 'nature' as a cost-effective, simple, ergonomics intervention strategy with positive physical and mental health benefits. It has the potential to be a transformational paradigm shift that is consistent with the mission of the ergonomics discipline as a whole.

Conclusions

This collection of papers from leading researchers and practitioners in our discipline represents a tantalising glimpse into a fascinating ergonomics future – or possible futures. Given the shifting nature of the systems in which ergonomists work, there is clearly a strong need to revisit the ergonomics unit of analysis and a groundswell of support for a broader, stronger, and more rigorous approach to systems thinking. There is a corresponding need, and opportunity, to enhance our ergonomics practices, with several important advances now afforded by new technology and the discipline's maturing state of knowledge in critical areas. Finally, there is an equally strong need to combine these paradigm shifts in theory, methods and practice with continued scrutiny of where ergonomics should be applied, how it should be applied, and to what ends. In other words, to keep asking ourselves 'what do we stand for?'. With this in mind, the collection of papers provides a series of important take-home messages. These include:

- The need to face up to the 'quantum effect' in ergonomics science and analyse systems at the appropriate level and with appropriate methods.
- That non-linear dynamical systems approaches offer a way of dealing with some of the true complexities of sociotechnical systems.
- That autonomous systems pose a significant threat to our way of life and we need to ensure they are implemented in an ethical manner.
- That work is required to develop ergonomics methods that are capable of modelling key systems thinking concepts such as normal performance, future accidents and system migration.
- That quantitative modelling on cognitive ergonomics is now able to predict performance and can be used to assess risk and design systems.
- That new modelling approaches can model normal performance and assess risk in complex sociotechnical systems.
- That the sheer quantity of contextual user data available from our digital work, social and home lives will provide a richer understanding of human, organisational and system behaviour.
- That neuroergonomics methods can be used for practical visual search tasks and measures of attention in applied settings.
- That the fatigue failure process could explain musculoskeletal disorders and reduce workplace injury.
- That connection with nature can offer transformational physical and mental health benefits for urban dwellers and office workers, and is an emerging ergonomics criterion.

It is hoped that readers of this special issue will be inspired by these new paradigms in ergonomics and seek to push the boundaries even further. The lifeblood of the science depends on continual evolvement and developments to take on the challenges we face in complex sociotechnical systems design and evaluation. Perhaps the most significant take-home message is that all these papers have demonstrated how theory maps onto practice. As such, perhaps the only remaining paradigm shift is for these ideas and concepts to be taken up more widely and the discipline advanced, until the next paradigm shift occurs.

References

Gallagher, S., and M. Schall, Jr. this issue. "Musculoskeletal Disorders as a Fatigue Failure Process." *Ergonomics*.

Gilbreth, F. B. 1912. *Primer of Scientific Management*. New York: Van Nostrand.

Green, D. M., and Swets J.A. 1966. *Signal Detection Theory and Psychophysics*. New York: Wiley.

Guastello, S. this issue. "Nonlinear Dynamical Systems for Theory and Research in Ergonomics." *Ergonomics*.

Hancock, P. this issue. "Imposing Limits on Autonomous Systems." *Ergonomics*.

Heinrich, H. W. 1931. *Industrial Accident Prevention: A Scientific Approach*. New York, NY: McGraw-Hill.

Hutchins, E. 1995. *Cognition in the Wild*. Cambridge, MA: MIT Press.

Leveson, N. 2004. *Engineering a Safer World: Systems Thinking Applied to Safety*. Cambridge, MA: MIT Press.

Mijovic, P., V. Kovic, M. De Vos, I. Macuzic, P. Todorovic, B. Jeremic, and I. Gligorjevic. this issue. "Towards Continuous and Real-time Attention Monitoring at Work." *Ergonomics*.

Moray, N., J. Groeger, and N. A. Stanton. this issue. "Quantitative Modelling in Cognitive Ergonomics." *Ergonomics*.

Perrow, C. [1984] 1999. Normal Accidents: Living with High Risk Technologies. Revised ed. Princeton, NJ: Princeton University Press.

Rasmussen, J. 1997. "Risk Management in a Dynamic Society: A Modelling Problem." *Safety Science* 27: 183–213.

Reason, J. 1990. *Human Error*. Cambridge: Cambridge University Press.

Richardson, M., M. Maspero, D. Golightly, D. Sheffield, V. Staples, and R. Lumber. this issue. "Nature: A New Paradigm for Well-being and Ergonomics." *Ergonomics*.

Salmon, P. M. 2016a. "Bigger, Bolder, Better? Methodological Issues in Science." *Theoretical Issues in Ergonomics Science* 17 (4): 337–344.

Salmon, P. M. 2016b. "Bridging the Gap between Research and Practice in Ergonomics Methods: Methodological Issues in Ergonomics Science Part II." *Theoretical Issues in Ergonomics Science*. 17 (5–6): 459–467.

Salmon, P., G. Walker, G. Read, N. Goode, and N. A. Stanton. this issue. "Fitting Methods to Paradigms: Are Ergonomics Methods Fit for Systems Thinking?" *Ergonomics*.

Sawyer, B., W. Karwowski, P. Xanthopoulos, and P. Hancock. this issue. "Detection of Error-related Negativity in Complex Visual Stimuli: A New Neuroergonomic Arrow in the Practitioner's Quiver." *Ergonomics*.

Sharples, S., and R. Houghton. this issue. "The Field Becomes the Laboratory? The Impact of the Contextual Digital Footprint on the Discipline of E/HF." *Ergonomics*.

Stanton, N. A. 2014. "Representing Distributed Cognition in Complex Systems: How a Submarine Returns to Periscope Depth." *Ergonomics* 57 (3): 403–418.

Stanton, N. A., A. Dunoyer, and A. Leatherland. 2011. "Detection of New in-path Targets by Drivers Using Stop & Go Adaptive Cruise Control." *Applied Ergonomics* 42 (4): 592–601.

Stanton, N. A., and C. Harvey. this issue. "Beyond Human Error Taxonomies in Assessment of Risk in Sociotechnical Systems: A New Paradigm with the EAST 'broken-links' Approach." *Ergonomics*.

Stanton, N. A., P. M. Salmon, L. A. Rafferty, G. H. Walker, C. Baber, and D. Jenkins. 2013. *Human Factors Methods: A Practical Guide for Engineering and Design*. 2nd ed. Aldershot: Ashgate.

Stanton, N. A., G. H. Walker, D. P. Jenkins, P. M. Salmon, K. Revell, and L. Rafferty. 2009. *Digitising Command and Control: Human Factors and Ergonomics Analysis of Mission Planning and Battlespace Management*. Aldershot: Ashgate.

Stanton, N. A., G. H. Walker, and L. J. Sorensen. 2012. "It's a Small World after All: Contrasting Hierarchical and Edge Networks in a Simulated Intelligence Analysis Task." *Ergonomics* 55 (3): 265–281.

Taylor, F. W. 1911. *The Principles of Scientific Management*. New York: Harper & Brothers.

Vicente, K. 1999. *Cognitive Work Analysis: Toward Safe, Productive, and Healthy Computer-based Work*. Mahwah, NJ: Lawrence Erlbaum Associates.

Walker, G., P. Salmon, M. Bedinger, and N. A. Stanton. this issue. "Quantum Ergonomics: Shifting the Paradigm of the Systems Agenda." *Ergonomics*.

Walker, G. H., N. A. Stanton, P. M. Salmon, D. P. Jenkins, and L. A. Rafferty. 2010. "Translating Concepts of Complexity to the Field of Ergonomics." *Ergonomics* 53 (10): 1175–1186.

Quantum ergonomics: shifting the paradigm of the systems agenda

Guy H. Walker, Paul M. Salmon, Melissa Bedinger and Neville A. Stanton

ABSTRACT

A paradigm is an accepted world view. If we do not continually question our paradigm then wider trends and movements will overtake the discipline leaving it ill adapted to future challenges. This Special Issue is an opportunity to keep systems thinking at the forefront of ergonomics theory and practice. Systems thinking prompts us to ask whether ergonomics, as a discipline, has been too timid? Too preoccupied with the resolution of immediate problems with industrial-age methods when, approaching fast, are developments which could render these operating assumptions an irrelevance. Practical case studies are presented to show how abstract systems problems can be tackled head-on to deliver highly innovative and cost-effective insights. The strategic direction of the discipline foregrounds high-quality systems problems. These are something the discipline is well able to respond to provided that the appropriate operating paradigms are selected.

Practitioner Summary: High-quality systems problems are the future of the discipline. How do we convert obtuse sounding systems concepts into practical interventions? In this paper, the essence of systems thinking is distilled and practical case studies used to demonstrate the benefits of this new paradigm.

Introduction

According to the Oxford English Dictionary, a paradigm is a 'conceptual or methodological model underlying the theories and practices of a science or discipline at a particular time; (hence) a generally accepted world view' (www.oed.com; 2015). But how often do we really scrutinise the paradigm we are operating under? For most people, indeed, for most scientific subjects, the view of the world may as well 'be' the world, so pervasive and unquestioned it is. Looking at paradigms, let alone changing them, is difficult yet we argue necessary (Stanton, Salmon and Walker, 2015; Walker 2016). Woods and Dekker (2000), Hancock (1997), Rasmussen 1997; De Greene (1980), Lee (2001), Guastello (2002), Walker, et al. (2010), Dul et al. (2012), Chung and Shorrock (2011) and many others are already questioning the existing paradigm and the need for it to adapt. Still others, like Rifkin (2014), Miller (2015), Marcovici (2014) and Kellmereit and Obodovski (2013) are going further still. Like Toffler in the 1980s (1980, 1981), they are suggesting this is a critical moment in history. A moment when infrastructures like the Internet are creating a form of globally connected nervous system, when more 'things' will be connected to the Internet than 'people' (hence the term

'Internet of Things') which, in turn, will give rise to enormous stores of 'big data' measured in zettabytes ($10^{\times 21}$) and growing all the time. Hitherto impossible ergonomic insights then become manifest in radically new ways (Drury 2015; Walker and Strathie 2016). In terms of paradigms, it is instructive to consider that for most of the post-war period the paradigm in the human sciences has been empirical falsification of the Popperian school, and the generalisability of findings derived from smaller samples to larger populations. This is the microscope through which we extract the majority of meaningful ergonomic insights and develop solutions to real-world problems. Could that be about to change? At least one social science journal has recently made it policy to ban null hypothesis significance testing from all future papers (Trafimow and Marks 2015); a bold and possibly significant move. It seems to indicate that in the new paradigm our entire population becomes our sample, and Big Data becomes the 'macroscope' (not 'microscope'; Rifkin 2014) through which radical new insights could be discovered. Ergonomists are already involved in projects and activities related directly to these challenges (e.g. Sharples et al. this issue) and this Special Issue on New Paradigms is therefore timely. Rather than representing problems that are scattered around the periphery of our discipline these deeply 'systemic' challenges, and the paradigm shifts necessary to cope with them, will become more rather than less central (Dul et al. 2012). This paper will put forward a selection of ergonomic problems emblematic of these paradigm shifts, illustrate some of the key issues in play and show the methods used to drive out practical insights. It is hoped the discussion will provoke deeper thought about our often unquestioned operating assumptions and how they may need to shift further in future.

The quantum world of ergonomics

Quantum physics – at the most basic level of explanation possible – is the idea that fundamental properties change at different scales of analysis. Key concepts include Heisenberg's uncertainty principle, in which the more one property is known, the less it is possible to know another property. Bohr's complementarity principle, and the duality that exists when items exhibit contradictory properties simultaneously. And Schrodinger's equation which helps to define the degrees of freedom that a system time evolves within. Stated in exceedingly crude terms, quantum physics is about understanding what these different states are, how jumps are made from one state to another, how quantum systems evolve over time and how nano and macro-scale system behaviours are related. Its influence, as a mode of thought, has expanded far beyond the narrow confines of Physics. It has shifted the wider scientific paradigm in terms of epistemology (making our knowledge of physical reality more probabilistic in nature) and ontology (creating the possibility of multiple valid positions on reality). By crude analogy to quantum physics, therefore, we can legitimately ask whether our own Ergonomic raw material, the 'atoms' and 'particles' of tasks, agents, functions and people (and the rest) can also simultaneously exhibit more than one property? The question then becomes what we choose to abstract from those situations. Different practitioners will observe different properties. Constraining one property experimentally may reduce our ability to measure others. A focus on the nanoscale of ergonomics problems may in turn lead to a loss of perspective on the macro-scale of entire sociotechnical systems. There follow three 'quantum case studies' which, between them, help to meet the paper's two main objectives. Firstly, to remind ourselves that ergonomics is well equipped to cope with systems challenges and they have been spoken of frequently in the past. Secondly, to reinforce the idea that systems approaches are practical and usable, not merely abstract or theoretical. Framing these two objectives within the notion of 'quantum ergonomics' provides a contemporary language for advancing an agenda for more systems thinking and more value to be extracted from ergonomic endeavours.

Catastrophic counter-intuitiveness

Ergonomics science is not quantum physics, but ergonomic problems can certainly exhibit what we might loosely label 'quantum behaviour'. That is, systems have many different realities depending on the level of analysis, and elements of the system can indeed exhibit 'complementarity', being simultaneously one thing and another. There is certainly a duality between nano and macro-scale phenomenon as the following study amply demonstrates.

The ergonomic case study in question centred on road freight and logistics, and the use of vehicle telematics to help reduce carbon emissions (see Walker and Manson 2014). Telematics works at the interface of road vehicles and the road network, helping to ensure the former make the most efficient and rational use of the latter. We know from the literature that the shape and topology of a network is a strong contingency factor in how it will perform (Leavitt 1951; Pugh et al. 1968; Watts and Strogatz 1998 etc.) yet behavioural issues like these are rarely embedded in studies of driver behaviour. It is far more common to extract the problem (the driver and the technology) from the network, place them (and them alone) in a driving simulator or other lab-based study for analysis, then reinsert them once the solution has been applied, all on the tacit assumption that the whole will be no more or less than the sum of its parts. This is a crude and perhaps unfair classification, but it is born out in the widespread engineering assumption that more technology, more route guidance and the greater the driver's knowledge of the wider traffic conditions, the better the network will perform (e.g. Nijkamp, Pepping, and Banister 1997). In other words, all it requires is for people in the system to behave in the correct way. A radical alternative is to use an agent-based approach called traffic microsimulation, create virtual models of entire towns and cities populate them with thousands of virtual vehicles each exhibiting ergonomically referenced behavioural patterns, run the models many times each and analyse the probabilistic behaviour of the entire traffic network as it evolves over time (Figure 1). The output from running these microsimulations is a set of raw data for each individual vehicle (out of anywhere between 3000 and 6000 total vehicles) as it progresses through the network. Microsimulation calculates the positional coordinates of each individual vehicle every half a second. This data can then be combined with specific vehicle 'drive cycles' to calculate fuel use and emissions, alongside a wide range of other performance

Figure 1. Agent-based traffic microsimulation software allows the effect of nano-scale individual driver behaviours to be analysed in terms of macro-level global effects on the network.

metrics. This constitutes 'big data' compared to the samples more typical under the empirical paradigm common in the ergonomics discipline.

In this study, four real-life towns and cities were modelled, each possessing a distinct network topology. Three types of virtual drivers were set free in the different networks; 50% without any route guidance, 25% with route guidance that was only acted on half of the time and 25% who followed their route guidance instructions explicitly. The simulation worked by communicating live network information to the virtual drivers (such as queue lengths, link delays, traffic densities) with the virtual drivers seeking more optimum routes through the network in the proportions given above (i.e. presented with this information and a route choice to be made, the virtual drivers would make a positive, optimising choice 0, 25 or 50% of the time).

In this case, optimum performance relates to network performance, and the optimum balance of journey times, emissions and costs. Stakeholders in this study assumed, like much of the literature, that the more route guidance that was provided (and the more it was adhered to) the better the network would perform. This was not the case in practice.

On some networks, vehicles with route guidance they followed strictly did no better than those drivers who had nothing. On other networks, the investment in more telematics, and enforcing its use, would have sharply diminishing returns (e.g. Figure 2). The detailed findings are contained in Walker and Manson (2014) but the overriding point is that highly rational empirically derived solutions at the individual level can translate into highly dysfunctional outcomes at the level of the total system. In other words, the nano-scale of individual behaviours diverges from the macro-scale of overall system behaviours, and moreover, this is not always captured using existing state of science approaches in ergonomics.

Everything is a network

Systems thinking is a framework for conceptualising or viewing the world as a set of interrelated elements (Hall and Fagen 1956; Carvajal 1983). There is some confusion about how this term is used because there are different types of system and they do not all mean the same thing. People refer to systems methods when they are, in fact, referring to a particular 'type' of system. It is possible to argue that all ergonomics methods are systems methods, because it is entirely possible to have 'sets of interrelated elements' that act as highly deterministic closed systems,

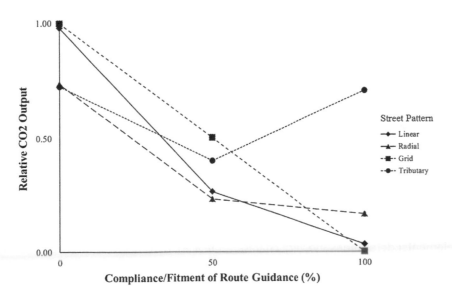

Figure 2. More route guidance at the nano-scale of individual vehicles is not always associated with lower CO2 emissions at the macro-scale of the entire traffic network: sometimes the reverse.

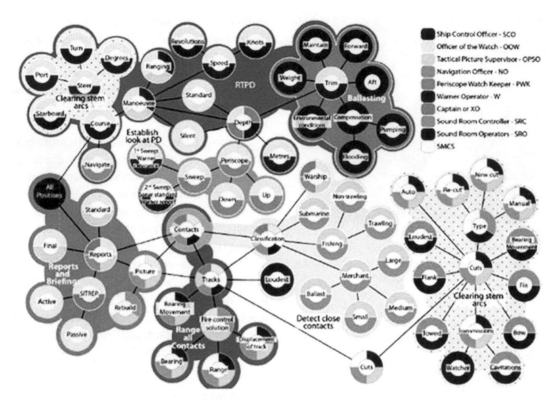

Figure 3. A composite network of combined task, social and knowledge networks representing the system involved in piloting a nuclear submarine to periscope depth.

ones in which the input, processing, output characteristics of each node are well understood, as are the properties and flows occurring on the links between them. Systems thinking, however, also refers to open systems, and not all systems methods cope equally well with open systems problems. One could argue that in common usage within ergonomics the term 'system' is really referring to 'open system' (Walker et al. 2010). An open system 'may attain (certain conditions presupposed) a time-independent state where the system remains constant as a whole … though there is a constant flow of the component materials. This is called a steady state' (Bertalanffy 1950, 23). Steady-state behaviour means that open systems can

> grow by processes of internal elaboration. They manage to achieve a steady state while doing work. They achieve a quasi-stationary equilibrium in which the enterprise as a whole remains constant, with a continuous 'throughput', despite a considerable range of external changes. (Trist 1978, 45).

This is why 'open systems methods', as distinct from a more general class of methods often labelled as 'systems-based', tend to foreground structure. ' … the set of relations [between system elements] determines the very character of the system … [and] … the structure of the system determines its function' (Ropohl 1999, 4). Metcalfe's Law reinforces this point: 'as the number of [parts in a system] increases linearly the potential 'value or effectiveness' of

the [system] increases exponentially' (Alberts, Garstka, and Stein 1999). Structure is a discoverable aspect of systems, even highly complex open-systems, and it can provide powerful clues as to how the system will behave. This feature can be revealed in the following case study examining how nuclear hunter-killer submarines return to periscope depth (Stanton 2014).

This is a system comprised of many interacting social and technical elements, one that requires a constant throughput of energy, information, actions and other inputs in order to remain as a coherent entity. One that is able to respond to its environment and change and adapt. Figure 3 illustrates a composite network of task, social and information networks extracted via a method called EAST (Event Analysis for Systemic Teamwork; Stanton & Bessell, 2014). From this structural representation alone, created via existing ergonomic methods such as Hierarchical Task Analysis (Annett et al. 1971), Social Network Analysis (e.g. Monge and Contractor 2003; Driskell and Mullen 2005) and semantic networks (e.g. Oden 1987), it was possible to discern exactly what was going on between the submarine's on-board sound and control rooms, and how submariners shared information related to the tasks they were performing.

Seemingly abstruse network representations like that shown in Figure 3 translate into significant practical outcomes. One of these is a full-scale submarine simulator facility (see Figure 4) in which alternative structural

Figure 4. Network representations which foreground the role of structure drove the design of this submarine control room simulator, which allows total flexibility in task, social and information structure to be analytically prototyped and tested.

configurations of task, social and information networks can be systematically tested with real submarine crews to examine their effects on performance, and the system remodelled again in network terms. These insights were simply not available or tractable according to the previous operating paradigms, which were either based around empirical cognitive psychology (highly resource intensive multi-measure, multi-factorial experiments) or loose, single case-study analyses (from which only loose and unsystematic insights could be derived). Instead, by focusing on the structural aspects of systems and viewing key features of performance as networks, a powerful new approach emerged. This, in turn, is leading directly to significant operational changes on new classes of hunter-killer submarine.

Dynamism and change

De Greene (1980) identifies 'major conceptual problems in the systems management of ergonomics research' and notes the particular difficulties that stem from the use of static models. Lee (2001), Guastello (2002) and Sharma (2006) also express concern over the use of static models of cognition and human/machine interaction. Woods and Dekker (2000), however, put it most forcefully:

> The quickening tempo of technology change and the expansion of technological possibilities has largely converted the traditional shortcuts for access to a design process (task analysis, guidelines, verification and validation studies, etc.) into oversimplification fallacies that retard understanding, innovation, and, ultimately, [ergonomic's] credibility. (272)

In Walker, Stanton, Revell et al. (2009) and Walker et al. (2010), we talked about emergence and the notion that simple, component-level behaviours can give rise to

disproportionate, nonlinear collective effects over time. We have also tried to demonstrate this feature with the road freight and logistics example shown above, whereby locally rational behaviours translated into globally irrational outcomes. The working definition put forward for the concept of emergence is 'phenomenon wherein complex, interesting high-level function is produced as a result of combining simple low-level mechanisms in simple ways' (Chalmers 1990, 2). Like the term 'system', emergence actually describes a range of phenomenon because as 'systems become more complex [...], self-organization appears at more than one level [...]. Such systems have multiple, hierarchical levels of self-organization, and calculation of system level emergent properties from the component level rapidly becomes intractable' (Halley and Winkler 2008, 12). In other words, there is a continuum of emergence and a judgement available to be made about where the ergonomics discipline generally operates in relation to the different types, as Table 1 shows:

It is possible to argue that Type 3 emergence is a key characteristic of the deeply perplexing Ergonomic problems which seem to be increasingly exposed by the fact that certain Ergonomic 'low hanging fruit' have been (and will continue to be) dealt with successfully using existing approaches. Dealing with Type 3 emergence, on the other hand, requires knowledge of individual system components (an individual, component-level view which features heavily in the discipline); knowledge of the dynamics of those components (achievable with some methods which acknowledge the role of time); states and configurations of the system (achievable with the kinds of network-based approaches shown above, and others); but also knowledge of the wider environment. This is a special, more conceptual form of environmental knowledge which needs extra clarification.

It might be more helpful to reframe the question of Type 3 emergence in more practical terms. Basically, how do you analyse systems with multiple actors, multiple relationships, multiple degrees of freedom, multiple states and multiple possible behaviours ... in a tractable, time- and cost-efficient manner? One option is to take the system apart and separate it into its constituent components. This is possible but the problems with doing so are twofold. Firstly, the more one isolates key variables and controls for other extraneous factors, the less the phenomena of interest looks and behaves like the 'real' multidimensional problem that is attempting to be solved. Experimental control, therefore, will more frequently find itself in conflict with ecological validity. Secondly, to empirically evaluate all possible states of a truly multidimensional system would likely need a sample size, due to the sheer number of conditions, manipulations and required statistical power that would exceed the human population on Earth. It is no wonder, then, that ergonomic analyses of these systems

Table 1. Types of emergence, the information needed in order to make a diagnosis of the system's behaviour, and tentative examples of existing methods and approaches at each level (definitions from Bar-Yam 2004).

Emergence	Type	Information needed in order to make a diagnosis of the system's collective behaviour	Examples of ergonomic methods
None	Deterministic	Knowledge of individual system components sufficient to fully explain global system behaviour.	Static, deterministic ergonomic models and methods such as Hierarchical Task Analysis (HTA)
Weak	Type 0		
	Type 1	As for Type 0 but with additional knowledge about the positions and dynamics of individual entities in a system, this being sufficient to describe the 'microscopic as well as macroscopic properties of the system' (Bar-Yam 2004c, 17).	Ergonomic models and methods that incorporate a timeline such as Operator Sequence Diagrams (OSD)
Strong	Type 2	As for Type 1 but with additional knowledge of possible states and configurations the system can adopt. 'the state of one part may determine (or be coupled to) the state of other parts' (Bar-Yam 2004c, 17).	Models incorporating nodes, links and layers, such as Work Domain Analysis (WDA) and System-Theoretic Accident Model and Processes (STAMP)
	Type 3	As for Type 2 but with additional knowledge of the environment that the system resides in. 'This is not contained in the conventional discussion of properties of a system as determined by the system itself' (Bar-Yam 2004c, 17).	Methods which use networks to model how systems move within a phase space, such as Event Analysis for Systemic Teamwork (EAST)

often shy away from what this multidimensionality really means, proceeding instead on the basis of more ad hoc, single-case study approaches which, nevertheless, are often still labelled 'experiments' (e.g. CCRP 2004).

An alternative approach to the conceptual problems posed by Type 3 emergence is to tackle it from the top-down (the system) instead of from the bottom-up (the components). Approaching the system top-down in terms of structure is highly beneficial, as the submarine case-study above shows, but it is possible to go further still. The central idea can be summed up by the simple cybernetic principle that: 'if all the variables are tightly coupled, and if you can truly manipulate one of them in all its freedoms, then you can indirectly control all of them' (Kelly 1994, 121). Thus, the behaviour of individual components are not measured in isolation, rather the product of those collective behaviours are measured as a whole. Phase spaces are a tractable ergonomic approach for doing this.

Phase spaces arise from the study of complex dynamical systems and the inadequacies of simple one-dimensional time-series analyses in capturing and representing their behaviour. Unlike simple time-series graphs, with time on the x-axis and one variable on the y-axis, phase spaces are a graphical way to analyse the time-varying interactions between multiple variables. Although the resulting representation is entirely abstract it can reveal fundamental properties about the system in question. Indeed, it is quite common for prominent graphical features to have some kind of physical analogue (in the famous Lorenz example it is fluid flow changing direction, for example). Unlike physical systems, many of the variables of interest within ergonomics problems are not easily or usefully reduced to a set of fundamental equations which can be plotted into a coordinate space with complete precision. This has been resolved to some extent in related approaches like Functional Holography (Baruchi and Ben-Jacob 2004; Baruchi et al. 2006). As in the case of

physical systems, whereby multidimensionality is reduced algebraically, here the multidimensionality can be reduced via Principal Components Analysis, or other similar means, based entirely on the structural properties of networks. By clustering individual components and having them form the principle axes of a phase space, the relationships between multiple dimensions of a system are preserved: the reduced set of dimensions still enables you to indirectly manipulate all others.

In our own research, we have used a dimensionally reduced organisational 'phase space' called the NATO SAS050 model of command and control (NATO 2007). We combined it with Social Network Analysis as a way to represent real command and control teams during a large-scale military operational field trial. To convey an idea of the scale and complexity of this study, and the challenges inherent in conducting an Ergonomic analysis in this domain, it can be noted that it took place (outside in poor weather conditions) in a 116-square-kilometre military training area, employed a fully functioning Brigade level field headquarters, geographically dispersed Battlegroup headquarters, 73 personnel, a simulated attack and over 2500 communication exchanges. It was also not possible to interfere with any of the normal activities being performed or to extract or apply more typical behavioural measures. Combining social network analysis with an organisational phase space represented a novel way to drive out insights the stakeholders were in critical need of.

A social network is 'a set of entities and actors [...] who have some type of relationship with one another' (Driskell and Mullen 2005, 58–1). Communications links between actors in the scenario were captured by the command and control system and via observation, thus meeting the stakeholder's stipulations governing our participation in this live exercise. Social network 'analysis' is about mathematically scrutinising those relationships to discern properties that are not necessarily apparent from visual

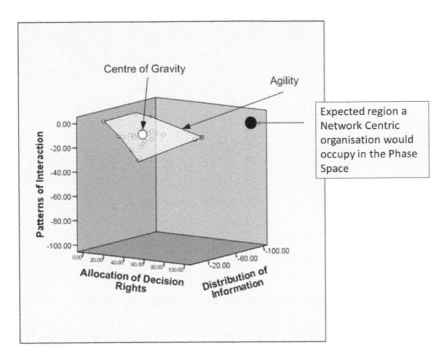

Figure 5. A large scale military operational field trial was analysed using social network analysis and plotted into a dimensionally reduced organisational phase space.
Note: It showed that the command and control organisation was not as agile or 'net-centric' as was originally thought and this, in turn, was born out in various measures of performance.

inspection of the network alone (e.g. Harary 1994). These properties include several that relate to the main dimensions of the NATO 'organisation model', or phase space, being used in this study. For example, it was established that sociometric status, a measure denoting the prominence of a node in the network based on its communications with other nodes, had good construct validity with the organisational property 'decision rights'. Network diameter, a measure denoting the maximum number of communication links needed to traverse the network from one side to the other, had good construct validity with the property 'patterns of interaction'. Network density, a measure of the average interconnectivity within the network, had good construct validity with the final axis called 'distribution of information' (see Walker et al. 2009). Mapping social network metrics to an organisational phase space meant that an organisation as complex and dynamic as this could be positioned in the organisation model (the phase space) based on the communications taking place within it, which were straightforward to collect. The ultimate goal of this activity was to show if the organisation's position in the phase space, after considerable financial investment in new equipment and procedures, was where it should be positioned. In this particular case it was not, as Figure 5 shows.

Different actors in this scenario needed to communicate with other actors at different times for different purposes. By analysing the social networks over successive periods of

time, it was possible to observe how their interconnection changed as a function of the task and the operating environment. These changes relate to a property called agility, or the 'ability to reconfigure structures rapidly [...] enabled by an information environment that allows rapid reconfiguration of the underlying network and knowledge bases' (Ferbrache 2005, 104). As the social networks change so the network metrics change. The range of different values these metrics describe over time says something important about how the sociotechnical system is reconfiguring itself, in particular how fast and by how much. When the changing values of the network metrics are plotted into a graph with the x-axis representing time and the y-axis representing the corresponding social network metric value, the resultant time-series resembles a complex waveform. Spectral analysis methods were then be used to unpick this complex waveform into a set of constituent frequencies which provided insights into organisational agility in a very direct manner (see Walker et al. 2010).

Figure 6 shows the results of this unpicking process in a representation called a periodogram. In this application, the x-axis shows the extent to which the social network is reconfiguring, the y-axis shows how quickly it is reconfiguring. In the example shown there is a strong 'fundamental frequency' of organisational reconfiguration, but weaker spikes of faster reconfigurations occurring at multiples of the fundamental frequency (these are called 'harmonics'). Patterns like these reveal something important about the

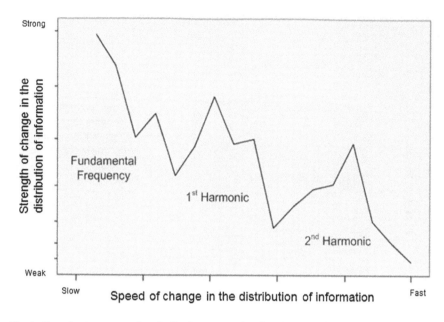

Figure 6. Periodogram illustrating the presence of periodic changes in the distribution of information (i.e. network density) within a live military command and control case study.
Note: A pattern is obtained that approximates to first, second and third order harmonic effects.

sociotechnical system's dynamics. Evidence of periodicity indicates that the organisation is repeatedly drawn into, or attracted, into specific locations of its phase space, revealing in turn specific configurations of that organisation which are more persistent than others. The lack of a pattern or periodicity is just as interesting. It suggests that the forces present in the environment which are driving the organisation's dynamic reconfigurations might be more chaotic in nature.

To be even more specific, clusters that form in defined areas of the phase space, and into which the organisation was repeatedly drawn (like those shown in Figure 5) can be termed 'fixed attractors'. Fixed attractors represent something analogous to equilibrium state(s) of the system or the dominant prevailing behaviours. This does not fully describe the dynamical behaviour of the real command and control organisation under analysis, however, because many of the other points did not fall into defined clusters at all and were instead scattered widely. The organisation was thus attracted to these other points in the phase space by forces whose underlying dynamics were not stable and deterministic, but unstable and chaotic. In the parlance of complex adaptive systems, these would be termed 'strange attractors'.

The presence of attractors in an organisational phase space suggests a much more fundamental source of organisational dynamics, linking in turn to a much more abstract idea of 'environment'. The fact sociotechnical systems of any kind are drawn or otherwise propelled into different regions of their phase spaces indicates that, like a ball rolling across a surface containing dips and hollows, this abstract environment possesses a defined 'causal texture' (e.g. Emery and Trist 1965). This is a classic sociotechnical systems concept. So is the idea that a sociotechnical system is itself part of the environment and through its actions able to influence its causal texture. We come back, then, to the idea that there are multiple interconnected pathways which are constantly moving and shifting … no predefined system layers … no predefined (static) holes in the 'Swiss Cheese' … and no possibility of detecting the things we really want to detect by focusing exclusively on the components. At least now, via phase spaces and innovative use of ergonomic methods, these features are able to become visibly manifest and open to inspection. In this case, the stakeholders were able to gain a much deeper understanding of what organisational agility actually meant, how to measure it and valuable insights into process improvements targeted at shifting their position in the phase space into more desirable regions, and the ability to measure and check those improvements ongoing.

Conclusions

The aim of this paper has been to take advantage of a unique opportunity (and Special Issue) to stand back and critically examine the ergonomic paradigm and the manner in which it may have to change. Strategically it has been written that the discipline will be required to work on more high-quality, high-impact systems problems (e.g. Dul et al. 2012) but this, in turn, requires that systems thinking remains at the forefront of ergonomics theory and practice. This does not mean a shift in the discipline's overriding philosophical stance of pragmatism. On the contrary, theoretical issues in ergonomics science are the route along which the next generation of methods and approaches will ultimately stem. The case studies chosen in this paper have also been deliberate: although founded on sometimes obtuse, highly conceptual theories all of them were carried out at the behest of industrial clients with thorny problems that had proven resistant to existing approaches. In other words, these 'quantum concepts' have all been practically applied and without exception provided useful, cost-effective insights for real ergonomic clients. Insights which, up until this point, had alluded them.

This paper's journey into the ergonomic paradigm has also taken us towards something nearer-term and more pressing. Picking up on a nascent theme from an earlier paper,

> there is a fundamental need to match our methods to the developing nature of our ergonomic problems. Quite often, the precipitating event which leads to the engagement of ergonomic services is a problem that existing tools and techniques cannot make progress on. Increasingly, we would argue, these problems contain significant amounts of emergence for which our simple and supposedly pragmatic methods have the most trouble dealing with. (Walker 2016, 456)

Matching methods to problems will inevitably mean a much stronger systems perspective being adopted. In this regard, and compared to some other fields, it could be argued that ergonomics may have been too timid. Too focused on the resolution of immediate problems by tools and techniques now ill-suited to new challenges. The answer is open to debate and will be for the discipline to decide, but journeying out into uncharted ergonomic territory we have been able to catch glimpses of what the discipline needs to do to remain relevant in the future. If we are not able to adapt to a new 'quantum world' of multidimensionality, collinearity, dynamism and complexity then other disciplines will do it for us. The warning from the future is that other disciplines are already revealing new ergonomic insights from the Internet of things and big data, simply because they have adapted more quickly than we have. Let us be bolder. There is an opportunity for ergonomics to become a leading rather than following discipline in new scientific and engineering paradigms. After all, these new paradigms are entirely human-constructed entities, and so too are the key future challenges.

Acknowledgement

We gratefully acknowledge the assistance of Dr Miranda Cornelissen, who was involved in the brainstorming sessions and aided with ideas and transcription.

Disclosure statement

No potential conflict of interest was reported by the authors.

References

Alberts, D. S., J. J. Garstka, and F. P. Stein. 1999. *Network Centric Warfare: Developing and Leveraging Information Superiority*. Washington, DC: CCRP.

Annett, J., K. D. Duncan, R. B. Stammers, and M. J. Gray. 1971. *Task Analysis*. London: HMSO.

Baruchi, I., and E. Ben-Jacob. 2004. "Functional Holography of Recorded Neuronal Networks Activity." *Neuroinformatics* 2 (3): 333–352.

Baruchi, I., D. Grossman, V. Volman, M. Shein, J. Hunter, V. Towle, and E. Ben-Jacob. 2006. "Functional Holography Analysis: Simplifying the Complexity of Dynamical Networks." *Chaos: An Interdisciplinary Journal of Nonlinear Science* 16: 015112.

Bar-Yam, Y. 2004. "Multiscale Complexity/entropy." *Advances in Complex Systems* 7: 47–63.

Bertalanffy, L. v. 1950. "The Theory of Open Systems in Physics and Biology." *Science* 111: 23–29.

Carvajal, R. 1983. "Systemic Netfields: The Systems' Paradigm Crises. Part I." *Human Relations* 36 (3): 227–245.

CCRP (Command and Control Research Program). 2004. *Code of Best Practice for C2 Assessment*. Washington, DC: CCRP.

Chalmers, D. J. 1990. Thoughts on Emergence [online]. Accessed March 13, 2015. http://consc.net/notes/emergence.html

Chung, A. Z. Q., and S. T. Shorrock. 2011. "The Research-practice Relationship in Ergonomics and Human Factors – Surveying and Bridging the Gap." *Ergonomics* 54 (5): 413–429.

De Greene, K. B. 1980. "Major conceptual problems in the systems management of human factors/ergonomics research." *Ergonomics* 23 (1): 3–11.

Driskell, J. E., and B. Mullen (2005). "Social Network Analysis." In *Handbook of Human Factors and Ergonomics Methods*, edited by N. A. Stanton, P. M. Salmon, G. H. Walker, C. Baber, and D. P. Jenkin, 58.1–58.6 Boca-Raton, FL: CRC.

Drury, C. 2015. "Human Factors/ergonomics Implications of Big Data Analytics: Chartered Institute of Ergonomics and Human Factors Annual Lecture." *Ergonomics* 58 (5): 659–673.

Dul, J., R. Bruder, P. Buckle, P. Carayon, P. Falzon, W. S. Marras, J. R. Wilson, and B. van der Doelen. 2012. "A Strategy for Human Factors/ergonomics: Developing the Discipline and Profession." *Ergonomics* 55 (4): 377–395.

Emery, F. E., and E. L. Trist. 1965. "The Causal Texture of Organisational Environments." *Human Relations* 18 (1): 21–32.

Ferbrache, D. 2005. "Network Enabled Capability: Concepts and Delivery." *Journal of Defence Science* 8 (3): 104–107.

Guastello, S. J. 2002. *Managing Emergent Phenomena: Nonlinear Dynamics in Work Organizations*. Mahwah, NJ: Lawrence Erlbaum Associates.

Hall, A. D., and R. E. Fagen. 1956. "Definition of System." *General Systems* I: 18–28.

Halley, J. D., and D. A. Winkler. 2008. "Classification of Emergence and Its Relation to Self-organization." *Complexity* 13 (5): 10–15.

Hancock, P. A. (1997). Essays on the Future of Human-machine Systems. Eden Prairie: MN: Banta.

Harary, F. 1994. *Graph Theory*. Reading, MA: Addison-Wesley.

IEA. (2015). What is Ergonomics: Definition and Domains of Ergonomics. http://www.iea.cc/whats/.

Kellmereit, D., and D. Obodovski. 2013. *The Silent Intelligence: The Internet of Things*. San Francisco, CA: DnD Ventures.

Kelly, K. 1994. *Out of Control: The New Biology of Machines, Social Systems, and the Economic World*. New York: Purseus.

Leavitt, H. J. 1951. "Some Effects of Certain Communication Patterns on Group Performance." *The Journal of Abnormal and Social Psychology* 46: 38–50.

Lee, J. D. 2001. "Emerging Challenges in Cognitive Ergonomics: Managing Swarms of Self-organizing Agent-based Automation." *Theoretical Issues in Ergonomics Science* 2 (3): 238–250.

Marcovici, M. 2014. *The Internet of Things*. Hamburg: Books on Demand.

Miller, M. 2015. *The Internet of Things: How Smart TVs, Smart Cars, Smart Homes, and Smart Cities Are Changing the World*. Indianapolis, IN: QUE.

Monge, P. R., and N. S. Contractor. 2003. *Theories of Communication Networks*. Oxford: Oxford University Press.

NATO. (2007). RTO-TR-SAS-050 Exploring New Command and Control Concepts and Capabilities. Report: RTO-TR-SAS-050 AC/323(SAS-050)TP/50. NATO. http://www.rta.nato.int/Pubs/RDP.asp?RDP=RTO-TR-SAS-050.

Nijkamp, P., G. Pepping, and D. Banister. 1997. *Telematics and Transport Behaviour*. Berlin: Springer-Verlag.

Oden, G. C. 1987. "Concept, Knowledge, and Thought." *Annual Review of Psychology* 38: 203–227.

Pugh, D. S., D. J. Hickson, C. R. Hinings, and C. Turner. 1968. "Dimensions of Organisation Structure." *Administrative Science Quarterly* 13 (1): 65–105.

Rasmussen, J. 1997. "Risk Management in a Dynamic Society: A Modelling Problem." *Safety Science* 27: 183–213.

Rifkin, J. 2014. *The Zero Marginal Cost Society*. New York: Palgrave Macmillan.

Ropohl, G. 1999. "Philosophy of Socio-technical Systems." *Society for Philosophy and Technology* 4 (3): 1–10.

Sharma, S. 2006. "An Exploratory Study of Chaos in Human-machine System Dynamics." *IEEE Transactions on Systems, Man, and Cybernetics-Part A: Systems and Humans* 36 (2): 319–326.

Stanton, N. A. 2014. "Representing Distributed Cognition in Complex Systems: How a Submarine Returns to Periscope Depth." *Ergonomics* 57 (3): 403–418.

Stanton, N. A., and K. Bessell. 2014. "How a Submarine Returns to Periscope Depth: Analysing Complex Socio-technical Systems Using Cognitive Work Analysis." *Applied Ergonomics* 45 (1): 110–125.

Stanton, N. A., P. M. Salmon, and G. H. Walker. 2015. "Let the Reader Decide: A Paradigm Shift for Situation Awareness in Sociotechnical Systems." *Journal of Cognitive Engineering and Decision Making.* 9 (1): 44–50.

Toffler, A. 1980. *The Third Wave*. London: Pan.

Toffler, A. 1981. *Future Shock: The Third Wave*. New York: Bantam.

Trafimow, D., and M. Marks. 2015. "Editorial." Basic and Applied Social Psychology 37 (1): 1–2.

Trist, E. L. 1978. "On Socio-technical Systems." In *Sociotechnical Systems: A Sourcebook*, edited by W. A. Pasmore and J. J. Sherwood, 32–51. San Diego, CA: University Associates.

Walker, G. H. 2016. "Fortune Favours the Bold." *Theoretical Issues in Ergonomics Science* 17 (4): 452–458.

Walker, G. H., and A. Manson. 2014. "Telematics, Urban Freight Logistics and Low Carbon Road Networks." *Journal of Transport Geography* 37: 74–81.

Walker, G. H., N. A. Stanton, K. Revell, L. Rafferty, P. M. Salmon, and D. P. Jenkins. 2009. "Measuring Dimensions of Command and Control Using Social Network Analysis: Extending the NATO SAS-050 Model." *International Journal of Command and Control* 3 (2): 1–47.

Walker, G. H., N. A. Stanton, P. M. Salmon, D. P. Jenkins, and L. Rafferty. 2010. "Translating Concepts of Complexity to the Field of Ergonomics." *Ergonomics* 53 (10): 1175–1186.

Walker, G. H., and A. Strathie. 2016. "Big Data and Ergonomics Methods: A New Paradigm for Tackling Strategic Transport Safety Risks." *Applied Ergonomics* 53: 298–311.

Watts, D. J., and S. H. Strogatz. 1998. "Collective Dynamics of 'Small-world' Networks." *Nature* 393 (6684): 440–442.

Woods, D. D., and S. Dekker. 2000. "Anticipating the Effects of Technological Change: A New Era of Dynamics for Human Factors." *Theoretical Issues in Ergonomics Science* 1 (3): 272–282.

Nonlinear dynamical systems for theory and research in ergonomics

Stephen J. Guastello

ABSTRACT

Nonlinear dynamical systems (NDS) theory offers new constructs, methods and explanations for phenomena that have in turn produced new paradigms of thinking within several disciplines of the behavioural sciences. This article explores the recent developments of NDS as a paradigm in ergonomics. The exposition includes its basic axioms, the primary constructs from elementary dynamics and so-called complexity theory, an overview of its methods, and growing areas of application within ergonomics. The applications considered here include: psychophysics, iconic displays, control theory, cognitive workload and fatigue, occupational accidents, resilience of systems, team coordination and synchronisation in systems. Although these applications make use of different subsets of NDS constructs, several of them share the general principles of the complex adaptive system.

Practitioner Summary: Nonlinear dynamical systems theory reframes problems in ergonomics that involve complex systems as they change over time. The leading applications to date include psychophysics, control theory, cognitive workload and fatigue, biomechanics, occupational accidents, resilience of systems, team coordination and synchronisation of system components.

1. Introduction

The growing complexity of sociotechnical systems has produced a need for a paradigm shift in the scientific development and practice of ergonomics (Karwowski 2012; Walker et al. 2010), as it has done across the range of social and life sciences (Allan and Varga 2007; Dore and Rosser 2007; Fleener and Merritt 2007; Guastello and Bond 2007; Ibanez 2007; Zausner 2007). A new scientific paradigm would produce new concepts for understanding phenomena, new questions to be asked, new methods to accompany the new questions and new explanations to be offered for phenomena. It would also improve on extant explanations for phenomena that might have been fragmented or elusive altogether by changing a perspective in a significant way. The next sections of this article expand on NDS' axioms, historical origins, core constructs, methodologies and some of its developing application areas in ergonomics. In each case, it is necessary to be concise so as not to lose the big picture of the transformation that is taking place.

1.1. Axioms

Nonlinear dynamical system (NDS) has four features that should be considered axiomatic. First, simple deterministic functions can produce events that are apparently random;

identifying the function can be challenging. The analysis of variability is at least as important as the analysis of means, which pervades the linear paradigm. Unlike the analysis of variance and the twenty-first century developments thereto, NDS is concerned with the amount and structure of the variability, the underlying processes that produce those patterns of variability, and the identification of system variables that could influence the outcomes within those structures.

Second, there are many types of change that systems can produce, not just one simple type of generic change. They are represented by numerous modelling structures such as attractors, bifurcations, chaos, fractals, catastrophes, entropy, self-organisation, emergence and synchronisation (Guastello and Liebovitch 2009; Sprott 2003). Temporal patterns are essentially the footprints of particular underlying dynamics. Except in rare circumstances time is an implicit variable, not an explicit independent variable. Instead, the core functions are framed such that a state of an agent at time-2 is a nonlinear function of the agent's state at time-1 and other influences from control variables; see the fourth axiom below.

Third, contrary to common belief, systems are not simply waiting in a state of equilibrium until they are perturbed by some force from outside the system. Rather, stabilities,

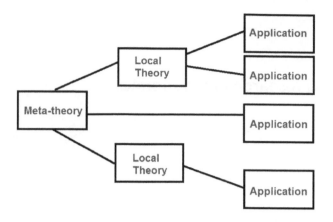

Figure 1. The hierarchical relationships among meta-theory, local theories, and applications.

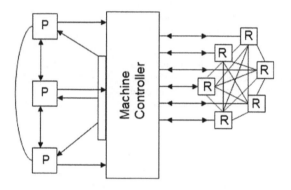

Figure 2. Configuration for the operation of swarming robots. Source: Reprinted from Guastello (2014a, 376) with permission of Taylor and Francis.

instability and other change dynamics are produced by the system as it behaves 'normally'. This is not to say that perturbations or random shocks cannot originate outside the system. Indeed they can do so, and perturbations can arise from inside the system as well. The researcher would then look for *control variables* that govern the reactivity of the system to perturbations.

Fourth, the concept of causality changes to concepts of *control* and *emergence*. In the more traditional thinking, Event A that occurs today causes Event B tomorrow. If A did not happen, neither would B, and that would be the end of the possible outcomes. In NDS, each agent is following a dynamic path that is strongly influenced by its previous state. Other variables and external events can deflect or redirect the path of the agent, either by a little or a lot. The effect of such variables is dependent on the previous state of the agent or system; small influences can produce large results and large influences can produce nothing at all, depending on the prior state of the agent or system. Furthermore, not all control variables play the same role in the dynamics.

Many problems that need to be solved, either in a theoretical or pragmatic context, cannot be reduced to single underlying causes. Rather, they are emergent products of complex system behaviours and internal interactions. Emergent phenomena involve both bottom-up and top-down processes. Although they often appear to 'come out of nowhere', there are known processes by which they can occur; one just needs to view a situation through the right lens. The new forms of research questions that proceed from the foregoing axioms bring about a need to develop appropriate research methods to answer those questions.

1.2. Historical roots

NDS is a general systems theory in the sense that its principles and models have found applications to a wide range of phenomena that might have appeared unrelated at first blush until their common intrinsic dynamics were discovered. It has made good use of some general systems concepts from the 1950s such as feedback loops and mathematical formalisms as the core elements of its applications (Guastello and Liebovitch 2009). NDS' primary origin was differential topology, and over the years the connections among its contributing constructs have grown. Although many of the ideas were developed in the context of problems in physics, any isomorphism between phenomena in physics and those of the life sciences reside in the mathematics, not in any necessary assumptions about physical processes. Nonetheless, good analogies often pay off well.

When NDS principles combine with the theoretical constructs within a particular application, they play the role of a meta-theory: Broad theoretical principles guide the organisation of constructs within the local theory, which in turn shapes the approach to studying a local phenomenon (Figure 1). There could be occasions where a phenomenon needs investigating, but a local theory has not been previously defined. Here, the meta-theory could shape the development of both.

Ergonomics has made productive use of systems thinking. The person–machine system is perhaps the most obvious systems idea (Meister 1977). The nature of the system has expanded in scope to include the immediate workspace, less immediate environments where related subsystems can reside, and multiple integrated person–machine systems, or swarming robot configurations (Figure 2). Teams, group dynamics and communications are have also risen in importance (Cooke, Duchon, et al. 2012; Gorman, Cooke, and Salas 2010).

Applications of NDS in ergonomics first appeared in the 1980s, starting with shift work and industrial production, physical workload, fatigue and occupational accidents (Guastello 1982, 1985, 1988; Guastello and McGee 1987).

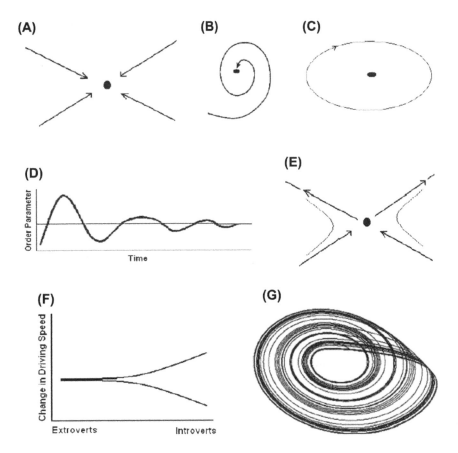

Figure 3. Gallery of elementary dynamics: (A) fixed point, (B) fixed point, spiral type, (C) oscillator or limit cycle, (D) dampened oscillator, (E) saddle, (F) bifurcation, (G) top-down view of the Rossler chaotic attractor.

A broader state-of-the-science compendium of psychological applications today would include neuroscience; psychophysics, sensation, perception and cognition; motivation and emotion, group dynamics, leadership, and collective intelligence; developmental, abnormal psychology and psychotherapy; and organisational behaviour and social networks (Dishion 2012; Dooley, Kiel, and Dietz 2013; Guastello 2009a; Guastello, Koopmans, and Pincus 2009); psychomotor coordination and control (Shelhamer 2009); creativity (Guastello and Fleener 2011), education (Stamovlasis and Koopmans 2014) and medical practice (Katerndahl 2010; Sturmberg and Martin 2013).

2. Primary NDS constructs

2.1. Attractors and bifurcations

The structures previously known as 'equilibria' are now understood as one of several basic forms of attractor. An attractor is a piece of space. When an object enters, it does not exist unless a substantial force is applied to it. The simplest attractor is the *fixed point*.

Oscillators, also known as *limit cycles*, are another type of attractor. They can be simple oscillators, dampened to a fixed point by means of a *control parameter*, or perturbed

in the opposite direction to become *aperiodic oscillators*. There is a gradual transition from aperiodic attractors to chaos.

Repellors are the opposite of attractors. Objects that veer too close to them are pushed outwards and can go anywhere. This property of an indeterminable final outcome is what makes repellors unstable. Fixed points and oscillators, in contrast, are stable. Chaotic attractors (described below) are also stable in spite of their popular association with unpredictability.

A *saddle* has mixed properties of an attractor and a repellor. Objects are drawn to it, but are pushed away once they arrived. A saddle is also unstable.

Bifurcations are splits in a dynamic field that can occur when an attractor changes from one type to another, or where different dynamics are occurring in juxtaposing pieces of space. Bifurcations can be as simple as a single point, or they may involve trajectories and patterns of instability that can become very complex. The example in Figure 3 is one of the simpler forms of bifurcation. This example from Cobb (1981) and (Zeeman 1977) shows the effect of alcohol on driving speed in a simulator. Extroverts maintain speed, whereas introverts show substantial changes in both directions. Very few observations fall within the space between the two curves.

2.2. Chaos and fractals

The essence of *chaos* is that seemingly random events can actually be explained by simple deterministic processes or equations (Kaplan and Glass 1995; Sprott 2003). Chaos has three hallmark properties: sensitivity to initial conditions, boundedness and unpredictability.

Sensitivity to initial conditions means that two points can start by being arbitrarily close together, but as the same function continues to iterate for both of them, the two points become increasingly farther apart; this is the well-known *butterfly effect* (Dooley 2009; Lorenz 1963). *Boundedness* means that, in spite of any volatility, the values of the measurement stay within a fixed range. *Unpredictability* is actually a matter of degree and sometimes an overstatement, depending on the context. What actually occurs is that it the predictability of a point from a previous one decays as the time interval becomes larger (Smith 2007). The ability to predict points changes in our favour if we know the nonlinear function in advance, which is seldom the case in ergonomics at the present time.

The basin, or outer rim, of a chaotic attractor has a fractal shape. A *fractal* is a geometric structure that has non-integer dimensionality and self-repeats as one zooms in or out from a target image. When viewed in their abstract forms, it is possible to see the entire fractal image at larger and smaller levels of scale. Although they have strong aesthetic appeal (Sprott 2004), the aesthetic properties of fractals are not particularly useful for ergonomics research. Fractal *structure,* however, can be analysed to interpret the ruggedness of a landscape (Mandelbrot 1983), the complexity of a time series (Guastello and Gregson 2011; Mandelbrot 1999), or the distribution of the sizes of events or objects after a self-organising process has taken place (Bak 1996). For this purpose, we would compute the *fractal dimension* of a time series of behavioural observations and examine how it varies across situations or from the individual, group or greater collective levels of analysis.

Fractal dimensions close to 0 denote fixed points. Dimensions around 1.0 indicate either a line or a perfect oscillator. Dimensions between 1 and 2 are in the *zone of self-organised criticality* (SOC), in which a system adopts a lower entropy structure after having been in a state of chaos, or far-from-equilibrium conditions for some time previously. In living systems, it reflects a healthy balance between the variability needed to make an adaptive response with the least amount of effort.

Levi flights are bursts of behaviour in an unexpected direction after the system has spent most of its time in a relatively steady state; they tend to occupy dimension ranges between 2 and 3. Chaos is usually associated with dimensions of 3.0 or more (Nicolis and Prigogine 1989), but there are some well-known chaotic attractors that have fractal dimensions less than 3.0.

2.3. Information and entropy

The entropy construct underwent some important developments since it was introduced in the late nineteenth century. Initially, it meant 'heat loss'. This definition led to the principle that systems will eventually dissipate heat and expire from 'heat death'. A century later, this generalisation turned out to be incorrect, when it was discovered that systems respond to high entropy conditions by self-organising (Haken 1984; Prigogine and Stengers 1984) in a way that establishes a low-entropy state.

The second perspective originated in the early twentieth century with statistical physics. It was not possible to target the location of individual molecules, but it was possible to define metrics of the average motion of the molecules. Shannon entropy was the third perspective: A system can take on any number of discrete states over time. It takes *information* to predict those states, and any variability for which information is not available to predict is considered entropy. Entropy and information add up to H_{MAX}, maximum information, which occurs when all the states of a system have equal probabilities of occurrence:

$$H_s = \Sigma_i \left[p_i \log_2(1/p_i) \right] \qquad (1)$$

where i is a system state and p is the probability of that state (Shannon 1948).

The NDS perspective on entropy, however, is that entropy is generated *by* a system as it changes behaviour over time (Nicolis and Prigogine 1989; Prigogine and Stengers 1984) and thus it has become commonplace to treat Shannon information and Shannon entropy as interchangeable quantities. Other computations of entropy have been developed for different types of NDS problems (Guastello and Gregson 2011).

2.4. Catastrophe models

Catastrophes are sudden changes in events; they are not necessarily bad or unwanted events as the word 'catastrophe' might suggest in English. Catastrophes involve combinations of attractors and bifurcations, and are operating in some self-organising events. According to the classification theorem (Thom 1975), all discontinuous changes of events can be described by seven elementary topological models. The models are hierarchical such that the simpler ones are embedded in the larger ones.

The *cusp* model (Figure 4) is the second-simplest in the series – just complex enough to be very interesting, unique and useful. There are two control parameters, *asymmetry* and *bifurcation*. To visualise the dynamics, start at the stable state on the left and follow the outer rim of the surface where bifurcation is high. If we change the value of the asymmetry parameter, nothing happens until it reaches a

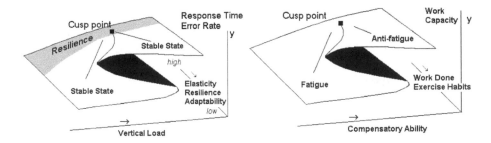

Figure 4. The cusp catastrophe response surface. These examples are labelled for the cognitive workload and fatigue problem.

critical point, at which we have a sudden change in behaviour: The *control point*, which indicates what behaviour is operating, flips to the upper sheet of the surface. A mirror image process occurs when shifting from the upper to the lower stable state. Notice that the upward and downward paths occur around different thresholds.

When the bifurcation effect is large, the discontinuous dynamics of the outer rim of the surface occur. When the bifurcation effect is low, change is relatively smooth. The darkened area between the stable states (attractors) is an area where few data points are likely to land, because a repellor is located there. The cusp point (a saddle) is the most instable location on the surface. The paths drawn between the cusp point and the stable states are *gradients*. Cusp models have been informative for problems involving cognitive workload and fatigue, biomechanics, and accident analysis and prevention; those applications are expanded further in Sections 4.3, 4.5, and 4.8.

2.5. Self-organisation and emergence

The group of NDS constructs that is often known as 'complexity theory' is concerned with self-organising phenomena and the effect of one subsystem behaviour on another. Self-organisation is sometimes known as 'order for free' because systems acquire patterns of behaviour without any input from outside sources. Three distinct models of self-organisation are considered next: the rugged landscape, (Kauffman 1993, 1995); the sand pile (Bak 1996); and synergetics (Haken 1984). The principle that they share is that systems self-organise in response to the flow of information from one subsystem to another.

For the *rugged landscape* scenario, imagine that a species of organism (or generic *agent*) is located on the top of a mountain in a comfortable ecological niche. The organisms have numerous individual differences in traits that are not relevant to survival. Then one day something happens and the organisms need to leave their old niche and find new ones on the rugged landscape, so they do. In some niches, they only need one or two traits to function effectively. For other possible niches, they need several traits. There will be more organisms living in a new 1-trait

environment, not as many in a 2-trait environment, and so on.

Figure 5 is a distribution of *K*, the number of traits required, and *N* the number of organisms exhibiting that many traits in the new environment. Notice, however, that there is a niche towards the right side of the distribution that looks more populated than its lower-*K* neighbours. A simple rule that predicts when clusters like these will occur has not been officially determined yet, but there is reason to suspect that it results from interactions among heterogeneous agents within the niche.

Kauffman's third parameter is *C*, complexity of interaction. A landscape is rugged to the extent that that the interactions among agents within a niche are frequent and varied in nature. If the interaction level is high, it becomes more difficult for new agents to assimilate into the niche.

Vidgen and Bull (2011) introduced two more parameters of complexity to the rugged landscape: *S*, the number of interacting species of agent, and *R*, the differential rates of change for the different species. Interactions among agents eventually produce change in the agents' behaviour and cognitive schemata. As a result they co-evolve. The antics of one species set the stage for the evolution of the others. They amplify each other's variation through interaction. Ergonomically, what we can observe is that person–machine systems morph as they interact with other systems. The rates of change for the interacting agents can be very different, and these differences alone can produce substantial variability in the final outcomes

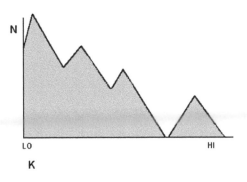

Figure 5. N|K distribution of traits in a rugged landscape.

of all the interactions (Guastello, Reiter, and Malon 2015; Matsumoto and Szidarovszky 2014).

For Bak's (1996) sand pile and avalanche model, imagine that we have a pile of sand, and new sand is slowly drizzled on top the pile. At first nothing happens, but suddenly the pile avalanches into distribution large and small piles. Importantly, the frequency distribution of large and small piles follows a *power law distribution*. A power law distribution is defined as

$$FREQ[X] = aX^b \qquad (2)$$

where X is the variable of interest (pile size), a is a scaling parameter and b is a shape parameter.

Different shapes are produced when b is negative versus positive. The distributions with negative b are of greater interest in NDS, however. Two examples appear in Figure 6. When b becomes more severely negative, the long tail of the distribution drops more sharply to the X axis. Self-organising phenomena of interest contain negative values of b. The $|b|$ is the fractal dimension for the process that presumably produced them. The widespread reports of the $1/f^b$ relationships led to the interpretation of fractal dimensions between 1.0 and 2.0 as being the *range of SOC*.

Synergetics captures two further aspects of self-organising systems. One is that the connections that occur between subsystems can have different temporal dynamics. For instance, if a chaotic *driver* feeds information to a periodic *slave*, the slave will, as a result, produce aperiodic output. This is just one simple example of analysing a system for driver–slave dynamics that are potentially complicated.

A second theme from synergetics is the phase shift that a system undergoes when it changes from one configuration of connections to another. For instance, a person might be experiencing a medical or psychological pathology that is unfortunately stable and prone to continue until there

is an intervention. The intervention takes some time and effort, but the system eventually breaks up its old form of organisation and adopts a new one (Heinzel, Tominschek, and Schiepek 2014; Leining, Strunk, and Mittlestadt 2013). This is also the phase shift form of emergence (Goldstein 2011). Many examples of phase shifts have been reported across disciplinary areas. The change in the system is akin to water turning to ice or to vapour, or vice versa. The challenge is to predict when the change will occur, which is what researchers who follow the synergetics reasoning often try to do. There is a sudden burst of entropy in the system just before the change takes place, which the researcher (therapist, operations engineer) would want to measure and monitor.

An important connection here is that the phase shift that occurs in self-organising phenomena *is* a cusp catastrophe function (Gilmore 1981). Researchers do not always describe it as such, but the equation they generally use to depict the process is the potential function for the cusp; the only difference is that sometimes the researchers hold the bifurcation variable constant, rather than treating it as a variable that is manipulated or measured in the study.

Figure 7 is a common expression of the phase shift. The little ball, which indicates the state the system is in, is stuck in a low-entropy well that represents an attractor. When sufficient energy or force is applied, the ball comes out of the well, and with just enough of a push it moves into the second well. In some situations we know what well the system is stuck in, but not necessarily the nature of the well it needs to visit next. The question of how to form a new attractor state is a challenge in its own right.

The essence of emergence is that, 'The whole is greater than the sum of its parts'. This maxim has been attributed to Aristotle (Gorman 2014), the Gestalt psychologists of perception, and Durkheim (Sawyer 2005). For Durkheim, a scientific study of sociology needed to study phenomena

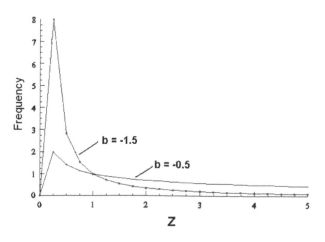

Figure 6. Two power law distributions with negative shape parameters.

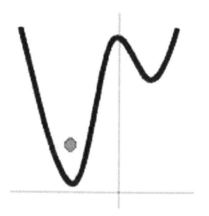

Figure 7. Two-well phase shift.

that could not be reduced to the psychology of individuals. The essential solution was as follows: The process starts with individuals who interact, do business and so on. After enough interactions, patterns form and become institutionalised or become institutions. When an institution forms, it has a top-down effect on the individuals such that any new individuals entering the system need to conform to the demands, which are hopefully rational, of the overarching system.

Emergent events can be light or strong. In the light version, the overarching structure forms but does not have a visible top-down effect. In the strong situations, there is a visible top-down effect. Two further types of emergence are frequently observed in live social systems. One is the phase shift dynamic. The second is the avalanche dynamic that produces $1/f^b$ relationships. Physical boundaries also place constraints on emerging phenomena. For instance, the social dynamics of a group of people can be shaped by the location of furniture in a room.

2.6. Synchronisation

The dynamics of synchronisation follow from those of self-organisation and emergence generally and driver–slave relationships more specifically. Whereas a driver-slave relationship is unidirectional, synchronisation phenomena are bi-directional or N-directional. A prototype illustration is synchronisation of a particular species of fireflies, as told by Strogatz (2003): In the early part of the evening the flies flash on and off, which is their means of communicating with each other, which they do at their own rates. As they start to interact, they pulse on and off in synchrony so that the whole forest lights up and turns off as if one were flipping a light switch. The common features among the many other examples are that synchronised systems contain two oscillators, a feedback loop between them, and a control parameter that speeds up the oscillation. When the speed reaches a critical level, *phase lock* sets in and the synchronised pulsing can be observed. Synchronisation has become an interesting component of team coordination dynamics, as discussed in Sections 4.9 and 4.10.

2.7. Agent-based Modelling

If a system contains thousands of agents that interact randomly, it becomes virtually impossible to calculate all their outcomes from those interactions at an individual level. Thus, computer programs known as agent-based models were developed to determine final system states – evolutionarily stable states – given the complexity of all the agents' interactions (Axelrod, 1981; Maynard-Smith, 1992). Agents interact according to specific rules defined

within the theory that is being programmed. Agent-based models are closely related to several other computational systems that illustrate self-organisation dynamics. Some of the principles are developed by Axelrod and Maynard-Smith surface in the theory of team coordination (Section 4.9).

2.8. The complex adaptive system

The concept of a *complex adaptive system* (CAS) can describe virtually any living system, although in the present context we are thinking primarily of multiple interacting person–machine systems. The CAS originated in biology (Gell-Mann 1984), and, in some circles, has become the new dominant model of organisational behaviour (Anderson 1999; Allen, Maguire, and McKelvey 2011; Dooley 1997, 2004). The perspective incorporates the NDS concepts that have been described already to study patterns of behaviour as they unfold over time: how the system recognises signals and events in the environment, harnesses its capabilities to make effective responses, changes its internal configurations as new adaptive responses require and interacts with the external environment.

The *schema* is the building blocks of the behaviour patterns, as seen in the development of psychomotor skill (Newell 1991). Schemata also bear a strong resemblance to perception-action sequences as defined in Gibson's (1979) ecological perspective on perception. Schemata also depict rules of interaction with other agents that exist within a system or with agents outside the system's boundaries (Dooley 1997). A team's schemata are often built from existing building blocks that are brought into the group when members arrive. Dominant patterns emerge from interactions among agents, and they often have a top-down influence on incoming agents. Although the individual schemata self-organise into one or more supervening mental models, there can be individual differences remaining that could provide enough entropy for further modification of the schemata.

Schemata change through mutation, recombination and acquisition of new ideas from outside sources. Change occurs in response to both changing environments and changing internal conditions. Indeed, there are numerous sources of entropy that could arise from changes in other agents in the environment, the capabilities of the incoming work force, technologies, products, markets and governmental regulations (Bailey 1994; Bigelow 1982; Guastello 2002). When schemata change, requisite variety, robustness and reliability are ideally enhanced (Dooley 1997). *Reliability* denotes error-free action in the usual sense. *Robustness* denotes the ability of the system to withstand unpredictable shock from the environment. *Requite variety* refers to Ashby's (1956) Law: For the effective control of a system, the complexity of the controller must be at least equal to the complexity of the system that is being controlled. *Complexity* in this context refers the number of system states, which are typically conceptualised as discrete outcomes.

The reliability and robustness of a system are not guaranteed. Schemata and the CAS as a whole undergo evolutionary and sometimes revolutionary change as their schemata self-organise further to enhance fitness and functionality. Increased levels of contradictions within the system's schemata are precursors of higher entropy levels in system performance, which are in turn precursors of an impending phase shift. This principle has been identified as symmetry-breaking (Prigogine and Stengers 1984; Sulis 2009).

A complex *adaptive* system remains poised 'on the edge of chaos' or far-from-equilibrium conditions ready to reorganise itself in response to new demands. The sequence of states through which it reorganises are relatively unpredictable, although at some point it should be possible to envision possible future scenarios and states; again some people are better at system forecasting than others (Heath 2002). A team's particular sequence of stages is often subjected to initial conditions which contribute to the global unpredictability of the system. The states of team organisation are irreversible once they have taken hold and stabilised (Dooley 1997). Although teams can redeploy old schemata, the effect is not the same because of the history that accumulated, events that occurred and time that has elapsed.

3. Nonlinear methods

The constructs described here are not simply enchanting metaphors. They are analytic, meaning that it is possible to assess empirically which dynamics are occurring when, how and under what conditions. There are equations representing these phenomena that can be computed with real data assessed statistically.

Nonlinear methods can be organised into three broad categories that are associated with the objectives of mathematicians, biologists and physicists, and social scientists (Guastello 2009a). The social scientists' perspective on data handling falls closest to the concerns for research in ergonomics. Their strategy arises from their history of developing linear models and separating them from 'error'. A nonlinear model, chaotic or otherwise, also needs to be tested statistically and separated from residual variance. That said, we can provide an overview of the analyses that have had the largest impact on nonlinear human science, some of which are fundamentally mathematical, and others derive from extensions of statistical theory. The interested reader should see Guastello and Gregson (2011) for expansions on each technique.

3.1. *Phase-space analysis*

A phase space diagram is a picture of a dynamical process that can tell us quite a bit about the behaviour of a system that follows a particular function. It is usually drawn as a plot of the change in X (ΔX) at time $t + 1$ as a function of X at time t – velocity versus position. The attractors shown in Figure 4 are phase space diagrams. There are a few other varieties of phase space diagram such as plots of two variables and the *state space grid* (Figure 8; Hollenstein 2007).

Phase-space analysis for real data has been problematic, however, because it does not provide any inferential statistics by itself, and the look of the diagram can be seriously affected by noise, and projection in an inadequate number of spatial dimensions. Problems of noise and spatial projection are generally handled through combinations of filters and analysis such as *false nearest neighbors* (Shelhamer 2007). Problems with statistical inference are typically handled a few different ways. One can start with non-statistical calculations such as the correlation dimension, the Lyapunov exponent, or the Hurst exponent on a time series (see the next sections of this article); from there one can compare the results of a single target time series against surrogate samples. If experimental conditions are involved, compare the various metrics found in each condition. Another class of alternatives is to use a statistical analogue of the metric.

3.2. *Correlation dimension and surrogate data analysis*

The *correlation dimension* is a computation of a fractal dimension for a time series. The 'correlation' in its name reflects that it is looking for long-range patterns of behaviour and measuring their structure (Grassberger and Procaccia 1983). The basis of the procedure is to treat the time series as a complicated line graph and cover it with

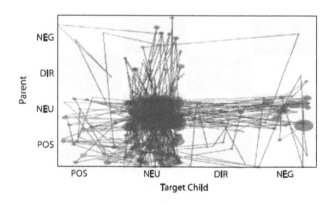

Figure 8. State space grid depicting a high-entropy and high conflict relationship, from Dishion et al. (2012, 346). Source: Reprinted with permission of the Society for Chaos Theory in Psychology and Life Sciences.

circles of radius r. Count the number of circles required, then change the radius and repeat many times. Then correlate the log of the number of circles required with the log of the radius. The result will be a line with a negative slope. The absolute value of the slope is the fractal dimension.

Some of the earliest NDS studies that used correlation dimensions attempted to draw a conclusion about the underlying dynamics from a single time series. On the one hand, researchers correctly observed that there was much more information in a time series compared to what was possible with conventional experimental designs. On the other hand, the problem of making generalisations from case studies still persists. The solution was to develop surrogate data by generating many time series by reshuffling the observations in the original time series. This step would preserve means and variances, but would disrupt the serial dependency among the observations. One then compares the correlation dimension (or other metric) from the target time series with those taken from the surrogate samples, and computes a one-sample z or t test to determine if the value from the target series could have occurred by chance.

3.3. Lyapunov exponent and entropy

In light of the large number of known chaotic systems, researchers in applied contexts usually focus on the generic properties of chaos and how it might be differentiated from other nonlinear dynamics. The test for chaos is currently performed with the Lyapunov exponent, which measures the amount of turbulence in a time series. It is calculated from sequential differences in values of the behavioural measurement and the extent to which the differences expand and contract. Larger values indicate faster information loss in a system as one attempts to predict the state of the system into many points into the future. There is also a statistical analogue of the Lyapunov exponent that can be calculated through nonlinear regression.

Chaotic time series contain both expansions and contractions. The Lyapunov exponent was first conceptualised as a spectrum of values calculated by taking a moving window of three observations and compiling them across the entire time series. The largest value in the spectrum is the most critical. If it is positive, the series is expanding and potentially chaotic. Furthermore, there would be many negative values corresponding to the contracting component. If the largest value is negative, however, the series is contracting only, and veering towards a fixed point. If the largest value is 0, then the system is a perfect oscillator. A slightly negative largest exponent indicates a dampening oscillator, which tends towards a fixed point eventually. A slightly positive largest exponent is aperiodic, and could be moving into more interesting regimes such as SOC, Levi flights and chaos.

The positive values of the Lyapunov exponent, λ, are more readily interpreted by converting it to a fractal dimension, $D_L = e^{\lambda t}$; where t = time units elapsed, which are usually set equal to 1.0. D_L is usually called a *Lyapunov dimension* when it is calculated in this manner. The interpretation of dimension values given λ in section 2.2 applies here.

3.4. Hurst exponent

The Hurst exponent, H, was first devised as a means of determining the stability of water levels in the Nile River. Water flows in and flows out, but does the net water level remain relatively stable or does it oscillate? To answer this question, a time-series variable, X with T observations is broken into several subseries of length n; n is arbitrary. Then for each subseries, the grand mean is subtracted from X, and the differences summed. The range, R, is the difference between the maximum and minimum values of the local means. Then rescaled range, $R/\sigma = (\pi n/2)^H$, where σ is the standard deviation of the entire series; $0 < H < 1$. Actual values of H may vary somewhat due to the choice of n and the time scale represented by the individual observations.

A value of $H = 0.5$ represents *Brownian motion*, meaning that the deflections in a time series can be upward or downward with equal probability on each step. Brownian motion has a Gaussian distribution and is nonstationary. Values that diverge from 0.5 in either direction are non-Markovian, meaning that there is memory in the system beyond the first step prior to an observation.

$H > 0.5$ denotes *persistence*. Deflections in a time series gravitate towards a fixed point if they are high enough. The autocorrelation of observations in a time series is positive. Values in the neighbourhood of 0.75 represent pink noise or self-organising systems. $H < 0.5$ denote anti-persistence; upward motions are followed by downward motions, and the autocorrelation is negative.

Negative values of H have also been reported (Guastello et al. 2014). When they occur, they are indicative of hysteresis around a bifurcation manifold, such as the one contained in a cusp catastrophe. Oscillations and hysteresis may appear similar, but topologically they are not.

3.5. Recurrence quantification analysis and plots

Although chaotic functions are said to produce non-repeating patterns, repeatability is really a matter of degree. Visualising what repeats and how often can provide new insights into dynamics of a process. A recurrence plot starts with a time series. The user specifies a *radius* of values that are considered to be similar enough and a *time lag* between observations that should be plotted. If the same value of $X(t)$ appears at $X(t + 1)$, a point is plotted.

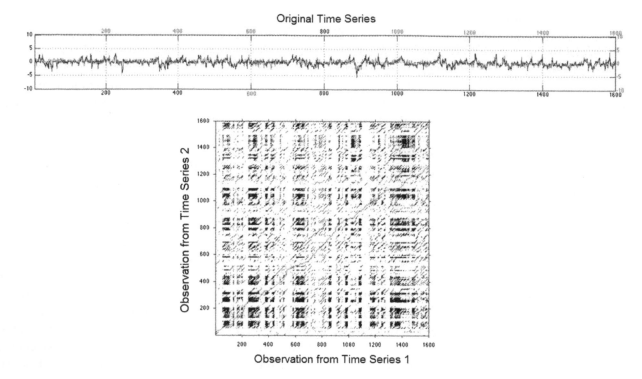

Figure 9. Example of a cross-recurrence plot generated from two galvanic skin responses times series.

The variable itself appears on the diagonal. One can also examine the *cross-recurrence* between two variables by substituting a $Y(t + 1)$ for $X(t + 1)$.

White noise would produce a plot containing a relatively solid array of dots with no patterning. An oscillator would produce diagonal stripes. More interesting deterministic functions would produce patterns; an example of a cross-recurrence pattern that was generated from two galvanic skin response series appears in Figure 9. One can then calculate metrics that distinguish one pattern from another such as the percentage of total recurrences, percentage of consecutive recurrences and the maximum length of the longest diagonal (shorter is more chaotic). A time series is interpreted as being more deterministic to the extent that more points line up on the central diagonal. The patterning of other points is often described as pink, or $1/f^b$, noise.

3.5. Nonlinear statistical theory

There is another genre of analysis that starts with a relatively firm hypothesis that a particular nonlinear function is inherent in the data. The function could be chaotic, reflect other types of dynamics or perhaps be flexible enough to discern the underlying dynamics from the results produced. Consistent with four goals of inferential statistics, the analyses are devised to predict points, identify the nonlinear dynamics in the data, report a measure of effect size and determine statistical significance of the parts of

the respective models. Other fundamental issues concern probability density functions, the structure of behavioural measurements and the stationarity of time series.

Students of the behavioural sciences in the late twentieth century were taught to divide their probability distributions into four categories: normal (Gaussian) distributions, those that would be normal when the sample sizes became larger, those that would be normal if transformations were employed and those that were so aberrant that non-parametric statistics were the only cure. In the NDS paradigm, however, one is encouraged to assume specific alternative distributions such as power law, exponential and complex exponential distributions. Differential functions can be rendered as a unique distribution in the exponential family, and the presence of an exponential distributions or a power law distribution in a time series variable denotes a nonlinear function contained therein.

The classic notion of a mental or behavioural measurement, following from Lord and Novick (1968) is that a measurement consists of two components – a true score and an error score. The true score is not really observable because of the presence of error. Errors are thought to be normally distributed with a mean of 0, and are uncorrelated with any other errors, true scores, or anything else. The two elements decompose, however, when the same observed measurement is put into a time series. The true score follows a function of some sort that could be linear but could also be nonlinear. Locally linear functions are often embedded in globally nonlinear functions to different degrees. The error score divides into two components: the classic independent error and dependent error. The independent error component is called IID, or *independently and identically distributed error* when it occurs in a time series.

Dependent error in a dynamical process might occur as follows: Imagine we have a measurement that is iterating through a process such that $X_2 = f(X_1)$. The function produces X_3, X_4, etc. in the same way. Now let a random shock of some sort intrude at X_4. At the moment the shock arrives, the error is IID. At the next iteration, however, $X_5 = f(X_4 + e)$; the error continues to iterate with the true score through the ensuing time series.

Dependent error also occurs when errors are autocorrelated or when complex lag effects are in play such that $X(t) = f[X(t-1), X(t-3)]$. Fortunately, there is a proof showing that the presence of dependent error in the residuals of a linear autocorrelation is a positive indicator of nonlinearity (Brock et al. 1996). Furthermore, it follows that, if a good-enough NDS function is specified for the iterations of the true scores, the dependent error is reduced to a nonsignificant level (Guastello 1995).

Specification of a good-enough nonlinear function requires that the dynamics for the phenomenon have been theorised in sufficient detail. Although there are no automatic procedures for finding the optimal nonlinear model for a problem, there are some good options that have some reasonable flexibility. Consider two examples:

$$\Delta z = \beta_0 + \beta_1 z_1^3 + \beta_2 B z_1^3 + \beta_3 A \tag{3}$$

$$Z_2 = \theta_1 B z_1 * \exp(\theta_2 z_1) + \theta_3 \tag{4}$$

where z is the dependent variable (order parameter) that is measured at two or more points in time, A and B are control (independent) variables, β_i are linear regression weights and θ_i are nonlinear regression weights. For models similar to Equations 3 and 4, data are prepared by setting a zero point and a standard calibration of scale for all variables in the model. Any error associated with approximating a differential function from a difference function becomes part of $[1-R^2]$.

Equation 3 is a cusp catastrophe executed with ordinary polynomial regression. There are also alternatives for evaluating catastrophe models by determining the extent to which the data fit a probability density function that is unique to the model. This method might be chosen in cases where the time-1 value of the order parameter z is 0 for all cases, or if the data-set has been collected to measure the behaviour at one point in time only (Guastello 2013; Guastello and Gregson 2011).

Equation 4 is part of a series of exponential models (Guastello 1995; 2011). In Equation 4, θ_2 corresponds to a

Lyapunov exponent; if θ_2 is positive and θ_3, the constant, is negative, then both expanding and contracting properties exist in the data-set. That combination of results in turn denotes chaos. One can then convert θ_2 into a Lyapunov dimension to interpret the dynamics further.

There is also a series of models for identifying oscillators and coupled oscillators (Boker and Graham 1998) that starts with the assumption that the time series contains an oscillator:

$$d^2 X(t)/dt^2 = \beta_1 (dX/dt) + \beta_2 X \tag{5}$$

The left-hand side of Equation 5 contains an acceleration of the time-series variable. The right-hand side contains its velocity and position. The model can be modified by adding components, such as a cubic term denoting a hard or soft spring or influences from another oscillator (Butner and Story 2011).

It should be emphasised that the statistical analyses are intended for *modelling, not curve-fitting*. Of course one wants to see statistical significance and large effect sizes, but the target nonlinear model should be compared against alternative theoretical models. The alternatives are often linear but not always so. The comparisons would involve not only effect size, but also whether the components of the model that represent the critical dynamics are all present.

3.6. Markov chains

Markov chain analysis originated outside the realm of nonlinear science, although it was intended for problems that are very similar to some found in NDS. The central idea is that the researcher is observing objects or people in a finite set of possible states. All objects have odds of moving from one state to another. The State X State matrix of odds is known as a *transition matrix*. The goal of the analyses is to establish the transition matrices and to predict the arrival of objects into states.

Some combinations of transitions result in fixed points, oscillators, bifurcations and chaotic outcomes (Gregson 2005, 2013; Merrill 2011). Thus, Markov chain analysis now falls within the scope of nonlinear dynamical analyses for categorical data.

3.7. Symbolic dynamics

Symbolic dynamics is an area of mathematics that finds patterns in a series of qualitative data. The elementary patterns themselves can be treated like qualitative states and analysed for higher order patterns. The class of techniques is ideally suited to analysing chaotic and related complex nonlinear dynamics, either in the form of continuously valued time-series or qualitative categorical data. Symbolic dynamics are particularly useful in situations where discontinuities, continuities, periodic functions and unnamable transients could co-exist within a relatively short time series.

For continuous data, events such as spikes and small or large uptrends and downtrends are coded nominally (eg with letter codes A, B, C, D, etc.) and then analysed. The computational procedures often include entropy metrics. For categorical data, the categories, which should be defined in a theoretically relevant way, are also given letter codes, and the computations are carried out the same way and produce the same metrics.

The algorithms that are available vary in their means for determining symbol sequences and the length of those sequences. The *orbital decomposition* procedure (Guastello, Hyde, and Odak 1998; Guastello, Peressini, and Bond 2011; Peressini and Guastello 2014), for instance, begins with the assumption that the time series *could* be chaotic, but chaos is not specifically required. According to a theorem by Newhouse, Ruelle, and Takens (1978), three coupled oscillators are minimally sufficient to product chaos. The central algorithm disentangles periodic patterns, and it is particularly good for empirically determining the optimal pattern length and isolating dominant patterns. It includes statistical tests for determining the extent to which the set of patterns isolated by the analysis deviate from chance levels of occurrence. Two of its application areas that are of interest to ergonomics are communication patterns in groups or teams (Guastello 2000) and task switching (Guastello et al. 2012).

4. Selected applications

Applications of NDS span *most* of the range of topics that are of current concern in ergonomics: psychophysics, visual displays, controls, neurocognitive processes, psychomotor functioning, stress and fatigue, accident analysis and prevention, and artificial life and complex systems. The state of the science on these matters is still evolving (Guastello 2014a). The selections that follow collectively emphasise a broad range of nonlinear principles.

4.1. Nonlinear psychophysics

Some of the more perplexing problems in psychophysics involve the detection of signals that appear and disappear irregularly as either the signal source moves through space or as the operator moves about. The prototype of the operator sitting in front of a display screen to which all the events are confined is only representative of a portion of real-world events. If the signal has multiple properties that need to be present, the detection challenge increases in complexity as well.

Nonlinear psychophysics (Gregson 1992, 2009) addresses this broad class of problems. Its centrepiece is the gamma function:

$$\Gamma: Y_{j+1} = -a(Y_j - 1)(Y_j - ie)(Y_j + ie) \qquad (6)$$

(Gregson 1992, 20). Equation 6 states that the strength of a response to a stimulus, Y, at time $j + 1$ is a function of the response at a previous point in time, j, a control parameter representing physical signal strength a, a situational control parameter e, and the imaginary number $\sqrt{-1}$. Response strength might be measured as a subjective numerical rating, although the preferred method would be to measure response time. The shorter response times will be required for stronger stimuli. It is noteworthy that as the signal strength becomes sufficiently weak, the pattern of responses over time becomes chaotic.

Equation 6 can be expanded to two response parameters that may be useful for modelling cross-modality responses, such as interpreting a hue of a coloured light from the perception of its brightness only, or studying the size–weight illusion. Two-dimensional outcome expansion is accomplished by substituting a real number, x, for i:

$$X_{j+1} = a(e^2 - e^2 X_j + X_j^2 - Y_j^2 - 3X_j Y_j^2 - X_j^3) \qquad (7)$$

$$Y_{j+1} = aY_j(-e^2 + 2X_j - 3X_j^2 + Y_j^2) \qquad (8)$$

(Gregson 1992, 44). Γ is further expandable into multidimensional inputs and outputs that were applied to real situations such as perceiving the taste of wine.

If one were to plot the perceived signal strength Y against ranges of a and e, one obtains not only the familiar ogive, but an escarpment of ogives with varying degrees of steepness (Gregson 2009). If one were to follow changes in Y over time as a and e vary, one can observe hysteresis around the underlying absolute threshold. Hysteresis around the threshold was known since Weber and Fechner; the solution in classical psychophysics was to use the average value of the two stimulus strengths that correspond to the sudden upswings and sudden downswings in response. Signal detection theory eliminated the hysteresis by randomising the stimuli that were presented to the observers. Randomised stimuli are probably not representative of most real-world events; chaotic or other deterministic flows of stimuli would be more likely.

Parameters a and e, therefore, are not restricted to constant values. They can be a nonlinear time-series variables as well. In those cases, the dynamics of X would be globally unstable as the respondent would go through periods of conscious and unconscious awareness of stimulus changes (Gregson 2013). The situation opens new frontiers in psychophysics research that would involve analysing properties of important stimulus sources, such as continuous camera feeds for vigilance tasks, and devising experiments in which humans respond to analogous stimulus flows with similar properties.

4.2. Nonlinear control theory

The effect of control actions in person–machine system interactions can be discrete, linear, exponentially growing or decaying, or oscillating. Control systems make use of feedback loops, velocities and accelerations. Thus, the stage has been set for a long time to make a smooth transition to nonlinear systems of the types considered here (Jagacinski and Flach 2003). For instance, multiple coupled oscillators can be overserved in rhythmic movements. One can further imagine the potential for chaotic conditions as three or more loosely coupled person–machine systems start to work together.

Modern control systems are growing in complexity, as evidenced by the use of multi-modal control systems and mode errors. How complex should a system be? According to Ashby's (1956) Law of Requisite Variety, a controller should be at least as complex as the system it needs to control. Some of the other applications considered next, however, indicate that complexity beyond requisite variety produces inefficiency.

4.3. Cognitive workload and fatigue

Increasing proportions of the work done each day by millions of people involve cognitive labour. Although computerisation can reduce work to some extent, it can generate new sources of fatigue and workload, particularly if people need to keep up with a fast flow of incoming data or task requests, or to keep up with automatic machine controls that seem to have a mind of their own. The negative effects of cognitive workload and fatigue on performance have been difficult to separate historically because they can both occur simultaneously, even though they might have different underlying dynamics. To complicate matters, coping, automaticity, practice and fatigue recovery have positive effects on performance over time simultaneously with increasing load and fatigue. For extensive background on the problems, see Hancock and Desmond (2001), Ackerman (2011), Matthews et al. (2012), and Hancock (2013). Operators can adapt to changing workloads in order to maintain desired performance levels, but after a point the potential for adaptation ends and catastrophic declines in performance occur (Hancock and Warm 1989). Operators can adapt to fatigue and boredom by switching tasks, but task switching incurs additional load on working memory (Guastello et al. 2012).

Furthermore, the performance decrement that occurs with fatigue can be relatively gradual or severe, depending on the level of workload involved.

The cusp catastrophe models for cognitive workload and fatigue (Figure 4) and their supporting research programme evolved in response to this nexus of difficulties. The overall objective is to separate the two processes, which have the same temporal dynamic structure but different contributing variables, using an integrated experimental design that tests them both in the same situation. The tasks that have been studied were chosen to capture an array of cognitive processes and to find what was generalisable about the control variables: an episodic memory task (Guastello et al. 2012), a pictorial memory task that required verbal retrieval cues (Guastello et al. 2012), perceptual-motor multitasking (Guastello et al. 2013), a vigilance dual task (Guastello et al. 2014) a financial decision-making task that captured both optimising and risk taking behaviour (Guastello 2016a) and an N-back task (Guastello et al. 2015).

Each of the workload models was paired with a fatigue model. The common feature of the experimental designs was to define starting and ending conditions that did not confound changes in workload with changes in duration of work. Some experimental procedures produced more of a workload effect than a fatigue effect and vice versa.

The model for cognitive workload invokes the concept of Euler buckling (Guastello 1985; Zeeman 1977). A piece of material that is subjected to sufficient amounts of stress in the form of repeated stretching will show a certain amount of deformity, or strain. Rigid materials break, whereas flexible materials rebound. In Figure 4, performance or response time would be the dependent variable, y. The amount of vertical weight is the asymmetry (a) parameter. The modulus of elasticity of the material is the bifurcation factor (b), with low elasticity located at the high end of the bifurcation axis. Psychological constructs that could be candidates as elasticity-rigidity variables should reflect adaptability versus rigidity.

Fatigue is defined as the loss of work capacity over time for both cognitive and physical labour (Dodge 1917; Guastello and McGee 1987). Depletion of work capacity is typically observed as a *work curve* that plots performance over time. Performance *usually* drops sharply under fatigue, but not everyone experiences a decline as result of the same expenditures, however. Some show an increase in physical strength akin to 'just getting warmed up', while others show consistently higher or lower performance levels for the duration of the work period. Fatigue is also accompanied by a higher degree of variability in performance. The cusp response surface (Figure 4) accounts for the full range of possible work curves. Work capacity displays the two stable states. Change in capacity is implied by change in performance. Psychological disengagement, or a drop in motivation, also contributes to performance decrements, although the drop in motivation is also symptomatic of physical or mental fatigue (Balagué et al. 2012; Guastello et al. 2012; Hockey 1997, 2011). At some point, the individual wants to stop working or switch to a different task.

The total quantity of work done between two measurement points would be the primary contributor to the bifurcation parameter: If the individual did not accomplish much in the time allotted, there would be little drain on work capacity. Those who accomplished more work could exhibit either positive or negative changes in work capacity.

The asymmetry parameter would be a compensatory strength measure. In the prototype example, labourers displayed changes in arm strength as a result of about two hours' worth of standard mill labour tasks, which primarily demanded arm strength. Leg strength, however, acted as a compensation factor for arm strength; those with greater leg strength experienced less fatigue in their arms (Guastello and McGee 1987). A similar effect is thought to occur in cognitive work.

The earliest efforts to identify compensatory abilities for the cognitive fatigue model stuck close to the physical labour prototype where the fatigue measure was a drop in capacity and the compensatory ability was a related capacity. Later efforts examined performance differences as the index of fatigue, and the abilities were not always so indirect, eg speeded arithmetic ability relative to the financial decisions task. Meanwhile, theoretically driven research was showing that working memory is part of fluid intelligence (Kane, Hambrick, and Conway 2005; Nusbaum and Silvia 2011). Thus, measures from the fluid domain such as anagrams and algebra flexibility were brought into the mix for testing. Field independence and algebra flexibility now seem to do double-duty as flexibility measures relative to workload and a compensatory ability relative to fatigue.

4.4. Elasticity and rigidity

The cognitive constructs that are part of the bifurcation variable for the workload model have been isolated as a separate subsection here because there is good reason to think that that they could be operable in other situations where the efficacy of a complex adaptive system would be involved. Variables that qualify as elasticity-rigidity constructs have some supporting rationale as bifurcation variables. One pole is associated with positive and negative discontinuities, whereas the other pole may be associated with gradual change or no change in performance at all. Some have contributed directly to the cusp models tested thus far, whereas others were only part of linear comparison models or indirectly related to workload through subjective ratings (Guastello et al. 2015). Five groups of constructs evolved as the project progressed:

(1) *Trait anxiety* can interfere with lucid decision-making, but it can also focus attention on details that others might miss. At present, anxiety only seems to be operative in contexts with interpersonal challenges or physical hazards. *Emotional intelligence* (EI) facilitates understanding of one's own emotions and the emotional messages from other people and forming appropriate actions in response. Low EI denotes rigidity in the form of indifference, which could be a buffer against stress effects. When stress gets too high, however, the system buckles and snaps in the form of poor decisions (Thompson 2010). *Frustration* can have a negative impact on performance making tough situations worse, but it can also spur the individual onward to work harder or differently.

(2) *Conscientiousness* is a trait that predisposes one to attention to details, rules and task orientation, and thus implies a type of rigidity. Flexibility or adaptiveness is not expected. *Work ethic* is thought to function in approximately the same way. *Impulsivity* is a facet of conscientiousness and reflects a tendency towards elasticity.

(3) One construct of *coping flexibility* is centred on emotional adjustments in the sense of long-term life issues (Kato 2012). More flexible people have a broader repertoire of coping strategies they can use. Another type of coping is oriented towards cognitive strategies such as planning, monitoring, decisiveness and inflexible responses to changing work situations (Cantwell and Moore 1996). So far the latter was found more closely related to cognitive workload dynamics.

(4) *Field independence* is a cognitive style that separates perceptions of a figure from a background. It was also proposed the field independent people use more of their working memory capacity (Pascual-Leone 1970). As such it worked well as a bifurcation variable in studies of problem-solving in chemistry (Stamovlasis and Tsaparlis 2012) and financial decision-making (Guastello 2016a).

(5) Other degrees of elasticity are inherent in the task structure such as whether operators can choose how to *sequence subtasks*.

4.5. Biomechanics of material handling

The buckling model for workload also provides a viable explanation for back injuries resulting from heavy lifting (Karwowski, Ostaszewski, and Zurada 1992). The load parameter is calibrated as kg/cm² of downward pressure on the spine as one lifts a heavy object from the floor or an elevated platform. Pressure is maximum at the moment when the individual stands upright with the object.

Elasticity is inherent in the material comprising the spinal column. Its capacity range can vary as a result of regular exercise, prior injury, brittleness of bones or the person's height relative to the lifting platform. Note the paradox here: Having the arm strength to lift a heavy object can be detrimental if the spine is too rigid to support the peak load. The flexibility in the spine as it moves during a lift has been determined to be chaotic (Khalaf, Karwowski, and Sapkota 2015).

A related problem is the stability of one's balance while carrying the load, particularly while walking across a narrow beam, rather than an ordinary floor. This is common but hazardous occurrence in the construction industry that could lead to falls from heights up to 5 m when individuals lose their balance. Gielo-Perczak, Karwowski, and Rodrick (2009) thus evaluated the natural variability in the centre of pressure that registers on a person's foot while carrying loads, using the Lyapunov exponent. They found greater volatility (deterministic chaos) in the centre of pressure when beams were narrower and loads were heavier. The increased variability is reflective of an adaptive response, but also a tendency towards instability when it becomes too great. The authors also remarked that a steady centre of pressure requires attentional resources that can be compromised by multitasking.

Surface electromyographs (EMG) can detect signs of muscle fatigue. As seen in other context, variability of the wave amplitudes is part of fatigue in addition to the decline in the capacity to perform. Rodrick and Karwowski (2006) hypothesised that the variability is an outgrowth of a deterministic nonlinear process. They compared Lyapunov exponents and fractal dimensions for EMG taken from participants in a lifting task and found that there were significant differences in both nonlinear metrics depending on the lifting posture. The task itself involved only differences in lifting posture and not external load. Future research could expand on these findings to establish load–fatigue relationships.

4.6. Degrees of freedom in movement and cognition

The principle of degrees of freedom underlies several types of self-organising dynamics. The idea was first introduced to explain the control of physical movement (Bernstein 1967; Turvey 1990). If all the movements of the marionette in Figure 10 were under control of an executive controller, it would take as many degrees of freedom to effect a movement as there are stings shown in the left diagram. If the system were self-organised, however, a much smaller number of degrees of freedom would be needed to create the same movement.

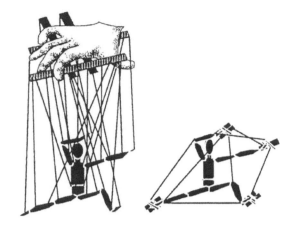

Figure 10. Degrees of freedom in the control of movement under complete executive control (left) and as a self-organised system (right). Source: Reprinted from Turvey (1990) with permission of the American Psychological Association.

The same idea appears to operate in some cognitive processes. When a task is first learned, several possible ways of performing are explored; some degrees are perceptual-motor, others are cognitive. Still others are inherent in the task itself, which might permit re-ordering or combining steps. Eventually, a path of least effort is discovered and adopted for continued use; this is *minimum entropy principle* (Hong 2010). The system self-organises with practice; the involvement in the executive function of working memory can be reduced, and task execution becomes more automatic (Lorist and Faber 2011). Efficiency conceived in this fashion reduces the variability of performance, but some variability needs to remain to facilitate further adaptations. Adaptation to work load could involve a reduction in executive control, which might not be beneficial.

The increase in performance variability and the performance decline signal two possible events. Either a strict minimum entropy effect ensues whereby the individual stops work, or a *redistribution* of degrees of freedom (Hong 2010) occurs so that the individual does something differently, uses different degrees of freedom configurations and gets a 'second wind'. Figure 11 depicts the performance of 105 people performing seven different tasks seven times each in four possible orders. All bends in the curve were statistically significant according to the trend analysis. Here we see the beginning of a performance drop at trial 4, a recovery and new peak at trial 6 and a second decline.

4.7. The performance-variability paradox

The minimum entropy principle holds that, as one learns to do a task, the learner finds ways of making physical and mental motions as efficient as possible with a minimum of wasted motion or decision time. Guastello et al. (2013) considered two conflicting influences on performance as it

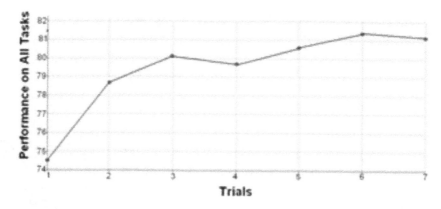

Figure 11. Performance on seven trials of seven tasks in four possible orders. Source: Reprinted from Guastello et al. (2013, 35) with permission from the Society for Chaos Theory in Psychology and Life Sciences.

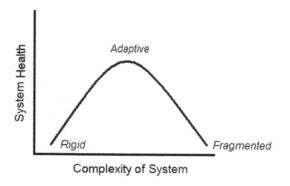

Figure 12. Hypothetical relationship between system variability or complexity and system health.

unfolds over time. First, 'Best' performers are also expected to be consistent performers, but some variability is needed to remain adaptable to chance events as they arise. Ashby's law of requisite variety is promoting the maintenance of high levels of internal complexity in order to response effectively to the demands of incoming stimuli. Variability is also necessary if performance capabilities are ever going to improve further. Second, *voluntarily* switching tasks can reduce fatigue, but it incurs a workload cost because of the added quantity of information we need to keep active in our working memories. System-driven switching, however, is likely to increase fatigue, (Hancock 2007).

When left to their own discretion, the participants in a study that involved seven perceptual-motor tasks gravitated towards four different task-switching strategies (Guastello et al. 2012): (a) *task first,* in which they completed all of one type before moving on to another; (b) *Set-first,* in which they completed set of (sub)tasks then repeated the sequence; (c) a *mixed strategy* that appeared to include patterns from task-first and set-first strategies; and (d) a *random strategy.* Given the way the study was designed, task-first produced the smallest number of task switches, and a relatively low level of entropy with respect to patterns of task choices. Set-first produced the smallest level

of entropy, but the largest number of switches. The analysis of performance results reflected a trade-off between the switches and entropy; the best performers minimised both to the extent possible. On the one hand, only about 50% of the participants engaged in one of the two efficient strategies. On the other hand, the exploration of the tasks was probably a motivating factor for some of the participants. The pattern extraction was performed by the orbital decomposition method. Abilities and elasticity-rigidity variables added to the prediction of performance in the multitasking experiment.

4.8. Optimum variability

The principles that produced the performance-variability paradox – Ashby's Law and minimum entropy – produce a closely related phenomenon, that of optimum variability. The difference is a shift in emphasis to an inverted-U relationship between the level of system complexity and the health or functionality of the system (Figure 12).

The principle of optimum variability originated with Goldberger (1991, 1997) and Cecen and Erkal (2009) who discovered that healthy heartbeats are chaotic over time, not rigidly periodic. Rather, a cardiac infarction is imminent if the electrocardiogram reveals strict periodicity. The principle generalises to a range of medical pathology problems (Bianciardi 2015; Vargas et al. 2015), psychomotor performance in sports and rehabilitation (Harrison and Stergiou 2015; Kiefer and Myer 2015; Morrison and Newell 2015) and cognition and work performance more generally (Corrêa et al. 2015; Navarro and Rueff-Lopes 2015). The condition of too much complexity has not been isolated in every instance. When it has been found, however, it is indicative of a poorly controlled system, such as psychomotor function after a patient has experienced a stroke. There are several ongoing research efforts to develop robotic tools that can force neuroplasticity towards and through neural pathways that remain undamaged (Guastello, Nathan, and Johnson, 2009).

The measurement of variability in the optimum variability studies varies by type of data and the options available at the time the research was conducted. Fractal dimensions and entropy metrics are often used (Schuldberg 2015). Although it could be tempting to compare standard deviations, the standard deviation does not capture any of the temporal dynamics in sequential observations, nor can it render a measurement of complexity that is scale-free, such as the fractal dimension. Scale-free metrics are independent of the actual standard deviation of the sample (Guastello and Gregson 2011).

4.8. Occupational accidents

Given the impact of stress on cognitive functioning, it is not unreasonable to surmise that stress has a substantial impact on industrial accident rates as well. If we all lived in rubber rooms, the consequences of stress would be limited. Ironically, the extant literatures on stress and environmental hazards have only considered one or the other influence on accidents (except perhaps on rare occasions). The two sources of influence were assembled into a cusp catastrophe model for the accident process in Guastello (1988, 1989), such that hazards contributed to the asymmetry parameter, and a variety of psychosocial variables, collectively labelled *operator load*, contributed to the bifurcation parameter. The cusp model for occupational accidents has been illustrated and compared for manufacturing, public transportation and health care settings (Guastello 2003).

Safety climate, which was first introduced by Zohar (1980), was part of the *operator load* parameter and tended towards lowering risk levels for an individual or group. Operator load (Guastello 1989) also included stress indicators, anxiety, beliefs about accident control, work-group size and work pace. Reason (1997) noted that work pace by itself can have what amounts to a hysteresis effect on accident rates: An organisation might make a concerted effort to reduce accidents, including adjusting work pace. Eventually, however, the organisation starts to demand higher production output, and, as a result, the organisation starts to zigzag up and down until the control point lands on the 'up' level of the cusp response surface.

The most recent study with the accident cusp (Guastello and Lynn 2014) responded to a meta-analysis (Clarke 2006) showing that safety climate had a generalisable relationship to safety behaviours but not to actual accident incidences or rates. Several things were thought to be missing from the simplistic correlation data: the cusp structure with hazards as the asymmetry parameter, the characterisation of safety climate as a bifurcation variable and anxiety as another bifurcation variable. Anxiety could have either a positive or negative impact on safe outcomes. It would

interfere with response time to emergency situations and interfere with clear decisions, or it could be symptomatic of hypervigilance for unsafe conditions.

The participants were 1,262 production employees of two steel-manufacturing facilities who completed a diagnostic survey that measured safety management (akin to safety climate), anxiety, and two types of hazard that were salient in that industry. The accident variable was also collected by a survey item in which the individual gave the number of OSHA-reportable accidents they had in the preceding three years. Because the survey was given at only one point in time, the static cusp model was used for the analysis.

Nonlinear regression analyses showed, for this industry, that the accident process could be explained by safety management, anxiety, hazards, and age and experience within the cusp structure ($R^2 = .72$). All parts of the cusp model were sustained. The alternative multiple linear regression with anxiety, safety management, hazards and age and experience was substantially less accurate ($R^2 = .08$). The results thus illustrated the advantages of framing accident problems through the nonlinear paradigm.

4.9. Resilience engineering

The concept of resilience in ergonomics (Hollnagel, Woods, and Leveson 2006; Sheridan 2008; Woods and Wreathall 2008) presents questions such as: How well can a system rebound from a threat or assault? Can it detect critical situations before they fully arrive? Building resilience into a system requires more than the analysis of accidents in hindsight or implementing a right-minded plan. It involves regular monitoring of the system for its proximity to critical situation, anticipating possible critical situations and taking action to adapt as necessary.

In this line of thinking, strain is an independent measure. In the low to moderate regions of strain, the outcome is a fairly consistent increase in the system's ability to meet demands of various types. Beyond a certain point, however, more strain is producing diminishing increments of capability in what the theorists call the 'extra region'. At the extreme of the extra region, the system faults. One would then look for a good place in the process to introduce an adaptation that would stretch the capability of the system.

The definition of a pre-emptive strategy requires a good sense of how a CAS operates. A CAS uses sensors that take the form of information gathering capabilities that can inform organisational members of changing events in the outside environment or internal operations. The most useful information inflows result in the system-level equivalent of situation awareness. Woods and Wreathall (2008) characterised this form of situation awareness as *calibration*, which is to know where the system is located along the stress-demand function in spite of changing circumstances.

A second feature of the CAS, which is not independent of the first, is to have a clear sense of how the system is organised for work and information flows. Here one should look for *functional resonance*, which is how the variability associated with one part of the process crosses to the adjacent parts of the process and travels through the system (Hollnagel 2012; Leonhardt et al. 2009). It would be then be possible to pinpoint non-resilient or non-adaptive features of the process. Non-resilient or non-adaptive features of the process often involve automated functions which, traditionally, permit little variation in the process and output and thus work very efficiently for a limited range of circumstances. One is thus looking for degrees of freedom within the system. A resilience event would occur when the human intervenes to make an adaptation that either extends or overrides an automatic process after detecting that a supercritical event was on the verge of occurring.

This construct of resilience corresponds to the rigidity-elasticity principle in the cusp model of mental workload (Figure 4), but it is now examined at level of a broader sociotechnical system. Rigidity, once again, has the benefit of controlling system variability up to a point, after which a decisive fault can be expected. Elasticity and resilience span the region of the surface that contains the cusp point, which is the most unstable point on the surface. It follows that too much flexibility, or resilience capability, can make a system unstable in the long run. Here one should look for features of the system that use too many degrees of freedom (or control) to do a job when fewer degrees of freedom would produce an equivalent result. There is an optimal level of variability associated with high levels of performance (Leonhardt et al. 2009), which is needed to generate an adaptive response.

4.9. Team coordination

Coordination occurs when two or more people take the same action or compatible actions at the same time or in the correct sequence. Whereas conventional thinking holds that successful coordination requires good communication and a shared mental model, the NDS perspective holds that coordination is fundamentally a nonverbal process, similar to biomechanical coordination, and that shared mental models can self-organise extemporaneously as situations unfold (Guastello 2009b; Guastello and Guastello 1998). The latter point was reflected in an example from Hurricane Katrina (Morris, Morris, and Jones 2007, 99) and in the concept of dynamic situation awareness (Chiappe, Strybel, and Vu 2015). Verbal communication does expedite the coordination process, nonetheless.

Different tasks can require different types of coordination, which are distinguished by the utilities for the group that are associated with different individual actions. Several types of coordination were defined in game theory. Although game theory pre-dated NDS as we know it today, its notions of equilibria that are associated with dominant strategies reflect attractors on some occasions and saddles on others. The connection with NDS consolidated more strongly with the works of Axelrod (1984) and Maynard-Smith (1982) who studied the long-run outcomes of repeated games, which are now known as *evolutionarily stable states*.

Axelrod and Maynard-Smith studied the Prisoners' Dilemma game, which required players to make a choice between a competitive and a cooperative option. The more recent game-theoretical studies investigated games that are strictly cooperative, particularly Intersection and Stag Hunt. An Intersection game is modelled after the four-way stop intersection. After approaching a four-way stop intersection with other cars coming from different directions, the goal is to figure out what rules of turn taking are in play for proceeding through the intersection, figuring out when one's turn comes up, and proceeding through the intersection at the correct moment. Experimental results show that Intersection coordination can occur when the participants are not allowed to talk to each other (Guastello and Bond 2007; Guastello and Guastello 1998; Guastello et al. 2005), including conditions where they only interact through making binary choices on a computer interface (Aruka 2011). Verbalisation does enhance performance, but it is not always necessary for effective performance (Guastello and Bond 2004; Guastello et al. 2005). This is fortunate because coordination among agents must be sustained during communication outages, which are going to occur in emergencies.

Coordination usually occurs as an implicit learning process. *Implicit learning* encompasses the procedures one learns while trying to learning something else explicitly; implicit learning often occurs unconsciously and transfers to new learning situations in much the same way as knowledge gained through explicit learning transfers to new situations (Seger 1994). The Intersection experiments illustrated implicit learning as a transfer of learning that occurred when the participants changed tasks; the players learned to coordinate, which is separate from learning the task itself. As with other forms of learning, coordinated performance is chaotic in its early stages and self-organised later on (Guastello and Guastello 1998; Guastello et al. 2005).

Figure 13 depicts the performance curves for groups engaged in a coordination-intensive task, 13 of which interacted nonverbally only and 13 were allowed to talk. The task was structured as a card game, which was played

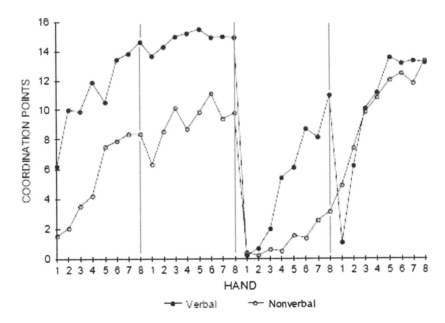

Figure 13. Learning curves for team coordination from Guastello and Bond 2007, 104). Source: Reprinted with permission of the Society for Chaos Theory in Psychology and Life Sciences.

in four rounds. The time series for the first round reflected a jagged learning curve. In the second round, the rule changed to a different rule of equal difficulty; the learning curve looked like a continuation of the first one. In the third round, the rule changed to one of greater difficulty. The fourth round reverted to the first rule. The data showed that coordination is learned and that it transfers from one version of the task to another.

Coordination within a group can withstand changes in personnel up to a point, but disruptions in coordination occur when too many group members change; the disruptions are not readily compensated by the availability of verbal channels (Guastello et al. 2005). Group performance can even be enhanced by exchanging personnel with members of other groups who were already trained and coordinated with different people on the same type of task (Gorman et al. 2006). Coordination occurs without any appointed leaders, but leaders do emerge from coordination-intensive situations (Guastello and Bond 2007) *even when the participants are not allowed to talk during the task activity.*

An unconfirmed hypothesis at this stage, however, is that verbalisation during coordination learning becomes increasingly valuable as the task becomes more complex. Verbalisation would include work structuring, role separation and agreements about optimal timing. Coordination learning would be *explicit* in such a context. These elements form the mental model, with the understanding that the model can modify or contain contingencies that can be enacted when the situation changes. The intuitive, implicit, or nonverbal components would still be present,

and probably more salient as well. Consider how the best efforts of participants to plan and coordinate are limited by people in the group who are 'slow to catch on' or 'just don't get it'.

In a *Stag Hunt game*, the players hypothetically choose between working with the group (hunting stag) and working on their own (hunting rabbits). The likelihood of a hunter joining the group depends on the group's efficacy. The performance of the group depends on the combined skill and efforts of all the hunters. The group outcome can tolerate some individual differences in contributions, but if one member makes too little effort, or too many mistakes, the outcome for the group is reduced to that of the least effective person (Crawford 1991; McCain 1992).

Guastello and Bond (2004) compared the economic game of Stag Hunt, a real stag hunt, an analogous game against a natural disaster, and an analogous game against a disaster that involved a sentient attacker. The key distinction is the dynamics of learning by both the humans and the prey or antagonist. The possible strategic options could be numerous, and the utilities associated with them vary each time the group faces a decision. It is seldom possible, therefore, to predict the evolutionarily stable states of an iterated Stag Hunt without a real-time experiment or simulation (Samuelson 1997). Iterated games of this type are essentially dynamic decisions, as Brehmer (2005) defined them.

Guastello and Bond (2004) operationalised Stag Hunt as a board game ('The Creature that Ate Sheboygan') in which an emergency response (ER) team played against a sentient attacker that was trying to destroy a city. The performance of the ER teams and the attackers were recorded over time. The principal findings were that (a) communication outages did not hinder the ER teams' performance, but attackers did better under those conditions; (b) the attackers improved their performance during consecutive games, which presented an added challenge to the ER teams, whereas ER teams showed both improvements and drops in performance in consecutive games; and (c) the number of ER team members who participated in a particular decision depended on the number of points scored by the attacker on the previous turn; ER team members tended to disengage when the going got tough.

In a later effort (Guastello 2010), the performance dynamics generated by ER teams of 4, 6, 9 or 12 players were analysed for Lyapunov exponents, which were interpreted as an index of adaptability. The comparison of groups sizes would determine whether the performance outcomes favoured smaller groups, which would meant that coordination was impaired, larger groups, which would mean that the larger groups contributed a critical mass of ideas needed for creative problem-solving, or no group effect, which would indicate that no group dynamic self-organised.

The Lyapunov exponent was useful for quantifying entropy and adaptation levels in the performance time series. ER teams operated in the middle self-organising range, suggesting that although they adapted, they stuck with the plans that appeared to have evolved. Attackers, in contrast, exhibited chaotic behaviour: a consistent path of destruction punctuated by opportunistic bursts into different regimes or directions (Lévy flights). Attackers can benefit from such tactics, but ER team members would probably benefit by being more predictable to each other. It is an open question as to whether the teams were as adaptive as they could reasonably be, or what new strategies they would need to deploy to improve.

4.10. Synchronisation

The interest in coordination dynamics has now morphed into an interest in synchronisation. Synchronisation in the behavioural sciences has acquired a few different but related meanings, some of which are not especially different from coordination as it has been discussed to this point. For present purposes, however, the construct of synchronisation that was introduced in Section 2.6 is the prototype construct. One manifestation of synchronisation is specific to psychomotor coordination. In the finger-tapping studies (Haken, Kelso, and Bunz 1985; Kelso 1995; Sleimen-Malkoun et al. 2010), the experimental participants tapped their fingers to a metronome that clicked at a fixed or controlled rate of speed in an alternating pattern of left-right-left-right. When the metronome speed

reached a critical point, the two fingers tapped simultaneously rather than alternating. The change from tapping out-of-phase to strictly in-phase is a *phase shift*, and the completely in-phase state is a *phase lock*.

Physiological synchronisation of EEGs, autonomic arousal and behavioural movements between people is thought to be an important component of work team coordination (De Vico Fallani et al. 2010; Delaherche et al. 2012; Guastello, Reiter, and Malon 2015; Henning, Boucsein, and Gil 2001; Salas et al. 2015; Stevens, Galloway, and Lamb 2014; Stevens et al. 2012, 2013) and other interpersonal dynamics. So far, it appears that the degree of coupling between two people is dependent on the task they are performing together, and probably other factors that are still under investigation. These ambiguities affect the lag lengths of the time series of physiological data streams, which in turn affect the rest of the statistical modelling procedure. In the case of a more-or-less ordinary conversation, the coupling was moderated by the empathy levels of the two parties (Guastello, Pincus, and Gunderson 2006; Marci et al. 2007).

The synchronisation links in the sample of dyads have been identified more often and with greater accuracy by Equation 9, compared to a linear alternative:

$$Z_2 = A \exp(Bz_1) + \exp(CP_1) \qquad (9)$$

In Equation 9, z is the normalised behaviour (autonomic arousal) of the target person at two successive points in time, P is the normalised behaviour of the partner at time-1, and A, B and C are nonlinear regression weights (Guastello, Pincus, and Gunderson 2006). Equation 5 can also be expanded to incorporate the influence of a coupled oscillator.

Nonlinear time-series modelling requires a strategy for determining appropriate lag lengths. If a measurement at time-2 is a function of itself at time-1 and a coupling effect from another source, how much real time is required to elapse between the two measurements in order to observe the coupling effect? Guastello, Reiter, and Malon (2015) examined four strategies for doing so. In the experiment, 73 undergraduates worked in pairs to perform a vigilance dual task for 90 min while galvanic skin responses (GSR) were recorded. Event rates on the vigilance task either increased or decreased without warning during the work period. Results based on minimum mutual information and a natural rate criterion supported a value of 20 s, whereas two other strategies were not calculable.

The next phase of the project that was based on the same experiment (Guastello 2016b) examined the properties of the linear autoregressive model, linear autoregressive model with a synchronisation component, the nonlinear autoregressive model (first addend of Equation 9) and the nonlinear autoregressive model with a synchronisation component (Equation 9). All models were more accurate at a lag of 20 s compared to 50 s (95 percentile) or customised lag lengths. Although the linear models were more accurate overall by a margin of 4–13% of variance accounted for, the nonlinear synchronisation parameters were more often related to psychological variables and performance. In particular, greater synchronisation was observed with the nonlinear model when the target event rate increased, compared to when it decreased, which was expected from the general theory of synchronisation. Nonlinear models were also more effective for uncovering inhibitory or dampening relationships between the co-workers as well as mutually excitatory relationships. Many aspects of the statistical structures of nonlinear time series that involve synchronisation still need to be investigated.

The adaptive value of high levels of physiological synchrony may require some qualification, however. Stevens, Galloway, and Lamb (2014) discovered that submarine navigation teams whose EEGs were less synchronised were more apt to make adaptive responses to their task when needed; this relationship flipped from negative to positive depending on whether the team was in a briefing, action, or debriefing segment of the simulation. Phase locks in human interaction are probably rare, although they could take the form of rigid conversation patterns in dysfunctional families (Gottman and Levinson 1986; Pincus 2001), which is quite the opposite of a highly functional work team. If synchronisation at the nonverbal level can produce desirable decisions at a more explicit level, it can also facilitate irrationality, particularly if stress and crowding are involved (Adamatzky 2005).

5. Concluding remarks

NDS qualifies as a paradigm of science by virtue of its novel constructs, methods for investigating them, new questions that it poses and the explanations of phenomena it can provide. It qualifies as a general systems theory, and a new paradigm in systems thinking, for the same reasons in addition to the applicability of its principles and methods to nonliving and living systems. The latter range from biophysics to macroeconomics, although the psychological content was perhaps most proximal to ergonomics thus far. Ergonomics, however, has been expanding into micro- and macro-systems, and constructs that provide some continuity among those levels of systems are needed (Karwowski 2012). The scaling principle that is inherent in fractal structures and $1/f^b$ relationships are particularly relevant to this concern.

Systems have also been growing in complexity in the sense of multiple interacting components and multiple outcomes from those interactions. Here, the NDS constructs of chaos and sensitivity to initial conditions,

entropy and emergence are well suited for describing and possibly predicting how a system could function (Walker et al. 2010). The principle of degrees of freedom was also introduced here for much the same reason. Add humans to the system, and the former notion of a system is upgraded to the CAS (Dooley 1997; Karwowski 2012). The next question, however, was whether NDS has actually taken root as a paradigm in ergonomics. To answer this question, several streams of research were specifically considered.

Nonlinear psychophysics by itself represents a paradigm shift of a proportion similar to the contribution of signal detection theory relative to classical psychophysics. It is now possible to interpret the double threshold associated with increasing and decreasing signal strengths as the result of a deterministic process. Rather than framing sensation phenomena as stochastic processes in a deterministic environment, nonlinear psychophysics frames the phenomenon as a deterministic process occurring in a stochastic environment (Gregson 2009, 115). Nonlinear psychophysics also accommodates multidimensional stimuli that change over time, and the response variable can be chaotic under some conditions.

Control theory has already built on constructs shared with NDS theory. It is only a matter to take the ideas a few step further to investigate the outcomes from a complex array of controllers working semi-autonomously. The new construct, relative to control theory, is Ashby's Law, which is now seen to work in counterpoint with other NDS constructs such as minimum entropy and degrees of freedom.

The NDS perspective on cognitive workload and fatigue has managed to untangle some perplexing questions that have haunted theorists for a century, and accounts for changes in performance under a variety of conditions. The central contribution was the use of two models, not just one, that separate what are often confounding processes. The dynamics of attractors, bifurcations and control variables are part of the catastrophe models. The studies currently on record show that, overall, about one-third of the variance in the performance can be attributed to the nonlinear structure itself (Guastello 2014b). Then, having separated the roles of asymmetry and bifurcation variables it was possible to embark on a productive search for variables that behaved in specific ways. The amount of work accomplished in a given amount of time, which is predicated on workload, feeds into the fatigue process as a bifurcation variable.

The elasticity-rigidity variables that were identified in the workload and fatigue research connect to another important group of questions regarding the resilience of systems. The elasticity-rigidity constructs need to be explored further at a team or larger level of a system. Similarly, the notion of degrees of freedom in fatigue could also explain spontaneous recovery from fatigue and self-organising cognitive processes more generally.

The construct of elasticity versus rigidity originated in material science, as did the first cusp catastrophe model for buckling phenomena. If one considers human and animal bodies as complex and composite materials, the basic cusp model for buckling stress is broadly applicable (Guastello 1985; Karwowski, Ostaszewski, and Zurada 1992). No one, since the beginning of experimental psychology in 1879 has determined what materials constitute the substance of mental structures. If anything, psychologists gave up trying to figure it out decades ago, and many would probably prefer that 'materials' be used in quotation marks. Nonetheless, elasticity and rigidity can be observed in psychological and complex systems; this observation again supports the general systems nature of NDS.

The type of fatigue that was studied in the experiments reported here examined performance fluctuations that occurred in laboratory experiments of about two hours duration and did not extend to the type of fatigue that is produced by sleep deprivation or disruptions of circadian cycles. A useful direction for future research would be to expand the workload and fatigue research to include the circadian influence as part of a more complex dynamical process.

The dual-task methodologies that were used in the early experiments on cognitive workload capacity were fashioned to absorb unused channel capacity with one task while evaluating performance on a target task (Kantowitz 1985). Contemporary work life has induced a small change in perspective such that both tasks are now important, and *multitasking* is a new verb in the workplace. The workload and fatigue studies encountered multitasking as another source of fatigue, and prompted studies on task switching which have some interesting temporal dynamics of their own. Symbolic dynamics and entropy functions were the primary nonlinear methods and constructs. The results converged with other streams of research to produce the performance-variability paradox, which results from the Law of Requisite Variety and minimum entropy principles.

An accident at face value is a discontinuous change of events. As such it looks amenable to analysis with the cusp catastrophe, and in fact it is. When this accident paradigm was first introduced (Guastello 1988, 1989), the objective was to go beyond the traditional single cause, chain-of-events, and other relatively simple models of the process to a model that represents the dynamics comprehensively, including the underlying nonlinearities. The nonlinear properties also describe a process by which situations move from sub-critical to super-critical. The original safety climate (Zohar 1980) remains firm in the single cause mentality. The NDS view is that a safety climate is comprised not only of managerial behaviours

and safe actions of operators, but also other environmental influences such as hazards, hazard perception, stress and anxiety levels. The foregoing elements self-organise into social climate that can explain group accident levels and also have a top-down influence on the experiences of individual operators (Guastello and Lynn 2014). The empirical value of this paradigm shift amounts to a much greater explanation of variance in accident outcomes in addition to an explicative model.

The objective of the resilience engineering is to prevent accidental injuries, deaths and other total system failures from happening by rethinking what it takes to make the system a complex adaptive system. Notions of elasticity and rigidity are also involved, although illustrative empirical analyses have not arrived yet. Nonetheless, the idea of system-wide resilience is consistent with NDS theory, and certainly did not originate anywhere else. Surprising events can occur when systems self-organise and situations emerge (McDaniel and Driebe 2005). If a system is sticking too automatically to its operating procedures that have worked well in the past, it can become blind-sided to critical novelties that require an adaptive response.

The NDS studies on group coordination introduced some radical thinking relative to the status quo of the research area: coordination is an implicitly learned response with chaotic and self-organising dynamics involved, it occurs at a non-verbal level although verbalisation adds value, mental models of task situations evolve and self-organise on the fly, and the internal utility structures of the tasks have an important effect on how these events might unfold. A decade later, there is finally a recognition that team mental models and situation awareness develop with experience and opportunities to adapt and interact (Chiappe, Strybel, and Vu 2015; Cooke et al. 2012, 2013; Gorman, Hessler et al. 2012; Likens et al. 2014). The latest developments in this area are expanding the concept of coordination to include synchronisation of body movements and neurocognitive events across people, and integrating the synchronisation levels with communication patterns (Guastello et al. 2016; Salas et al. 2015; Stevens et al. 2012, 2013; Stevens, Galloway, and Lamb 2014). Thus, to answer the central question, one can paraphrase an old piece of folk wisdom: If it looks like a paradigm, walks like a paradigm, and quacks like a paradigm, it is probably a paradigm.

Disclosure statement

No potential conflict of interest was reported by the author.

References

Ackerman, P. L., ed. 2011. *Cognitive Fatigue*. Washington, DC: American Psychological Association.

Adamatzky, A. 2005. *Dynamics of Crowd-Minds: Patterns of Irrationality in Emotions, Beliefs and Actions.* Singapore: World Scientific.

Allan, P. M., and L. Varga. 2007. "Complexity: The Co-Evolution of Epistemiology, Axiology, and Ontology." *Nonlinear Dynamics, Psychology, and Life Sciences* 11: 19–50.

Allen, P., S. Maguire, and B. McKelvey, eds. 2011. *The Sage Handbook for Complexity Management.* Thousand Oaks, CA: Sage.

Anderson, P. 1999. "Complexity theory and organization science." *Organization Science* 10: 216–233.

Aruka, Y. 2011. "Avatamsaka Game Structure and Experiment on the Web." In *Complexities of Production and Interacting Human Behaviour*, edited by Y. Aruka, 203–222. Heidelberg: Physica-Verlag.

Ashby, W. R. 1956. *Introduction to Cybernetics.* New York: Wiley.

Axelrod, R. 1984. *The Evolution of Cooperation.* New York: Basic Books.

Bailey, K. D. 1994. "Talcott Parsons, Social Entropy Theory, and Living Systems Theory." *Behavioral Science* 39: 25–45.

Bak, P. 1996. *How Nature Works: The Science of Self-Organized Criticality.* New York: Springer-Verlag/Copernicus.

Balagué, N., R. Hristovski, D. Agagonés, and G. Tenebaum. 2012. "Nonlinear Model of Attention Focus during Accumulated Effort." *Psychology of Sport and Exercise* 13: 591–597.

Bankes, S., and R. Lempert. 2004. "Robust Reasoning with Agent-Based Modeling." *Nonlinear Dynamics, Psychology, and Life Sciences* 8: 259–278.

Bernstein, N. 1967. *The Coordination and Regulation of Movements.* Oxford: Pergamon.

Bianciardi, G. 2015. "Differential Diagnosis: Shape and Function, Fractal Tools in the Pathology Lab." *Nonlinear Dynamics, Psychology, and Life Sciences* 19: 437–464.

Bigelow, J. 1982. "A Catastrophe Model of Organizational Change." *Behavioral Science* 27: 26–42.

Boker, S. M., and J. Graham. 1998. "A Dynamical Systems Analysis of Adolescent Substance Abuse." *Multivariate Behavioral Research* 33: 479–507.

Brehmer, B. 2005. "Micro-Worlds and the Circular Relation between People and Their Environment." *Theoretical Issues in Ergonomic Science* 6: 73–93.

Brock, W. A., W. Dechert, J. Scheinkman, and B. LeBaron. 1996. "A Test for Independence Based on the Correlation Dimension." *Economic Reviews* 15: 197–235.

Butner, J., and N. Story. 2011. "Oscillators with Differential Equations." In *Nonlinear Dynamical Systems Analysis for the Behavioral Sciences Using Real Data*, edited by S. J. Guastello and R. A. M. Gregson, 367–400. Boca Raton, FL: CRC Press.

Cantwell, R. H., and P. J. Moore. 1996. "The Development of Measures of Individual Differences in Self-Regulatory Control and Their Relationship to Academic Performance." *Contemporary Educational Psychology* 21: 500–517.

Cecen, A. A., and C. Erkal. 2009. "The Long March: From Monofractals to Endogenous Multifractality in Heart Rate Variability Analysis." *Nonlinear Dynamics, Psychology, and Life Sciences* 13: 181–206.

Chiappe, D., T. Z. Strybel, and K.-P. L. Vu. 2015. "A Situated Approach to the Understanding of Dynamic Situations." *Journal of Cognitive Engineering and Decision Making* 9: 33–43.

Clarke, S. 2006. "The Relationship between Safety Climate and Safety Performance: A Meta-Analytic Review." *Journal of Occupational Health Psychology* 11: 315–327.

Cobb, L. 1981. "Parameter Estimation for the Cusp Catastrophe Model." *Behavioral Science* 26: 75–78.

Cooke, N. J., A. Duchon, J. C. Gorman, J. Keyton, and A. Miller. 2012. "Preface to the Special Section on Methods for the Analysis of Communication." *Human Factors* 54: 485–488.

Cooke, N. J., J. C. Gorman, C. W. Myers, and J. L. Duran. 2012. "Theoretical Underpinning of Interactive Team Cognition." In *Theories of Team Cognition: Cross-Disciplinary Perspectives*, edited by E. Salas, S. M. Fiore and M. P. Letsky, 187–207. New York: Routledge.

Cooke, N. J., J. C. Gorman, C. W. Myers, and J. L. Duran. 2013. "Interactive Team Cognition." *Cognitive Science* 37: 255–285.

Corrêa, U. C., R. N. Benda, D. L. de Oliveira, H. Ugrinowitsch, A. M. Freudenheim, and G. Tani. 2015. "Different Faces of Variability in the Adaptive Process of Motor Skill Learning." *Nonlinear Dynamics, Psychology, and Life Sciences* 19: 465–488.

Crawford, V. P. 1991. "An 'Evolutionary Interpretation' of Van Huyk, Batalio, and Beil's Experimental Results on Coordination." *Games and Economic Behavior* 3: 25–59.

De Vico Fallani, F., V. Nicosia, R. Sinatra, L. Astolfi, F. Cincotti, et al. 2010. "Defecting or Not Defecting: How to Read Human Behavior during Cooperative Games by EEG Measurements." *PLoS One* 5 (12): e14187.

Delaherche, E., M. Chetouani, A. Mahdhaoui, C. Saint-Georges, S. Viaux, and D. Cohen. 2012. "Interpersonal Synchrony: A Survey of Evaluation Methods across Disciplines." *IEEE Transactions on Affective Computing* 3 (3): 1–20.

Dishion, T. J. 2012. "Relationship Dynamics in the Development of Psychopathology: Introduction to the Special Issue." *Nonlinear Dynamics, Psychology, and Life Sciences* 16: 237–242.

Dishion, T. J., M. Forgatch, M. Van Ryzin, and C. Winter. 2012. "The Nonlinear Dynamics of Family Problem Solving in Adolescence: The Predictive Validity of a Peaceful Resolution Attractor." *Nonlinear Dynamics, Psychology, and Life Sciences* 16: 331–352.

Dodge, R. L. 1917. "The Laws of Relative Fatigue." *Psychological Review* 24: 89–113.

Dooley, K. J. 1997. "A Complex Adaptive Systems Model of Organization Change." *Nonlinear Dynamics, Psychology, and Life Sciences* 1: 69–97.

Dooley, K. J. 2004. "Complexity Science Models of Organizational Change and Innovation." In *Handbook of Organizational Change and Innovation*, edited by M. S. Poole and A. H. Van de Ven, 354–373. New York: Oxford University Press.

Dooley, K. J. 2009. "The Butterfly Effect of the 'Butterfly Effect'." *Nonlinear Dynamics, Psychology, and Life Sciences* 13: 279–288.

Dooley, K. J., L. D. Kiel, and A. S. Dietz. 2013. "Introduction to the Special Issue on Nonlinear Organizational Dynamics." *Nonlinear Dynamics, Psychology, and Life Sciences* 17: 1–2.

Dore, M. H. I., and J. B. Rosser Jr. 2007. "Do Linear Dynamics in Economics Amount to a Kuhnian Paradigm?" *Nonlinear Dynamics, Psychology, and Life Sciences* 11: 119–148.

Fleener, M. J., and M. L. Merritt. 2007. "Paradigms Lost?" *Nonlinear Dynamics, Psychology, and Life Sciences* 11: 1–18.

Gell-Mann, M. 1984. *The Quark and the Jaguar.* New York: W. H. Freeman.

Gibson, J. J. 1979. *The Ecological Approach to Visual Perception.* Boston, MA: Houghton Mifflin.

Gilmore, R. 1981. *Catastrophe Theory for Scientists and Engineers.* New York: Wiley.

Goldberger, A. 1991. "Is Normal Heartbeat Chaotic or Homeostatic?" *News in Physiological Science* 6: 87–91.

Goldberger, A. 1997. "Fractal Variability versus Pathologic Periodicity: Complexity Loss and Stereotypy in Disease." *Perspectives in Biology and Medicine* 40: 543–561.

Goldstein, J. 2011. "Emergence in complex systems." In *The Sage handbook of complexity and management*, edited by P. Allen, S. Maguire and B. McKelvey, 65–78. Thousand Oaks, CA: Sage.

Gorman, J. C. 2014. "Team Coordination and Dynamics: Two Central Issues." *Current Directions in Psychological Science* 23: 355–360.

Gorman, J. C., N. J. Cooke, H. K. Pedersen, J. Winner, D. Andrews, and P. G. Amazeen. 2006. "Changing Team Composition after a Break: Building Adaptive Command-and-Control Teams". *Proceedings of the Human Factors and Ergonomics Society, 50th Annual Meeting* (pp. 487–491). Baltimore, MD: Human Factors and Ergonomics Society.

Gorman, J. C., N. J. Cooke, and E. Salas. 2010. "Preface to the Special Issue on Collaboration, Coordination, and Adaptation in Complex Sociotechnical Settings." *Human Factors* 52: 147–161.

Gorman, J. C., N. J. Cooke, P. G. Amazeen, and S. Fouse. 2012. "Measuring Patterns in Team Interaction Sequences Using a Discrete Recurrence Approach." *Human Factors* 54: 503–517.

Gorman, J. C., E. E. Hessler, P. G. Amazeen, N. J. Cooke, and S. M. Shope. 2012. "Dynamical Analysis in Real Time: Detecting Perturbations to Team Communication." *Ergonomics* 55: 825–839.

Gottman, J. M., and R. W. Levinson. 1986. "Assessing the Role of Emotion in Marriage." *Behavioral Assessment* 8: 31–48.

Grassberger, P., and I. Procaccia. 1983. "Characterization of Strange Attractors." *Physical Review Letters* 50: 346–349.

Gregson, R. A. M. 1992. *N–Dimensional Nonlinear Psychophysics*. Hillsdale, NJ: Lawrence Erlbaum Associates.

Gregson, R. A. M. 2005. "Identifying Ill-Behaved Nonlinear Processing without Metrics: Use of Symbolic Dynamics." *Nonlinear Dynamics, Psychology, and Life Sciences* 9: 479–503.

Gregson, R. A. M. 2009. "Psychophysics." In *Chaos and Complexity in Psychology: The Theory of Nonlinear Dynamical Systems*, edited by S. J. Guastello, M. Koopmans and D. Pincus, 108–131. New York: Cambridge University Press.

Gregson, R. A. M. 2013. "Symmetry-Breaking, Grouped Images and Multistability with Transient Unconsciousness." *Nonlinear Dynamics, Psychology, and Life Sciences* 17: 325–344.

Guastello, S. J. 1982. "Color Matching and Shift Work: An Industrial Application of the Cusp-Difference Equation." *Behavioral Science* 27: 131–139.

Guastello, S. J. 1985. "Euler Buckling in a Wheelbarrow Obstacle Course: A Catastrophe with Complex Lag." *Behavioral Science* 30: 204–212.

Guastello, S. J. 1988. "Catastrophe Modeling of the Accident Process: Organizational Subunit Size." *Psychological Bulletin* 103: 246–255.

Guastello, S. J. 1989. "Catastrophe Modeling of the Accident Process: Evaluation of an Accident Reduction Program Using the Occupational Hazards Survey." *Accident Analysis and Prevention* 21: 61–77.

Guastello, S. J. 1995. *Chaos, Catastrophe, and Human Affairs: Applications of Nonlinear Dynamics to Work, Organizations, and Social Evolution*. Mahwah, NJ: Lawrence Erlbaum.

Guastello, S. J. 2000. "Symbolic Dynamic Patterns of Written Exchange: Hierarchical Structures in an Electronic Problem Solving Group." *Nonlinear Dynamics, Psychology, and Life Sciences* 4: 169–187.

Guastello, S. J. 2002. *Managing Emergent Phenomena: Nonlinear Dynamics in Work Organizations*. Mahwah, NJ: Lawrence Erlbaum Associates.

Guastello, S. J. 2003. "Nonlinear Dynamics, Complex Systems, and Occupational Accidents." *Human Factors in Manufacturing* 13: 293–304.

Guastello, S. J., D. E. Nathan, and M. J. Johnson. 2009. "Attractor and Lyapunov models for reach and grasp movements with application to robot-assited therapy." *Nonlinear Dynamics, Psychology, and Life Sciences* 13: 99–121.

Guastello, S. J. 2009a. "Chaos as a Psychological Construct: Historical Roots, Principal Findings, and Current Growth Directions." *Nonlinear Dynamics, Psychology, and Life Sciences* 13: 289–310.

Guastello, S. J. 2009b. "Group Dynamics: Adaptability, Coordination, and Leadership Emergence." In *Chaos and Complexity in Psychology: Theory of Nonlinear Dynamical Systems*, edited by S. J. Guastello, M. Koopmans and D. Pincus, 402–433. New York: Cambridge University Press.

Guastello, S. J. 2010. "Nonlinear Dynamics of Team Performance and Adaptability in Emergency Response." *Human Factors* 52: 162–172.

Guastello, S. J. 2013. "Catastrophe Theory and Its Applications to I/O Psychology." In *Frontiers of Methodology in Organizational Research*, edited by J. M. Cortina and R. Landis, 29–61. New York: Routledge/Society for Industrial and Organizational Psychology.

Guastello, S. J. 2014a. *Human Factors Engineering and Ergonomics: A Systems Approach*. 2nd ed. Boca Raton, FL: CRC Press.

Guastello, S. J. 2014b. "Catastrophe Models for Cognitive Workload and Fatigue: Memory Functions, Multitasking, Vigilance, Financial Decisions and Risk." *Proceedings of the Human Factors and Ergonomics Society* 58: 908–912.

Guastello, S. J., D. E. Marra, C. Perna, J. Castro, M. Gomez, and A. F. Peressini. 2016. "Physiological synchronization in emergency response teams: Subjective workload, drivers and empaths." *Nonlinear Dynamics, Psychology, and Life Sciences* 20: 223–270.

Guastello, S. J., ed. 2016a. *Cognitive Workload and Fatigue in Financial Decision Making*. Tokyo: Springer.

Guastello, S. J. 2016b. "Physiological Synchronization in a Vigilance Dual Task." *Nonlinear Dynamics, Psychology, and Life Sciences* 20: 49–80.

Guastello, S. J., and R. W. Bond. 2004. "Coordination Learning in Stag Hunt Games with Application to Emergency Management." *Nonlinear Dynamics, Psychology, and Life Sciences* 8: 345–374.

Guastello, S. J., and R. W. Bond Jr. 2007. "The Emergence of Leadership in Coordination-Intensive Groups." *Nonlinear Dynamics, Psychology, and Life Sciences* 11: 91–118.

Guastello, S. J., and M. J. Fleener. 2011. "Chaos, Complexity, and Creative Behavior." *Nonlinear Dynamics, Psychology, and Life Sciences* 15: 143–144.

Guastello, S. J., and R. A. M. Gregson, eds. 2011. *Nonlinear Dynamical Systems Analysis for the Behavioral Sciences Using Real Data*. Boca Raton, FL: CRC Press/Taylor and Francis.

Guastello, S. J., and D. D. Guastello. 1998. "Origins of Coordination and Team Effectiveness: A Perspective from Game Theory and Nonlinear Dynamics." *Journal of Applied Psychology* 83: 423–437.

Guastello, S. J., and L. S. Liebovitch. 2009. "Introduction to Nonlinear Dynamics and Complexity." In *Chaos and Complexity in Psychology: Theory of Nonlinear Dynamical Systems*, edited by S. J. Guastello, M. Koopmans, and D. Pincus, 1–40. New York: Cambridge University Press.

Guastello, S. J., and M. Lynn. 2014. "Catastrophe Model of the Accident Process, Safety Climate, and Anxiety." *Nonlinear Dynamics, Psychology, and Life Science* 18: 177–198.

Guastello, S. J., and D. W. McGee. 1987. "Mathematical Modeling of Fatigue in Physically Demanding Jobs." *Journal of Mathematical Psychology* 31: 248–269.

Guastello, S. J., T. Hyde, and M. Odak. 1998. "Symbolic Dynamic Patterns of Verbal Exchange in a Creative Problem Solving Group." *Nonlinear Dynamics, Psychology, and Life Sciences* 2: 35–58.

Guastello, S. J., B. Bock, P. Caldwell, and R. W. Bond Jr. 2005. "Origins of Group Coordination: Nonlinear Dynamics and the Role of Verbalization." *Nonlinear Dynamics, Psychology, and Life Sciences* 9: 175–208.

Guastello, S. J., D. Pincus, and P. R. Gunderson. 2006. "Electrodermal Arousal between Participants in a Conversation: Nonlinear Dynamics for Linkage Effects." *Nonlinear Dynamics, Psychology, and Life Sciences* 10: 365–399.

Guastello, S. J., M. Koopmans, and D. Pincus, eds. 2009. *Chaos and Complexity in Psychology: Theory of Nonlinear Dynamical Systems*. New York: Cambridge University Press.

Guastello, S. J., A. F. Peressini, and R. W. Bond Jr. 2011. "Orbital Decomposition for Ill-Behaved Event Sequences: Transients and Superordinate Structures." *Nonlinear Dynamics, Psychology, and Life Sciences* 15: 465–476.

Guastello, S. J., H. Boeh, M. Schimmels, H. Gorin, S. Huschen, E. Davis, N. E. Peters, M. Fabisch, and K. Poston. 2012. "Cusp Catastrophe Models for Cognitive Workload and Fatigue in a Verbally Cued Pictorial Memory Task." *Human Factors* 54: 811–825.

Guastello, S. J., H. Boeh, C. Shumaker, and M. Schimmels. 2012. "Catastrophe Models for Cognitive Workload and Fatigue." *Theoretical Issues in Ergonomics Science* 13: 586–602.

Guastello, S. J., H. Gorin, S. Huschen, N. E. Peters, M. Fabisch, and K. Poston. 2012. "New Paradigm for Task Switching Strategies While Performing Multiple Tasks: Entropy and Symbolic Dynamics Analysis of Voluntary Patterns." *Nonlinear Dynamics, Psychology, and Life Sciences* 16: 471–497.

Guastello, S. J., H. Boeh, H. Gorin, S. Huschen, N. E. Peters, M. Fabisch, and K. Poston. 2013. "Cusp Catastrophe Models for Cognitive Workload and Fatigue: A Comparison of Seven Task Types." *Nonlinear Dynamics, Psychology, and Life Sciences* 17: 23–47.

Guastello, S. J., H. Gorin, S. Huschen, N. E. Peters, M. Fabisch, K. Poston, and K. Weinberger. 2013. "The Minimum Entropy Principle and Task Performance." *Nonlinear Dynamics, Psychology, and Life Sciences* 17: 405–424.

Guastello, S. J., M. Malon, P. Timm, K. Weinberger, H. Gorin, M. Fabisch, and K. Poston. 2014. "Catastrophe Models for Cognitive Workload and Fatigue in a Vigilance Dual Task." *Human Factors* 56: 737–751.

Guastello, S. J., K. Reiter, A. Shircel, P. Timm, M. Malon, and M. Fabisch. 2014. "The Performance-Variability Paradox, Financial Decision Making, and the Curious Case of Negative Hurst Exponents." *Nonlinear Dynamics, Psychology, and Life Sciences* 14: 297–328.

Guastello, S. J., K. Reiter, and M. Malon. 2015. "Estimating Appropriate Lag Length for Synchronized Physiological Time Series: The Electrodermal Response." *Nonlinear Dynamics, Psychology, and Life Sciences* 19: 285–312.

Guastello, S. J., K. Reiter, M. Malon, P. Timm, A. Shircel, and J. Shaline. 2015. "Catastrophe Models for Cognitive Workload and Fatigue in N-Back Tasks." *Nonlinear Dynamics, Psychology, and Life Sciences* 19: 173–200.

Guastello, S. J., A. Shircel, M. Malon, and P. Timm. 2015. "Individual Differences in the Experience of Cognitive Workload." *Theoretical Issues in Ergonomics Science* 16: 20–52. doi: http://dx.doi.org/10.1080/1463922X.2013.869371.

Haken, H. 1984. *The Science of Structure: Synergetics*. New York: Van Nostrand Reinhold.

Haken, H., J. A. S. Kelso, and H. Bunz. 1985. "A Theoretical Model of Phase Transition in Human Hand Movements." *Biological Cybernetics* 51 (347): 356.

Hancock, P. A. 2007. "On the Process of Automation Transition in Multitask Human-Machine Systems." *IEEE Transactions on Systems, Man, and Cybernetics – Part a: Systems and Humans* 37: 586–598.

Hancock, P. A. 2013. "In Search of Vigilance: The Problem of Iatrogenically Created Psychological Phenomena." *American Psychologist* 68: 97–109.

Hancock, P. A., and P. A. Desmond, eds. 2001. *Stress, Workload, and Fatigue*. Mahwah, NJ: Lawrence Erlbaum Associates.

Hancock, P. A., and J. S. Warm. 1989. "A Dynamic Model of Stress and Sustained Attention." *Human Factors* 31: 519–537.

Harrison, S. J., and N. Stergiou. 2015. "Complex Adaptive Behavior in Dexterous Action." *Nonlinear Dynamics, Psychology, and Life Sciences* 19: 345–394.

Heath, R. A. 2002. "Can People Predict Chaotic Sequences?" *Nonlinear Dynamics, Psychology, and Life Sciences* 6: 37–54.

Heinzel, S., I. Tominschek, and G. S. Schiepek. 2014. "Dynamic Patterns in Psychotherapy: Discontinuous Changes and Critical Instabilities during the Treatment of Obsessive Compulsive Disorder." *Nonlinear Dynamics, Psychology, and Life Sciences* 18: 155–176.

Henning, R. A., W. Boucsein, and M. C. Gil. 2001. "Social-Physiological Compliance as a Determinant of Team Performance." *International Journal of Psychophysiology* 40: 221–232.

Hockey, G. R. J. 1997. "Compensatory Control in the Regulation of Human Performance under Stress and High Workload: A Cognitive-Energetical Framework." *Biological Psychology* 45: 73–93.

Hockey, G. R. J. 2011. "A Motivational Control Theory of Cognitive Fatigue." In *Cognitive Fatigue*, edited by P. Ackerman, 167–187. Washington, DC: American Psychological Association.

Hollenstein, T. 2007. "State Space Grids: Analyzing Dynamics across Development." *International Journal of Behavioral Development* 31: 384–396.

Hollnagel, E. 2012. *FRAM: The Functional Resonance Analysis Method*. Burlington, VT: Ashgate.

Hollnagel, E., D. D. Woods, and N. Leveson, eds. 2006. *Resilience Engineering*. Burlington, VT: Ashgate.

Hong, S. L. 2010. "The Entropy Conservation Principle: Applications in Ergonomics and Human Factors." *Nonlinear Dynamics, Psychology, and Life Sciences* 14: 291–315.

Ibanez, A. 2007. "Complexity and Cognition: A Meta-Theoretical Analysis of the Mind as a Topological Dynamical System." *Nonlinear Dynamics, Psychology, and Life Sciences* 11: 51–90.

Jagacinski, R. J., and J. M. Flach. 2003. *Control Theory for Humans*. Mahwah, NJ: Lawrence Erlbaum Associates.

Kane, M. J., D. Z. Hambrick, and A. R. A. Conway. 2005. "Working Memory Capacity and Fluid Intelligence Are Strongly Related Constructs: Comment on Ackerman, Beier, and Boyle (2005)." *Psychological Bulletin* 131: 66–71.

Kantowitz, B. H. 1985. "Channels and Stages in Human Information Processing: A Limited Analysis of Theory and Methodology." *Journal of Mathematical Psychology* 29: 135–174.

Kaplan, D., and L. Glass. 1995. *Understanding Nonlinear Dynamics*. New York: Springer-Verlag.

Karwowski, W., K. Ostaszewski, and J. Zurada. 1992. "Applications of the Catastrophe Theory in Modeling the Risk of Low Back Injury in Manual Lifting Tasks." *La Travail Humain*, (in English) 55: 259–275.

Katerndahl, D. 2010. "Cracking the Linear Lens." *Nonlinear Dynamics, Psychology, and Life Sciences* 14: 349–352.

Kato, T. 2012. "Development of the Coping Flexibility Scale: Evidence for the Coping Flexibility Hypothesis." *Journal of Counseling Psychology* 59: 262–273.

Kauffman, S. A. 1993. *Origins of Order: Self-Organization and Selection in Evolution*. New York: Oxford University Press.

Kauffman, S. A. 1995. *At Home in the Universe: The Search for Laws of Self-Organization and Complexity*. New York: Oxford University Press.

Kelso, J. A. S. 1995. *Dynamic Patterns: The Self-Organization of Brain and Behavior*. Cambridge, MA: MIT Press.

Khalaf, T., W. Karwowski, and N. Sapkota. 2015. "A Nonlinear Dynamics of Trunk Kinematics during Manual Lifting Tasks." *Work: A Journal of Prevention, Assessment and Rehabilitation* 51: 423–437.

Kiefer, A. W., and G. D. Myer. 2015. "Training the Antifragile Athlete: A Preliminary Analysis of Neuromuscular Training Effects on Muscle Activation Dynamics." *Nonlinear Dynamics, Psychology, and Life Sciences* 19: 489–510.

Leining, A., G. Strunk, and E. Mittlestadt. 2013. "Phase Transitions between Lower and High Level Management Learning in times of Crisis: An Experimental Study Based on Synergetics." *Nonlinear Dynamics, Psychology, and Life Sciences* 11: 517–542.

Leonhardt, J., Macchi, L., Hollnagel, E., and Kirwan, B. 2009. *A White Paper on Resilience Engineering for ATM*. Eurocontrol. Accessed March 9, 2011. http://www.eurocontrol.int/esp/gallery/content/public/library

Likens, A. D., P. G. Amazeen, R. Stevens, T. Galloway, and J. C. Gorman. 2014. "Neural Signatures of Team Coordination Are Revealed by Multifractal Analysis." *Social Neuroscience* 9: 219–234.

Lord, F. M., and M. R. Novick. 1968. *Statistical Theories of Mental Test Scores*. Reading, MA: Addison-Wesley.

Lorenz, E. N. 1963. "Deterministic Nonperiodic Flow." *Journal of Atmospheric Sciences* 20: 130–141.

Lorist, M. M., and L. G. Faber. 2011. "Consideration of the Influence of Mental Fatigue on Controlled and Automatic Cognitive Processes." In *Cognitive Fatigue*, edited by P. Ackerman, 105–126. Washington DC: American Psychological Association.

Mandelbrot, B. B. 1983. *The Fractal Geometry of Nature*. New York: Freeman.

Mandelbrot, B. B. 1999. *Multifractals and 1/f Noise*. New York: Springer.

Marci, C. D., J. Ham, E. Moran, and S. P. Orr. 2007. "Physiologic Correlates of Perceived Therapist Empathy and Social-Emotional Process during Psychotherapy." *Journal of Nervous and Mental Disease* 195: 103–111.

Matsumoto, A., and F. Szidarovszky. 2014, March. "Learning in Monopolies with Delayed Price Information". Paper presented to the 6th International Nonlinear Science Conference, Nijmegen, Netherlands.

Matthews, G., P. A. Desmond, C. Neubauer, and P. A. Hancock, eds. 2012. *The Handbook of Operator Fatigue*. Burlington, VT: Ashgate.

Maynard-Smith, J. 1982. *Evolution and the Theory of Games*. Cambridge, UK: Cambridge University Press.

McCain, R. A. 1992. "Heuristic Coordination Games: Rational Action Equilibrium and Objective Social Constraints in a Linguistic Conception of Rationality." *Social Science Information* 31: 711–734.

McDaniel Jr., R. R., and D. J. Driebe, eds. 2005. *Uncertainty and Surprise in Complex Systems*. New York: Springer-Verlag.

Meister, D. 1977. "Implications of the System Concept for Human Factors Research Methodology." *Proceedings of the Human Factors Society* 21: 453–456.

Merrill, S. J. 2011. "Markov Chains for Identifying Nonlinear Dynamics." In *Nonlinear Dynamical Systems Analysis for the Behavioral Sciences Using Real Data*, edited by S. J. Guastello and R. A. M. Gregson, 401–424. Boca Raton, FL: CRC Press.

Morris, J. C., E. D. Morris, and D. M. Jones. 2007. "Reaching for the Philosopher's Stone: Contingent Coordination and the Military's Response to Hurricane Katrina." *Public Administration Review* 67: 94–106.

Morrison, S., and K. M. Newell. 2015. "Dimension and Complexity in Human Movement and Posture." *Nonlinear Dynamics, Psychology, and Life Sciences* 19: 395–418.

Navarro, J., and P. Rueff-Lopes. 2015. "Healthy Variability in Organizational Behavior: Empirical Evidence and New Steps for Future Research." *Nonlinear Dynamics, Psychology, and Life Sciences* 19: 529–552.

Newell, K. M. 1991. "Motor Skill Acquisition." *Annual Review of Psychology* 42: 213–237.

Newhouse, S., D. Ruelle, and F. Takens. 1978. "Occurrence of Strange Attractors: An Axiom near Quasi-Periodic Flows on T^m, $M \geq 3$." *Communications in Mathematical Physics* 64: 35–40.

Nicolis, G., and I. Prigogine. 1989. *Exploring Complexity*. New York: Freeman.

Nusbaum, E. C., and P. J. Silvia. 2011. "Are Intelligence and Creativity Really So Different? Fluid Intelligence, Executive Processes, and Strategy Use in Divergent Thinking." *Intelligence* 39: 36–45.

Pascual-Leone, J. 1970. "A Mathematical Model for the Transition Rule in Piaget's Developmental Stages." *Acta Psychologica* 32: 301–345.

Peressini, A. F., and S. J. Guastello. 2014. *Orbital Decomposition: A Short User's Guide to ORBDE V2.4*. [Software]. Accessed May 1, 2014. http://www.societyforchaostheory.org/resources/, Menu 4.

Pincus, D. 2001. "A Framework and Methodology for the Study of Nonlinear, Self-Organizing Family Dynamics." *Nonlinear Dynamics, Psychology and Life Sciences* 5: 139–173.

Prigogine, I., and I. Stengers. 1984. *Order out of Chaos: Man's New Dialog with Nature*. New York: Bantam.

Reason, J. 1997. *Managing the Risks of Organizational Accidents*. Brookfield, VT: Ashgate.

Rodrick, D., and W. Karwowski. 2006. "Nonlinear Dynamical Behavior of Surface Electromyographical Signals of Biceps Muscle under Two Simulated Static Work Postures." *Nonlinear Dynamics, Psychology, and Life Sciences* 10: 21–35.

Salas, E., R. Stevens, J. Gorman, N. J. Cooke, S. J. Guastello, and A. A. von Davier. 2015. "What Will Quantitative Measures of Teamwork Look like in 10 Years?" *Proceedings of the Human Factors and Ergonomics Society* 59: 235–239.

Samuelson, L. 1997. *Evolutionary Games and Equilibrium Selection*. Cambridge, MA: MIT Press.

Sawyer, R. K. 2005. *Social Emergence: Societies as Complex Systems*. New York: Cambridge University Press.

Schuldberg, D. 2015. "What is Optimum Variability?" *Nonlinear Dynamics, Psychology, and Life Sciences* 14: 553–568.

Seger, C. A. 1994. "Implicit Learning." *Psychological Bulletin* 115: 163–196.

Shannon, C. E. 1948. "A mathematical theory of communication." *Bell System Technical Journal* 27: 379–423.

Shelhamer, M. 2007. *Nonlinear Dynamics in Physiology: A State-Space Approach*. Singapore: World Scientific.

Shelhamer, M. 2009. "Introduction to the Special Issue on Psychomotor Coordination and Control." *Nonlinear Dynamics, Psychology, and Life Sciences* 13: 1–2.

Sheridan, T. B. 2008. "Risk, Human Error, and System Resilience: Fundamental Ideas." *Human Factors* 50: 418–426.

Sleimen-Malkoun, R., J. J. Temprado, V. K. Jirsa, and E. Berton. 2010. "New Directions Offered by the Dynamical Systems Approach to Bimanual Coordination for Therapeutic Intervention and Research in Stroke." *Nonlinear Dynamics, Psychology, and Life Sciences* 14: 435–462.

Smith, L. A. 2007. *Chaos: A Very Short Introduction*. New York: Oxford University Press.

Sprott, J. C. 2003. *Chaos and Time Series Analysis*. New York: Oxford University Press.

Sprott, J. C. 2004. "Can a Monkey with a Computer Create Art?" *Nonlinear Dynamics, Psychology, and Life Sciences* 8: 103–114.

Stamovlasis, D., and M. Koopmans. 2014. "Editorial Introduction: Education is a Dynamical System." *Nonlinear Dynamics, Psychology, and Life Sciences* 18: 1–4.

Stamovlasis, D., and G. Tsaparlis. 2012. "Applying Catastrophe Theory to an Information-Processing Model of Problem Solving in Science Education." *Science Education* 96: 392–410.

Stevens, R. H., T. L. Galloway, P. Wang, and C. Berka. 2012. "Cognitive Neurophysiologic Synchronies: What Can They Contribute to the Study of Teamwork?" *Human Factors* 54: 489–502.

Stevens, R., J. C. Gorman, P. Amazeen, A. Likens, and T. Galloway. 2013. "The Organizational Neurodynamics of Teams." *Nonlinear Dynamics, Psychology, and Life Sciences* 17: 67–86.

Stevens, R. H., T. L. Galloway, and C. Lamb. 2014. "Submarine Navigation Team Resilience: Linking EEG and Behavioral Models." *Proceedings of the Human Factors and Ergonomics Society* 58: 245–249.

Strogatz, S. 2003. *Sync: The Emerging Science of Spontaneous Order*. New York: Hyperion.

Sturmberg, J. P., and C. M. Martin, eds. 2013. *Handbook of Systems and Complexity in Health*. New York: Springer.

Sulis, W. 2009. "Collective Intelligence: Observations and Models." In *Chaos and Complexity in Psychology: Theory of Nonlinear Dynamical Systems*, edited by S. J. Guastello, M. Koopmans and D. Pincus, 41–72. New York: Cambridge University Press.

Thom, R. 1975. *Structural Stability and Morphegenesis*. New York: Benjamin-Addison-Wesley.

Thompson, H. L. 2010. *The Stress Effect: Why Smart Leaders Make Dumb Decisions – And What to Do about It*. San Francisco: Jossey-Bass.

Turvey, M. T. 1990. "Coordination." *American Psychologist* 45: 938–953.

Vargas, B., D. Cuesta-Frau, R. Ruis-Esteban, E. Cirugeda, and M. Varela. 2015. ""What Can Biosignal Entropy Tell Us about Health and Disease? Applications in Some Clinical Fields." *Nonlinear Dynamics, Psychology, and Life Sciences* 19: 419–436.

Vidgen, R., and L. Bull. 2011. "Application of Kauffman's Coevolutionary NKCS Model to Management and Organization Studies." In *The Sage Handbook for Complexity Management*, edited by P. A. Allen, S. Maguire and B. McKelvey, 201–219. Thousand Oaks, CA: Sage.

Walker, G. H., N. A. Stanton, P. M. Salmon, D. P. Jenkins, and L. Rafferty. 2010. "Translating Concepts of Complexity to the Field of Ergonomics." *Ergonomics* 53: 1175–1186.

Woods, D. D., and J. Wreathall. 2008. "Stress-Strain Plots as a Basis for Assessing System Resilience." In *Resilience Engineering: Remaining Sensitive to the Possibility of Failure*, edited by E. Hollnagel, C. P. Nemeth and S. W. A. Dekker, 143–158. Aldershot, UK: Ashgate.

Zausner, T. 2007. "Process and Meaning: Nonlinear Dynamics and Psychology in Visual Art." *Nonlinear Dynamics, Psychology, and Life Sciences* 11: 149–166.

Zeeman, E. C. 1977. *Catastrophe Theory: Selected Papers 1972–1977*. Reading, MA: Addison-Wesley.

Zohar, D. 1980. "Safety Climate in Industrial Organizations: Theoretical and Applied Implications." *Journal of Applied Psychology* 65: 96–102.

Fitting methods to paradigms: are ergonomics methods fit for systems thinking?

Paul M. Salmon, Guy H. Walker, Gemma J. M. Read, Natassia Goode and Neville A. Stanton

ABSTRACT

The issues being tackled within ergonomics problem spaces are shifting. Although existing paradigms appear relevant for modern day systems, it is worth questioning whether our methods are. This paper asks whether the complexities of systems thinking, a currently ubiquitous ergonomics paradigm, are outpacing the capabilities of our methodological toolkit. This is achieved through examining the contemporary ergonomics problem space and the extent to which ergonomics methods can meet the challenges posed. Specifically, five key areas within the ergonomics paradigm of systems thinking are focused on: normal performance as a cause of accidents, accident prediction, system migration, systems concepts and ergonomics in design. The methods available for pursuing each line of inquiry are discussed, along with their ability to respond to key requirements. In doing so, a series of new methodological requirements and capabilities are identified. It is argued that further methodological development is required to provide researchers and practitioners with appropriate tools to explore both contemporary and future problems.

Practitioner Summary: Ergonomics methods are the cornerstone of our discipline. This paper examines whether our current methodological toolkit is fit for purpose given the changing nature of ergonomics problems. The findings provide key research and practice requirements for methodological development.

Introduction

Structured methods provide the foundation for our discipline (Stanton et al. 2013). Within the realm of cognitive ergonomics, researchers and practitioners have a wide range of methods available for studying aspects of operator, team and system performance. In the individual operator context, these include methods such as cognitive task analysis (e.g. Klein, Calderwood and Clinton-Cirocco, 1986), workload assessment (Hart and Staveland 1988), situation awareness measurement (e.g. Endsley 1995) and error identification techniques (e.g. Shorrock and Kirwan 2002). For teams, they include teamwork assessment (e.g. Burke 2004), analysis of communications (e.g. Houghton et al. 2006) and team workload assessment (e.g. Helton, Funke, and Knott 2014). More recently, methods such as Accimap (Svedung and Rasmussen 2002), the Event Analysis of Systemic Teamwork (EAST, Stanton et al. 2013), the MacroErgonomic Analysis and Design method (MEAD; Kleiner 2006), the Functional Resonance Analysis Method (FRAM; Hollnagel 2012) and Cognitive Work Analysis (CWA; Vicente 1999) are being applied to analyse overall systems and their emergent behaviours.

There is no doubt then that ergonomists have access to a diverse methodological toolkit; however, the systems in which ergonomists operate are becoming increasingly complex and technology driven (Dekker, Hancock, and Wilkin 2013; Grote, Weyer, and Stanton 2014; Walker et al. this issue; Woods and Dekker 2000). Whilst work systems have arguably been complex since the dawn of the discipline, a shift towards the systems thinking paradigm, along with increasing levels of technology and complexity, is beginning to expose the reductionist tendencies of many ergonomics methods. Indeed, an examination of recent papers published in this journal leaves no doubt that the issues currently being tackled are stretching the capabilities of our methods (e.g. Cornelissen et al. 2013; Stanton 2014; Trapsilawati et al. 2015; Walker, Stanton, Salmon and Jenkins, 2010; Young et al. 2015). Example, issues include emergence, resilience, performance variability, distributed cognition and even complexity itself. These issues (or lack of them) are to be found, increasingly prominently, in modern day catastrophes (a major focus of ergonomics research and practice). Typically, these have numerous contributory factors stretching over multiple

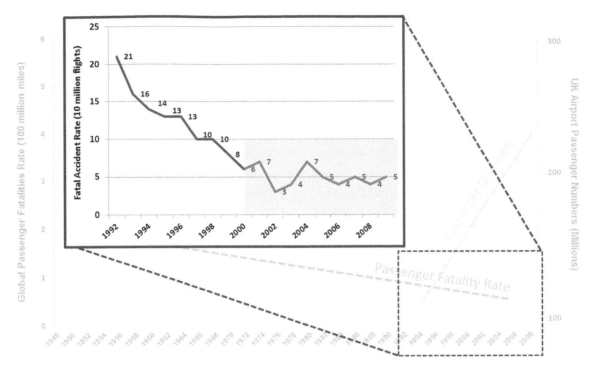

Figure 1. The pattern of global passenger fatalities per 10 million passenger miles on scheduled commercial air transport since 1993. The graph shows that the precipitous drop in fatality rates since 1945 has, since 2003, levelled off.Source: European Aviation Safety Authority(2010). Source: European Aviation Safety Authority (2010).

people, technologies, organisations, environments and time. With these complex problems in mind, it is dangerous to assume that our methods remain fit for purpose simply because we continue to use them.

On top of this is the fact that the issues themselves do not appear to be improving as they once were. The statistics across common ergonomics application areas make for sobering reading. In Australia, for example, over 600,000 workers are injured per year with an estimated annual cost of over $60 billion dollars (Safework Australia 2012a, 2012b). In areas such as road transport, road collisions take the lives of well over 1000 Australians per year and cause approximately 50,000 to be admitted to hospital (Bradley and Harrison 2008). Even in domains where the level of regulation and control is much higher, such as the aviation industry, there were still 221 serious incidents and around 5500 incidents reported to the Australian Transport Safety Bureau in 2013 (ATSB 2013). In the rail industry, there were 350 fatalities and 923 serious personal injuries across Australia between 2002 and 2012 (ATSB 2012). For all of the good ergonomics research and practice achieves, the outcomes that we seek to prevent are still occurring, and in large numbers. This is not of course entirely down to ergonomists and our methods, and clearly the methods being applied are having a beneficial effect; otherwise, there would not be a business or safety case for continued investment in them. Other issues play a role including the dissemination of ergonomics applications, the integration

of ergonomics practices within system design and operation practices, and the gap that exists between ergonomics research and practice (Underwood and Waterson 2013). Equally though, the numbers suggest that there are likely problems that are proving resistant to ergonomics methods. Why is this? Are our methods tackling (successfully) only the deterministic parts of problems, leaving the underlying systemic issues unaddressed? At the very least, it is legitimate to question the validity of our existing ergonomics toolkit. Not least because validity is often assumed but seldom tested (Stanton and Young 1999).

For ergonomics to remain relevant, it is imperative that our methods can cope with the problem spaces in which researchers and practitioners work and indeed with the paradigms that are driving this work. We are not alone in expressing these concerns. Dekker (2014), Leveson (2011), Salmon et al. (2011) and Walker et al. (this issue) present an, at times, alarming picture of the complexities of modern day systems and the extent to which they are rapidly outpacing the capabilities of our methodological toolkit. In addition, new concepts introduced to better deal with the increasingly complex nature of modern day systems, such as Safety II (Hollnagel et al. 2013), require appropriate methodological support.

The aim of this paper is to explore this by examining the contemporary ergonomics problem space and the extent to which ergonomics methods can meet the challenges posed. We discuss five key areas within the highly popular contemporary ergonomics paradigm of systems thinking, when applied to accident analysis and prevention activities, and examine the ability of existing ergonomics methods to respond to them: normal performance as a cause of accidents (Dekker 2011; Leveson 2004; Rasmussen 1997), accident prediction (Salmon et al. 2014a), system migration (Rasmussen 1997), systems concepts (Hutchins 1995; Stanton et al. 2006) and ergonomics in design (Read et al. 2015). Where our methods are deemed to be lacking, new methodological requirements and capabilities are identified. Whilst we acknowledge that other disciplines and areas of safety science (e.g. resilience, safety II) may possess alternative methodologies that fulfil some of the requirements discussed, the focus of this article is specifically on the ergonomics methods used by ergonomics researchers and practitioners. In line with the topic of the special issue, we see methodological extension, development and integration as an omnipresent issue for our discipline.

Normal performance as a cause of accidents

Systems thinking and its methodological implications

Significant progress has been made in understanding accident causation in safety critical systems. Systems models, in particular, are now widely accepted (Leveson 2004; Rasmussen 1997) and there are a range of methods that enable accidents to be analysed from this perspective (e.g. Hollnagel 2012; Leveson 2004; Svedung and Rasmussen 2002). This approach has a long legacy in safety science, from the foundational work of Heinrich (1931) through to the evolution of a number of more recent accident causation models and analysis methods (e.g. Leveson 2004; Perrow 1984; Rasmussen 1997; Reason 1990). Accidents are now widely acknowledged to be systems phenomena, just as safety is (Dekker 2011; Hollnagel 2004). Both safety and accidents therefore are emergent properties arising from non-linear interactions between multiple components distributed across a complex web of human and machine agents and interventions (e.g. Leveson 2004).

It is precisely this form of thinking, and the evolution of it, that brings the methods we use into question. Despite the great progress in safety performance that has been made in most safety critical sectors since the Second World War, significant trauma still occurs and in some areas progress may be slowing. Whilst Figure 1 shows a plateauing effect in commercial air transport, data shows a similar trend in other problem areas such as rail level crossings (Evans 2011). Moreover, in areas such as road transport where the intensity of operations is increasing, the global burden is increasing and is projected to increase significantly (WHO 2014). Leveson (2011) argues that little progress is now being made and suggests that one reason

is that our methods do not fully uncover the underlying causes of accidents. Part of the issue may be that the evolution in accident causation models is not reflected in current accident analysis methods. Another issue that could conceivably play a part is the well-documented research-practice gap, whereby practitioners continue to use older methodologies that do not reflect contemporary models of accident causation (Salmon et al. Forthcoming; Underwood and Waterson 2013).

One of the fundamental advances provided by state of the art models centres around the idea that the behaviours underpinning accidents do not necessarily have to be errors, failures or violations (e.g. Dekker 2011; Leveson 2004; Rasmussen 1997). As Dekker (2011) points out, systems thinking is about how accidents can happen when no parts are broken (Perrow 1984). This provides an advance over popular models that tend to subscribe to the idea that failure leads to failure (e.g. Reason 1990). Reason did note that latent conditions can emerge from normal decisions and actions, but to take this idea much further is to describe two key tenets. First, normal performance plays a role in accident causation, and second, accidents arise from the very same behaviours and processes that create safety. In his recent drift into failure model, Dekker (2011) argues that the seeds for failure can be found in 'normal, day-to-day processes' (99) and are often driven by goals conflicts and production pressures. These normal behaviours include workarounds, improvisations and adaptations (Dekker 2011), but may also just be normal work behaviours routinely undertaken to get the job done. It is only with hindsight and a limited investigation methodology that these normal behaviours are treated as failures. Both safety boundaries and behaviour can drift: what is safe today may not be safe tomorrow. It is notable in the Kegworth aviation accident in the UK, the pilots shut down of the left engine would have been the correct action on the previous generation of aircraft for which they were most familiar (Griffin, Young, and Stanton 2010; Plant and Stanton 2012).

Theoretical advances such as this have important implications for the methodologies applied to understand accidents. We require appropriate methodologies that reflect how contemporary models think about accident causation. The tenets described above provide an interesting shift in the requirements for accident analysis methodologies. Dekker (2014) argues that practitioners should not look for the known problems that appear in incident reporting data or safety management systems. Instead, he argues, that the focus should be in the places where there are no problems, in other words, normal work. In addition, the burgeoning concept of Safety II and I (Hollnagel et al. 2013) argues that safety management needs to move away from attempting to ensure that as little as possible goes wrong to ensuring that as much as possible goes right. A key part of this involves understanding performance when it went right as well as when it went wrong.

This raises critical questions – do our accident analysis methodologies have the capability to incorporate normal performance into their descriptions of accidents? Do we currently incorporate normal performance into accident analyses? And if we do, are we misclassifying it as errors, failures and inadequacies? Further, should we be investigating and analysing accidents at all or putting those efforts into auditing everyday work providing an opportunity to continuously understand and manage performance variability without waiting for major accidents to occur? If so, do we have appropriate methods to support this?

Accident analysis methods

According to the literature, the most popular accident analysis methods are Accimap (Rasmussen 1997), Systems Theoretic Accident Model and Process (STAMP) (Leveson 2004) and Human Factors Analysis and Classification System (HFACS) (Wiegmann and Shappell 2003). Accimap accompanies Rasmussen's now popular risk management framework and is used to describe accidents in terms of contributory factors and the relationships between them. This enables a comprehensive representation of the network of contributory factors involved. It does this by decomposing systems into six levels across which analysts place the decisions and actions that enabled the accident in question to occur (although the method is flexible in that the number of levels can be adjusted based on the system in question). Interactions between the decisions and actions are subsequently mapped onto the diagram to show the relationships between contributory factors within and across the six levels. A notable feature of Accimap is that it does not provide analysts with taxonomies of failure modes; rather, analysts have the freedom to incorporate any factor deemed to have played a role in the accident in question.

The STAMP method views accidents as resulting from the inadequate control of safety-related constraints (Leveson 2004), arguing that they occur when component failures, external disturbances and/or inappropriate interactions between systems components are not controlled (Leveson 2004, 2011). STAMP uses a 'control structure' modelling technique to describe complex systems and the control relationships that exist between components at the different levels. A taxonomy of control failures is then used to classify the failures in control and feedback mechanisms that played a role in the incident under analysis. An additional component of STAMP involves using systems dynamics modelling to analyse system degradation over time. This enables the interaction of control failures to be demonstrated along with their effects on performance.

Although not based on contemporary models of accident causation, the HFACS (Wiegmann and Shappell 2003) remains highly popular (e.g. Daramola 2014; Mosaly et al. 2014). HFACS is a taxonomy-based approach that provides analysts with taxonomies of error and failure modes across four system levels based on Reasons Swiss cheese model of organizational accidents: unsafe acts, preconditions for unsafe acts, unsafe supervision and organizational influences. Although developed originally for use in analysing aviation incidents, the method has subsequently been redeveloped for use in other areas including: mining (Lenné et al. 2012), maritime (Chauvin et al. 2013), rail (Baysari, McIntosh, and Wilson 2008) and health care (ElBardissi et al. 2007). Later versions of the method have extended the levels to incorporate an 'external influences' level which considers failures outside of organisations such as legislation gaps, design flaws and administration oversights (Chen et al., 2013).

Accident analysis methods and normal performance

A notable shortfall of the latter two methods is their focus on abnormal behaviours or failures. Both HFACS and STAMP provide taxonomies of error and failure modes that are used to classify the behaviours involved in accident scenarios, which in turn means that there is little scope for analysts to include behaviours other than those deemed to have been failures of some sort. There is no opportunity for analysts to incorporate normal behaviours in their descriptions of accidents – they have to force fit events into one of the error or failure modes provided. The output is a judgement on what errors or failures combined to create the accident under analysis. Whilst this is inappropriate given current knowledge on accident causation, a worrying consequence may be that the normal behaviours that contribute to accidents are not picked up during accident analysis efforts. This may impact accident prevention activities by providing a false sense of security that nothing else is involved and thus nothing needs fixing (apart from error producing human operators). A more sinister implication is that organisations who apply methods such as HFACS may not develop a sufficient understanding of accidents to prevent them. Although the aviation sector routinely monitors normal performance through flight data monitoring systems, arguably they do not run analyses of the role of normal performance in air crashes. Extending methods such as HFACS and STAMP to incorporate analyses of normal performance in accidents is therefore a pressing requirement. The benefits include developing a more holistic view of accident causation that is not entirely based on understanding errors and failures and understanding how normal behaviours lead to system failure.

Accimap, on the other hand, does not use a taxonomy of failure or error modes and so enables analysts to incorporate normal performance and to show its relationship with other behaviours. There is freedom for analysts to include any form of behaviour in the network of contributory factors. Despite this, Accimap descriptions still tend to incorporate many contributory factors prefixed with descriptors such as 'failure to', 'lack of' or ending with 'error'. A pressing question here then is the extent to which the failures described in Accimap analyses actually represent failure or are in fact normal behaviours. Salmon, Walker, and Stanton (2015) recently examined a sub-set of their own analyses and found examples where contributory factors originally described as failures could be reclassified as normal performance.

Another important line of inquiry is the extent to which researchers and practitioners understand the need to incorporate 'normal' behaviours in accident analyses. A downside of Accimap's flexibility is that there are no prompts for analysts to look beyond failures. A step-by-step procedure specifying this would be beneficial as would investigation techniques that prompt investigators to look beyond failures. For example, the form of questioning might be: (a) 'what behaviours would you reasonably expect to see given this context and these set of features', and (b) 'are those expected behaviours the ones you actually want to see'.

A related issue is that methods such as Accimap are not typically used to assess performance in which accidents were avoided. The need to monitor and understand performance that went right, as opposed to just performance that went wrong (Hollnagel et al. 2013), has been strongly argued for by proponents of resilience engineering and Safety II (e.g. Hollnagel et al. 2013). Notably, big data capabilities will enable this, and sectors such as aviation do monitor aspects of flight performance. In addition, many ergonomics methods exist for examining performance generally, such as Hierarchical Task Analysis (HTA; Stanton, 2006), the EAST (Stanton et al. 2013) and CWA (Vicente 1999), and also for examining variance in performance such as the MEAD (Kleiner 2006). Despite this, the focus of such applications is more often than not on theoretical development, the impact of introducing new procedures, training programmes or devices, rather than accident causation. Apart from Trotter, Salmon, and Lenné (2014), who used Accimap to examine the Apollo 13 incidents, to the authors' knowledge there are no other published applications in which performance not resulting in an accident of some sort is examined via methods such as Accimap. Despite its origins in accident analysis, it is these authors' opinion that the method provides a useful framework for examining both normal performance and events where an adverse outcome was avoided.

The conclusion then is that there is room for improvement in our accident analysis methods, both in terms of their structure and the guidance on how to use them. Not all state-of-the-art methods are consistent with our current understanding of accident causation. Further, even for the methods that are, it is questionable whether they are being used in a manner consistent with contemporary models of accident causation. This paradox represents a key issue for ergonomics researchers and practitioners and for safety science generally. On the one hand, there is now a widespread understanding that the role of normal performance in accidents is apparent and needs to be understood (Dekker 2011, 2014; Leveson 2011; Rasmussen 1997). On the other hand, accident analysis efforts, regardless of domain, do not seem to be dealing particularly well with this feature. This means our understanding of accidents may be incomplete. Worse, the countermeasures we recommend are based on incomplete analyses and doomed to fail. Dekker (2014) points out that we need to look where there are no holes; equally, we need methods that do not dig holes or take us down them.

Accident prediction

Forecasting accidents before they occur has been labelled the final frontier for ergonomics (Moray 2008; Salmon et al. 2011; Stanton and Stammers 2008). Although there have been various attempts at developing accident prediction models (e.g. Deublein et al. 2013), most are statistical models that are unable to identify and describe how behaviours across overall sociotechnical systems might combine to create failure scenarios. Other predictive methods are available, such as those that can be used to predict the kinds of 'human errors' that lead to accidents (see Stanton et al. 2013). Indeed, some of these methods have been shown to achieve acceptable levels of reliability and validity (e.g. Stanton et al. 2009). The problem is that they predict what is likely the last behaviour in a long and complex network of interacting and emergent behaviours occurring across various parts of the system. They predict consequences not causes and do not identify the network of contributory factors that might co-occur to create accidents. Whilst it is of course useful to examine what erroneous behaviours are created by the systems emergent properties, accident prevention efforts are better served by looking at the interactions that occur before the human operator makes the error. In short, it is the entire accident scenario, including interacting factors and emergent behaviours, that are important for understanding how to prevent accidents.

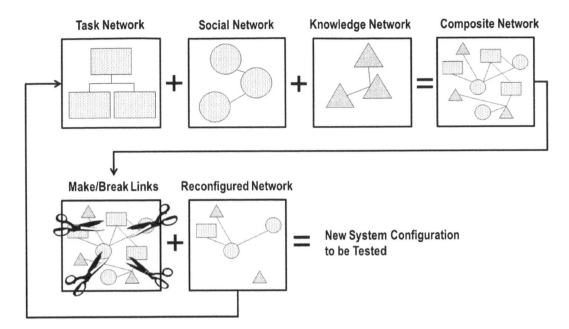

Figure 2. The EAST models sociotechnical systems by combining task, social and knowledge networks into composite networks which can be systematically degraded into all possible configurations.

A systems approach to prediction

A key requirement, then, is a prediction method that is underpinned by systems thinking, or at least by the same tenets that our accident causation models are. Error prediction methodologies can be thought of as reductionist (although they are not entirely reductionist as they do focus on human–machine interactions). Reductionist approaches, those which rely on taking the system apart in order to understand the components, then reassembling the components back into the complete system (on the tacit assumption that the whole cannot be greater or less than the sum of its parts) do not allow us to detect the emergent properties associated with the types of risk issues upon which we wish to make progress (see Walker et al. 2009). Systems approaches do. One means by which they can enable forecasting and prediction is to consider the causal texture of the systems environment, and the system's movement through that environment.

As discussed, the systems approach has become popular in part because of various systems analysis methodologies that can, to some extent at least, do this (e.g. Rasmussen 1997). These methods, for example Accimap (Rasmussen 1997), are becoming increasingly popular for accident investigation purposes. A major limitation of these methodologies is that, so far, they have not been used in a pro-active manner: organisations are effectively waiting for loss events to occur before they can work on prevention strategies. The lack of data resulting from improved safety trends combined with greater operational intensity and risk exposure means that, if anything, loss events are more likely to be large-scale and unexpected,

meaning that 'learning from disasters' is becoming increasingly dubious from an ethical perspective. The need for systems-based prediction approaches is discussed extensively in the literature (e.g. Moray 2008; Salmon et al. 2011; Stanton and Stammers 2008), but as yet a credible approach has yet to emerge.

Existing ergonomics methods and accident prediction

Encouragingly, there are methods available that could be used to predict accidents (Salmon et al. 2011). Systems analysis methods such as EAST (Stanton et al. 2013), CWA (Vicente 1999) and FRAM (Hollnagel 2012) all describe systems, interactions and their resulting emergent behaviours. There is no reason why these approaches cannot be used to predict emergent states such as accidents and some of them are being tested for this through exploratory work. We will discuss EAST here, but similar arguments have been made for CWA (Salmon et al. 2014a) and FRAM (Hollnagel 2012).

EAST provides an integrated suite of methods for analysing the inter-related task, social and information networks underlying the performance of sociotechnical systems (Stanton et al. 2013). Task networks are used to describe the goals and tasks that are performed within a system (i.e. which agents, both human and non-human, do what). Social networks are used to analyse the organisation of the system and the communications taking place between agents (i.e. who/what interacts and communicates with who/what during tasks). Information networks show how information and knowledge are distributed

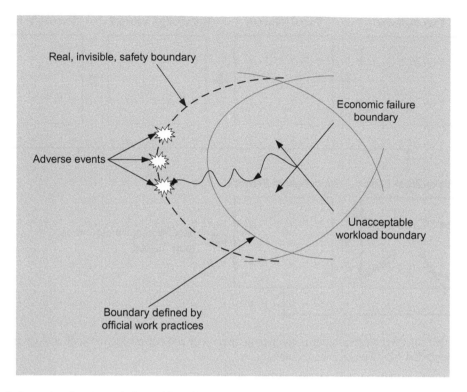

Figure 3. Rasmussen's dynamic safety space (adapted from Rasmussen 1997).

across different agents within the system (i.e. who/what knows what at different points in time).

Since its inception, the EAST method has been applied by its developers to retrospectively understand performance in many domains ranging from maritime (Stanton 2014; Stanton et al. 2006) and air traffic control (Walker et al.2010) to road transport (Salmon et al. 2014b) and railway maintenance (Walker et al. 2006). The primary limitation of this body of work, of course, is that the analyses were based on events that had happened.

In response, Stanton, Harris, and Starr (2014) have conducted initial pilot research to test the utility of EAST when used to model system performance in a predictive capacity. This involved adding and breaking links and nodes within the networks to explore different system states. The effect of this make/break link process is to create 'short circuits', 'long circuits' or 'no circuits', all of which put systems into new configurations (see Figure 2). For example, adding and breaking links within the networks reveals instances where a human operator may or may not be aware of a particular piece of information, where a task will or will not be fulfilled by a human operator or piece of technology, or where a required communication may or may not be made. By these means, it is possible to model the majority of possible accident pathways in a given system model under a wide range of different permutations. On the other hand, adding and breaking links may reveal aspects of normal performance that move towards the boundary of safe operations (Rasmussen 1997). This systems level focus

encourages a different approach to forecasting resilience: to examine persistent and emergent patterns that arise even though the boundaries of the system cannot always be fully known. Further testing of EAST in this capacity is currently being undertaken by the authors and it is recommended that similar testing involving methods such as CWA and FRAM are also undertaken.

Migration towards safety boundaries

As discussed, Rasmussen's risk management framework (Rasmussen 1997; see Figure 3) is becoming one of the most popular safety and risk management models of our time with an increasing set of applications (e.g. Goode et al. 2014; Salmon et al. 2014c; Underwood and Waterson 2014). In his seminal article, Rasmussen outlined the concept of migration based on the 'Brownian movements' of gas molecules, describing how organisations shift towards and away from safety and performance boundaries due to various constraints including financial, production and performance pressures. According to Rasmussen, there is a boundary of economic failure: these are the financial constraints on a system that influence behaviour towards greater cost efficiencies. There is also a boundary of unacceptable workload: these are the pressures experienced by people and equipment in the system as they try to meet economic and financial objectives. The boundary of economic failure creates a pressure towards greater efficiency, which works in opposition to a similar pressure against excessive workload. As systems involve human as well as technical elements, and because humans are able to adapt situations to suit their own needs and preferences, these pressures inevitably introduce variations in behaviour that are not explicitly designed and can lead to increasingly emergent system behaviours, both good and bad (Clegg 2000; Qureshi 2007). Over time, this adaptive behaviour can cause the system to cross safety boundaries and accidents to happen (Qureshi 2007; Rasmussen 1997). The key, then, is to detect in advance (a) where those boundaries are and (b) where the system is travelling in relation to them.

Mapping migration

Unfortunately, there are few ergonomics methodologies that are capable of describing where boundaries are situated and whereabouts organisations may be in relation to them. Further, the ability to dynamically track an organisation's migration is not readily supported by current ergonomics methods. Whilst information on so-called lagging indicators such as accidents and near misses can provide an indication of proximity to a safety boundary, there is an absence of ergonomics methods that use leading indicators to track organisational performance and safety. This is a key methodological requirement for the future.

Systems concepts

Systems thinking in accident analysis and prevention raises important questions around the nature of ergonomics concepts that are commonly cited as causal factors in accident investigation reports. In particular, why these concepts are not examined through a systems lens is pertinent to this discussion. Encouragingly, some ergonomics concepts are beginning to be examined in this manner. For example, cognition (Hutchins 1995) and situation awareness (Stanton et al. 2006) are two notable areas in which a systems approach has proved successful. Indeed, it is argued that advances such as these are now needed for Ergonomics to fulfil its role of supporting sociotechnical systems design and analysis. In accident analysis and prevention efforts, for example, the still prevalent but widely derided human error-driven old view on accident causation supports identification of errors and failures at the individual level. In relation to situation awareness, Salmon, Walker, and Stanton (2015) describe how a systems approach enables loss of situation awareness to be appropriately cited as a causal factor in accidents, whereas an individual approach (e.g. Endsley 1995) raises moral and ethical concerns and is incongruent with the systems approach to accident causation (Dekker, 2015).

So why are other ergonomics concepts not being examined through a systems lens? Notably, both Hutchins distributed cognition model and Stanton et al.'s (2006) distributed situation awareness model came equipped with an appropriate methodology supporting a shift towards the system as the unit of analysis. Indeed, the keen uptake of both models could not have been achieved without appropriate data collection and analysis methods. In the case of situation awareness, for example, at the time no other methodology was available to support examination at a systems level (Salmon et al. 2006), hence, the authors set about developing one. Without the resulting EAST framework, many recent distributed situation awareness applications could not have taken place (Salmon et al. 2014b; Stanton 2014; Walker et al. 2010).

A lack of systems methods

Although other popular ergonomics constructs appear suited to examination through a systems lens, it is apparent that we do not possess the methods to support this or even a systems of systems view on performance (Siemieniuch and Sinclair 2014). It could be that concepts will remain misunderstood or only partly understood at an individual level simply because we do not have appropriate methods to study them through a systems lens. Moreover, a continuing focus on concepts at an individual operator level will support the blame culture in accident analysis (Dekker 2015; Salmon, Walker, and Stanton 2015).

Mental workload (see Young et al. 2015) is a case in point. Whilst it has predominantly been thought of as an individual operator concept, it is increasingly being examined at the team level (Helton, Funke, and Knott 2014), and there is no reason why it cannot be considered at a systems level. Just as Stanton et al. (2006) describe how situation awareness is an emergent property of systems and is distributed across operators, mental workload can be also thought of in this way. Moreover, similar to the transactions in awareness described by Stanton and colleagues, transactions in mental workload between operators are readily apparent in sociotechnical systems. An example of this is the workload shedding by Air Traffic Controllers, dividing sectors up as the air traffic increases (Walker et al. 2010).

But do we have the methods to explore this? The answer, unfortunately, is not yet. There are many methods available to support the assessment of individual operator workload, including the highly popular NASA TLX (Hart and Staveland 1988) and many similar subjective rating scales (see Stanton et al. 2013). In addition, there are other individual focused methods, such as psychophysiological measures. Further, methodologies that support the assessment of team mental workload are emerging (Helton, Funke, and Knott 2014), although these are not without their problems. Unfortunately, methods that can consider mental workload at a systems level do not yet exist. Such a method, something akin to 'distributed mental workload' assessment would need to consider the workload of multiple actors, how interactions between actors shape each other's workload, how different levels of workload dynamically shift throughout task performance and, further, what wider systemic factors constrain or facilitate workload. In addition, under a systems view workload across system levels, and even non-human actors, should be considered.

The general lack of systems methods is a significant barrier to advancing many ergonomics concepts to a systems level. Although systems analysis methods and macro-ergonomics methods exist (See Stanton et al. 2005), the majority of popular methods are focused on individuals or teams. This is exemplified by Stanton et al.'s (2013) recent compilation of human factors methods. Of over 100 methods described, less than 10 have the ability to take the overall system as the unit of analysis. Of course, given the focus on humans and the theories that emerged at the dawn of our discipline, this is not surprising; however, this does not mean it should remain acceptable.

Ergonomics methods in design

Ergonomics has a key role to play in the design of new technologies, interfaces, training programs, procedures and indeed overall systems. Accordingly, there are many theories and methods within our discipline that have been applied in design efforts across many domains. Notably, many design-related applications have involved the use of ergonomics methods to evaluate and refine design concepts and prototypes (e.g. HTA, Stanton, 2006) or to specify design requirements which are then used to inform a design process of some sort (e.g. Endsley, Bolte, and Jones 2003). A criticism of many ergonomics methods is that they do not directly contribute to design – that is, they are not used by designers to design. Rather, the outputs are used to inform and/or evaluate and refine designs. This may be one reason why ergonomists are often seen as trouble shooters, called in to solve design flaws once technologies and systems have been implemented and flaws have been identified.

The requirement to shift the emphasis on ergonomics to the front end of the design life cycle is well known and has many suitors (Norros 2014). However, it requires ergonomics methods that can be used by designers and design teams to design, or at least that can be integrated with design approaches. It also requires ergonomics researchers and practitioners to take the lead to facilitate design

efforts, which surely would be a paradigm shift in the way complex systems are engineered.

A direct contribution to design?

Whilst a significant shift is required, it is notable that many of our methods are suited to directly and indirectly informing design. Rather than develop new methods per se, the pressing requirement seems to be the development of processes to bridge the gap between analysis and design. One such approach that aims to provide some traction in this area is the CWA design toolkit (CWA-DT; Read et al. 2015).

The CWA-DT intends to assist CWA users to identify design insights from the application of the framework and to use these insights within a participatory design paradigm. It promotes the collaborative involvement of experts (i.e. ergonomics professionals, designers and engineers), stakeholders (i.e. company representatives, supervisors, unions) and end users (i.e. workers or consumers) to solve design problems, based on insights gained through CWA.

Underlying the CWA-DT is both the design philosophy of CWA (i.e. 'let the worker finish the design') and the related sociotechnical systems theory approach which aims to design organisations and systems that have the capacity to adapt and respond to changes and disturbances in the environment (Cherns 1976; Clegg 2000; Trist and Bamforth 1951; Walker et al. 2009). Consequently, the CWA-DT includes design tools and methods that encourage consideration of the values underlying the sociotechnical systems approach (and indeed underpinning ergonomics more generally). These include the notion of humans as assets or adaptive decision-makers, rather than error-prone liabilities; of technology being designed as a tool to assist humans to achieve their goals, rather than implemented because of assumed efficiency or cost-savings; and design to promote the quality of life or wellbeing for end users. Further, the consideration of sociotechnical design principles, such as minimal critical specification, boundary management and joint design of social and technical elements, intends to achieve the design of systems that can operate within their safety and performance boundaries both on implementation and in an on-going fashion through continual monitoring and re-design.

Initial applications indicate that this design approach shows promise (e.g. Read, Salmon & Lenné, 2015), however time will tell if the approach is successfully taken up within the CWA and cognitive engineering field. In addition, there is no reason why the approach cannot be used in conjunction with the outputs of other systems ergonomics methods such as EAST, HTA and Accimap.

Conclusions

This paper examined the contemporary ergonomics paradigm of systems thinking in accident analysis and prevention and extent to which ergonomics methods can meet the challenges posed. Whilst we acknowledge that other disciplines and emerging areas within safety science (e.g. resilience, macroergonomics, safety II) possess methodologies that may be suited to some of the issues discussed, for the subset of ergonomics methods considered, it is apparent that there are key methodological gaps. Although ergonomists have a range of methods at our disposal, the popular paradigm of systems thinking may have extended the line of inquiry beyond their capabilities. The discussion has suggested that:

- accident analysis methods, though high on explanatory power, do not describe accident causation in a manner that is congruent with contemporary models;
- despite there being a range of appropriate candidate methods, we currently do not have a method that supports the prediction of accidents;
- the migration of systems towards and away from safety boundaries has not yet been dealt with by ergonomics methods;
- various ergonomics constructs may be suited to systems level analysis; however, there is a lack of ergonomics-based systems methods to support this; and
- despite its critical role in the design process, few ergonomics methods are actually used by designers to design.

It was not the authors' intention to paint a picture of doom and gloom, and is certainly not our intention to suggest that ergonomics is no longer relevant. Far from it in fact. Rarther it was our intention to raise the debate around our methods and their fitness for purpose given the shifting ergonomics problem space. Just as science and paradigms do not stand still, ergonomics methods should not, and indeed cannot. Encouragingly, the discussion has revealed instances where seemingly appropriate methods already exist, or where research is underway to develop the methods required. It is also noted that methodologies that fulfil some of the requirements discussed exist in other disciplines and areas of safety science. Further methodological development related to the research areas discussed is urged, as is methodological research and development in ergonomics generally. Our existing paradigms demand it, and future paradigms will also. If we desire ergonomics to maintain its relevance, we cannot rest on our methodological laurels.

Acknowledgement

Paul Salmon's contribution to this article was funded through his Australian Research Council Future Fellowship (FT140100681). We would also like to thank the two anonymous reviewers for their insightful and helpful comments on earlier drafts of this article.

Disclosure statement

No potential conflict of interest was reported by the authors.

Funding

This work was supported by the Australian Research Council [grant number FT140100681].

References

ATSB. 2012. "Australian Rail Safety Occurrence Data 1 July 2002 to 30 June 2012." Report number RR-2012-010. Canberra, ACT.

ATSB. 2013. "Aviation Occurrence Statistics 2004 to 2013." Accessed February 20, 2015. http://www.atsb.gov.au/publications/2014/ar-2014-084.aspx

Baysari, M. T., A. S. McIntosh, and J. R. Wilson. 2008. "Understanding the Human Factors Contribution to Railway Accidents and Incidents in Australia." *Accident Analysis and Prevention* 40 (5): 1750–1757.

Bradley, C. E., and J. E. Harrison. 2008. *Hospital Separations due to Injury and Poisoning, Australia 2004–05*. Canberra: AIHW Communications, Media and Marketing Unit.

Burke, S. C. 2004. "Team Task Analysis." In *Handbook of Human Factors Methods*, edited by N. A. Stanton, et al, 56–1. Boca Raton, FL: CRC Press.

Chauvin, C., S. Lardjane, G. Morel, J.-P. Clostermann, and B. Langard. 2013. "Human and Organisational Factors in Maritime Accidents: Analysis of Collisions at Sea Using the HFACS." *Accident Analysis & Prevention* 59: 26–37.

Cherns, A. 1976. "The Principles of Sociotechnical Design." *Human Relations* 29: 783–792.

Clegg, C. W. 2000. "Sociotechnical Principles for System Design." *Applied Ergonomics* 31: 463–477.

Cornelissen, M., P. M. Salmon, R. McClure, and N. A. Stanton. 2013. "Using Cognitive Work Analysis and the Strategies Analysis Diagram to Understand Variability in Road User Behaviour at Intersections." *Ergonomics* 56 (5): 764–780.

Daramola, A. Y. 2014. "An Investigation of Air Accidents in Nigeria Using the Human Factors Analysis and Classification System (HFACS) Framework." *Journal of Air Transport Management* 35: 39–50.

Dekker, S. 2011. *Drift into Failure: From Hunting Broken Components to Understanding Complex Systems*. Aldershot: Ashgate.

Dekker, S. 2014. *The Field Guide to Understanding Human Error*. 3rd ed. Aldershot: Ashgate.

Dekker, S. 2015. "The danger of losing situation awareness." *Cognition, Technology and Work* 17 (2): 159–161.

Dekker, S. W. A., P. A. Hancock, and P. Wilkin. 2013. "Ergonomics and Sustainability: Towards an Embrace of Complexity and Emergence." *Ergonomics* 56 (3): 357–364.

Deublein, M., M. Schubert, B. T. Adey, J. Köhler, and M. H. Faber. 2013. "Prediction of Road Accidents: A Bayesian Hierarchical Approach." *Accident Analysis & Prevention* 51: 274–291.

ElBardissi, A. W., D. A. Wiegmann, J. A. Dearani, R. C. Daly, and T. M. Sundt. 2007. "Application of the Human Factors Analysis and Classification System Methodology to the Cardiovascular Surgery Operating Room." *The Annals of Thoracic Surgery* 83: 1412–1419.

Endsley, M. R. 1995. "Toward a Theory of Situation Awareness in Dynamic Systems." *Human Factors: The Journal of the Human Factors and Ergonomics Society* 37: 32–64.

Endsley, M. R., B. Bolte, and D. G. Jones. 2003. *Designing for Situation Awareness*. London: Taylor & Francis.

European Aviation Safety Authority. 2010. *Annual Safety Review 2010*. Cologne: EASA.

Evans, A. W. 2011. "Fatal Accidents at Railway Level Crossings in Great Britain 1946–2009." *Accident Analysis & Prevention* 43 (5): 1837–1845

Goode, N., P. M. Salmon, M. G. Lenné, and P. Hillard. 2014. "Systems Thinking Applied to Safety during Manual Handling Tasks in the Transport and Storage Industry." *Accident Analysis & Prevention* 68: 181–191.

Griffin, T. G. C., M. S. Young, and N. A. Stanton. 2010. "Investigating Accident Causation through Information Network Modelling." *Ergonomics* 53 (2): 198–210.

Grote, G., J. Weyer, and N. A. Stanton. 2014. "Beyond Human-centred Automation–Concepts for Human-machine Interaction in Multi-layered Networks." *Ergonomics* 57 (3): 289–294.

Hart, S. G., and L. E. Staveland. 1988. "Development of NASA-TLX (Task Load Index): Results of Empirical and Theoretical Research." In *Human Mental Workload*, edited by P. A. Hancock and N. Meshkati, 139–183. Amsterdam: North Holland Press.

Heinrich, H. W. 1931. *Industrial Accident Prevention: A Scientific Approach*. New York, NY: McGraw-Hill.

Helton, W. S., G. J. Funke, and B. A. Knott. 2014. "Measuring Workload in Collaborative Contexts: Trait versus State Perspectives." *Human Factors: The Journal of the Human Factors and Ergonomics Society* 56 (2): 322–332.

Hollnagel, E. 2004. *Barriers and Accident Prevention*. Aldershot: Ashgate.

Hollnagel, E. 2012. *FRAM: The Functional Resonance Analysis Method: Modelling Complex Socio-technical Systems*. Aldershot: Ashgate.

Hollnagel, E., J. Leonhardt, T. Licu, and S. Shorrock. 2013. "From Safety-I to Safety-II: A White Paper." *Eurocontrol*. Accessed July 21, 2015. https://www.eurocontrol.int/sites/default/files/content/documents/nm/safety/safety_whitepaper_sept_2013-web.pdf

Houghton, R. J., C. Baber, R. McMaster, N. A. Stanton, P. Salmon, R. Stewart, and G. Walker. 2006. "Command and Control in Emergency Services Operations: A Social Network Analysis." *Ergonomics* 49 (12–13): 1204–1225.

Hutchins, E. 1995. *Cognition in the Wild*. Cambridge, MA: MIT Press.

Klein, G. A., R. Calderwood, and A. Clinton-Cirocco. 1986. "Rapid Decision Making on the Fireground." Proceedings of the 30th Annual Human Factors Society conference, 576–580. Dayton, OH: Human Factors Society.

Kleiner, B. M. 2006. "Macroergonomics: Analysis and Design of Work Systems." *Applied Ergonomics* 37 (1): 81–89.

Lenné, M. G., P. M. Salmon, C. C. Liu, and M. Trotter. 2012. "A Systems Approach to Accident Causation in Mining: An Application of the HFACS Method." *Accident Analysis and Prevention* 48: 111–117.

Leveson, N. G. 2004. "A New Accident Model for Engineering Safer Systems." *Safety Science* 42 (4): 237–270.

Leveson, N. G. 2011. "Applying Systems Thinking to Analyze and Learn from Events." *Safety Science* 49: 55–64.

Moray, N. 2008. "The Good, the Bad, and the Future: On the Archaeology of Ergonomics." *Human Factors: The Journal of the Human Factors and Ergonomics Society* 50 (3): 411–417.

Mosaly, P. R, L. Mazur, S. M. Miller, M. J. Eblan, A. Falchook, G. H. Goldin, and L. B Marks. 2014. "Assessing the Applicability and Reliability of the Human Factors Analysis and Classification System (HFACS) to the Analysis of Good Catches in Radiation Oncology." *International Journal of Radiation Oncology* 90 (1): Supplement, S750–S751.

Norros, L. 2014. "Developing Human Factors/Ergonomics as a Design Discipline." *Applied Ergonomics* 45 (1): 61–71.

Perrow, C. 1984. *Normal Accidents: Living with High-Risk Technologies* New York, NY: Basic Books.

Plant, K. L., and N. A. Stanton. 2012. "Why Did the Pilots Shut down the Wrong Engine? Explaining Errors in Context Using Schema Theory and the Perceptual Cycle Model." *Safety Science* 50 (2): 300–315.

Qureshi, Z. H. 2007. "A Review of Accident Modelling Approaches for Complex Critical Sociotechnical Systems." *Conferences in Research and Practice in Information Technology Series*. Vol.336 *Paper presented at the 12th Australian Workshop on Safety Related Programmable Systems (SCS'07)*, University of South Australia, Adelaide.

Rasmussen, J. 1997. "Risk Management in a Dynamic Society: A Modelling Problem." *Safety Science* 27 (2–3): 183–213.

Read, G. J. M., P. M. Salmon, M. G. Lenné, and D. P Jenkins. 2015. "Designing a Ticket to Ride with the Cognitive Work Analysis Design Toolkit." *Ergonomics* 58 (8): 1266–1286.

Reason, J. 1990. *Human Error*. New York, NY: Cambridge University Press.

Reason, J. 2008. *The Human Contribution – Unsafe Acts, Accidents and Heroic Recoveries*. Ashgate.

Safework Australia. 2012a. "Australian Work-Related Injury Experience by Sex and Age, 2009–10." Accessed October 21, 2013. http://www.safeworkaustralia.gov.au/sites/swa/about/publications/pages/austwri_bysexage2009-10

Safework Australia. 2012b. "The Cost of Work-Related Injury and Illness for Australian Employers, Workers and the Community: 2008–09." Accessed October 21, 2013. http://www.safeworkaustralia.gov.au/sites/swa/statistics/cost-injury-illness/pages/cost-injury-illness

Salmon, P. M., N. Stanton, G. Walker, and D. Green. 2006. "Situation Awareness Measurement: A Review of Applicability for C4i Environments." *Applied Ergonomics* 37 (2): 225–238.

Salmon, P. M., N. A. Stanton, M. Lenné, D. P. Jenkins, L. A. Rafferty, and G. H. Walker. 2011. *Human Factors Methods and Accident Analysis*. Aldershot: Ashgate.

Salmon, P. M., M. G. Lenne, G. Read, G. H. Walker, and N. A. Stanton. 2014a. "Pathways to Failure? Using Work Domain Analysis to Predict Accidents in Complex Systems." In *Proceedings of the 5th International Conference on Applied Human Factors and Ergonomics AHFE 2014*, edited by T. Ahram, W. Karwowski, and T. Marek, 258–266. Kraków, Poland, July 19–23, 2014.

Salmon, P. M., M. G. Lenne, G. H. Walker, N. A. Stanton, and A. Filtness. 2014b. "Using the Event Analysis of Systemic Teamwork (EAST) to Explore Conflicts between Different

Road User Groups When Making Right Hand Turns at Urban Intersections." *Ergonomics* 57 (11): 1628–1642.

Salmon, P. M., N. Goode, M. G. Lenné, E. Cassell, and C. Finch. 2014c. "Injury Causation in the Great Outdoors: A Systems Analysis of Led Outdoor Activity Injury Incidents." *Accident Analysis and Prevention* 63: 111–120.

Salmon, P. M., G. H. Walker, and N. A. Stanton. 2015. "Broken Components versus Broken Systems: Why It is Systems Not People That Lose Situation Awareness." *Cognition, Technology and Work* 17: 179–183.

Salmon, P. M., N. Goode, N. Taylor, C. Dallat, C. Finch, and M. G. Lenne. (Forthcoming). "Rasmussen's Legacy in the Great Outdoors: A New Incident Reporting and Learning System for Led Outdoor Activities." *Applied Ergonomics*. Accepted for publication July 14, 2015

Chen, Shih-Tzung, Alan Wall, Philip Davies, Zaili Yang, Jin Wang, and Yu-Hsin Chou. 2013. "A Human and Organisational Factors (HOFs) analysis method for marine casualties using HFACS-Maritime Accidents (HFACS-MA)." *Safety Science* 60: 105–114.

Shorrock, S., and B. Kirwan. 2002. "Development and Application of a Human Error Identification Tool for Air Traffic Control." *Applied Ergonomics* 33 (4): 319–336.

Siemieniuch, C. E., and M. A. Sinclair. 2014. "Extending Systems Ergonomics Thinking to Accommodate the Socio-technical Issues of Systems of Systems." *Applied Ergonomics* 45 (1): 85–98.

Stanton, N. A. 2014. "Representing Distributed Cognition in Complex Systems: How a Submarine Returns to Periscope Depth." *Ergonomics* 57 (3): 403–418.

Stanton, N. A., and R. B. Stammers. 2008. "Bartlett and the Future of Ergonomics." *Ergonomics* 51 (1): 1–13.

Stanton, N. A., and M. Young. 1999. "What Price Ergonomics?" *Nature* 399: 197–198.

Stanton, N. A., A. Hedge, E. Salas, H. Hendrick, and K. Brookhaus. 2005. *Handbook of Human Factors and Ergonomics Methods*. London: Taylor & Francis.

Stanton, N. A., R. Stewart, D. Harris, R. J. Houghton, C. Baber, R. McMaster, P. M. Salmon, G. Hoyle, G. H. Walker, M. S. Young, M. Linsell, R. Dymott, and D. Green. 2006. "Distributed Situation Awareness in Dynamic Systems: Theoretical Development and Application of an Ergonomics Methodology." *Ergonomics* 49: 1288–1311.

Stanton, N. A., P. M. Salmon, D. Harris, J. Demagalski, A. Marshall, M. S. Young, S. W. A. Dekker, and T. Waldmann. 2009. "Predicting Pilot Error: Testing a New Methodology and a Multi-methods and Analysts Approach." *Applied Ergonomics* 40 (3): 464–471.

Stanton, N. A., P. M. Salmon, L. Rafferty, C. Baber, G. H. Walker, and D. P. Jenkins. 2013. *Human Factors Methods: A Practical Guide for Engineering and Design*. 2nd ed. Aldershot: Ashgate.

Stanton, N. A., D. Harris, and A Starr. 2014. "Modelling and Analysis of Single Pilot Operations in Commercial Aviation." HCI-Aero, International Conference on Human-Computer Interaction in Aerospace, Silicon Valley, California, USA.

Svedung, I., and J. Rasmussen. 2002. "Graphic Representation of Accidentscenarios: Mapping System Structure and the Causation of Accidents." *Safety Science* 40: 397–417.

Trapsilawati, F., X. Qu, C. D. Wickens, and C.-H. Chen. 2015. "Human Factors Assessment of Conflict Resolution Aid Reliability and Time Pressure in Future Air Traffic Control." *Ergonomics* 58 (6): 897–908.

Trist, E. L., and K. W. Bamforth. 1951. "Some Social and Psychological Consequences of the Longxwall Method of Coal-getting: An Examination of the Psychological Situation and Defences of a Work Group in Relation to the Social Structure and Technological Content of the Work System." *Human Relations* 4: 3–38.

Trotter, M., P. M. Salmon, and M. G. Lenné. 2014. "Impromaps: Applying Rasmussen's Risk Management Framework to Improvisation Incidents." *Safety Science* 64: 60–70.

Underwood, P., and P. Waterson. 2013. "Examining the Gap between Research and Practice." *Accident Analysis & Prevention* 55: 154–164.

Underwood, P., and P. Waterson. 2014. "Systems Thinking, the Swiss Cheese Model and Accident Analysis: A Comparative Systemic Analysis of the Grayrigg Train Derailment Using the ATSB." *AcciMap and STAMP Models, Accident Analysis & Prevention* 68: 75–94.

Vicente, K. J. 1999. *Cognitive Work Analysis: Toward Safe, Productive, and Healthy Computer-Based Work*. Mahwah, NJ: Lawrence Erlbaum Associates.

Walker, G. H., H. Gibson, N. A. Stanton, C. Baber, P. M. Salmon, and D. Green. 2006. "Event Analysis of Systemic Teamwork (EAST): a Novel Integration of Ergonomics Methods to Analyse C4i Activity." *Ergonomics* 49: 1345–1369.

Walker, G. H., N. A. Stanton, P. S. Salmon, and D. P. Jenkins. 2009. *Command and Control: The Sociotechnical Perspective*. Aldershot: Ashgate.

Walker, G. H., N. A. Stanton, P. M. Salmon, D. P. Jenkins, and L. Rafferty. 2010. "Translating concepts of complexity to the field of ergonomics." *Ergonomics* 53 (10): 1175–1186.

Walker, G. H., N. A. Stanton, C. Baber, L. Wells, H. Gibson, P. M. Salmon, and D. P. Jenkins. 2010. "From Ethnography to the EAST Method: A Tractable Approach for Representing Distributed Cognition in Air Traffic Control." *Ergonomics* 53 (2): 184–197.

Walker, G. H., P. M. Salmon, M. Bedinger, M. Cornelissen, N. A. Stanton. This issue. "Quantum Ergonomics: New Paradigms Explored to Their Limits." *Ergonomics*. Submitted March 23, 2015.

Wiegmann, D. A., and S. A. Shappell. 2003. *A Human Error Approach to Aviation Accident Analysis. The Human Factors Analysis and Classification System*. Burlington, VT: Ashgate.

Woods, D. D., and S. Dekker. 2000. "Anticipating the Effects of Technological Change: A New Era of Dynamics for Human Factors." *Theoretical Issues in Ergonomics Science* 1 (3): 272–282.

World Health Organisation. 2014. The Global Burden of Disease from Motorized Road Transport. Accessed July 2, 2015. http://siteresources.worldbank.org/INTTOPGLOROASAF/Resources/IHME_T4H_FINAL_TO_WORLD_BANK-compressed.pdf

Young, M. S., K. A. Brookhuis, C. D. Wickens, and P. A. Hancock. 2015. "State of Science: Mental Workload in Ergonomics." *Ergonomics* 58 (1): 1–17.

Quantitative modelling in cognitive ergonomics: predicting signals passed at danger*

Neville Moray, John Groeger and Neville Stanton🅙

ABSTRACT

This paper shows how to combine field observations, experimental data and mathematical modelling to produce quantitative explanations and predictions of complex events in human–machine interaction. As an example, we consider a major railway accident. In 1999, a commuter train passed a red signal near Ladbroke Grove, UK, into the path of an express. We use the Public Inquiry Report, 'black box' data, and accident and engineering reports to construct a case history of the accident. We show how to combine field data with mathematical modelling to estimate the probability that the driver observed and identified the state of the signals, and checked their status. Our methodology can explain the SPAD ('Signal Passed At Danger'), generate recommendations about signal design and placement and provide quantitative guidance for the design of safer railway systems' speed limits and the location of signals.

Practitioner Summary: Detailed ergonomic analysis of railway signals and rail infrastructure reveals problems of signal identification at this location. A record of driver eye movements measures attention, from which a quantitative model for out signal placement and permitted speeds can be derived. The paper is an example of how to combine field data, basic research and mathematical modelling to solve ergonomic design problems.

Introduction

When we analyse complex human–machine interactions in real-life situations such as accidents, the reports are typically narrative accounts together with data from such sources as black boxes. These are related to data from site surveys, task analyses, ergonomics handbooks and databases. Quantitative measures are usually statistical means, and we try to explain events in relation to population statistics. In this paper, we show how a much more detailed quantitative methodology is possible. We use the field study and narrative reports to identify cognitive processes that are relevant to the events. We then use empirical observations and experiments to estimate how cognitive activity puts quantitative limits on human performance. We develop mathematical models for the interaction of cognition and the environment, even in dynamic situations. By combining field data, empirical data and mathematical modelling, we can account for the events that occurred at the level of real-time performance of an individual worker, and make highly specific design recommendations to improve safety and efficiency in the human–machine systems' performance. The approach we recommend can in principle be applied to any 'real life' situation, as distinct from mere laboratory studies, but the particular empirical studies needed and the choice of mathematical models will of course vary with the situation investigated. As an example, we consider a major railway accident.

On the 5 October, 1999, a commuter train left Paddington Station, UK, heading west ('down') driven by Michael Hodder, a fully qualified but relatively inexperienced train driver. Less than three minutes later, near Ladbroke Grove station, the train crossed several tracks from right to left, under the control of signals displayed on overhead gantries straddling the tracks. The tracks to be crossed included 'up' (towards London) and 'down' (away from London) lines for high-speed express trains, and on those tracks expresses could be travelling at up to 100 miles per hour (160 kph). Other tracks carried local or commuter trains. Hodder should have stopped his train as he crossed the lines because a signal associated with his track showed a red aspect ('R'). The record shows that his train initially slowed, but then accelerated through the red light resulting in a Signal Passed At Danger (SPAD).

---

*Distances and speeds are given both in Imperial and Metric units since there was no uniformity in the sources on which the paper is based. For the sources of data, see the Acknowledgement section of this paper. The paper analyses the 1999 Ladbroke Grove accident in detail, but the emphasis is less on the state of the railway infrastructure, at that time than on the methodology used. Considerable changes have been made to the rail infrastructure since 1999.

Just over half a minute later, an upbound express arrived at high speed and a catastrophic crash occurred.

A diagrammatic summary of these events is shown in Figure 1. Note that the figure is not to scale. Further details of the engineered system characteristics and infrastructure are given in the text.

The Public Inquiry (*The Ladbroke Grove Rail Inquiry*, Cullen and Lord 2000) sought input about human factors in order to understand why Hodder had passed the light at red. The first two authors were retained as expert witnesses by the rail Unions. The following paper is based mainly on their investigations at the time of the enquiry and later studies and analysis. The third author was retained as an expert witness by Network Rail and has reported

analyses of the incident previously (Stanton and Baber 2008; Stanton and Walker 2011).

Signals at R shown as white stars on gantries. Signal at Y shown as double ring. Dashed line is route of Hodder's train.

The 'Black Box' data from the train were recovered and are displayed in Figure 2. The main events are coded as follows:

a = Speed control notch 1 selected
b = Speed control notch 2 selected
c = Speed control notch 3 selected
d = Speed control notch 4 selected
e = Speed control notch 7 selected
f = SN43 AWS horn activated – green aspect – line 4 indicated
g = SN63 AWS horn activated – double yellow aspect
h = Speed control notch 0 selected + brake level 1 applied
i = brake level 1 released
j = SN87 AWS horn activated – single yellow aspect – position line junction indicator number 1
k = Speed control notch 5 selected
l = SN109 AWS horn activated – red aspect
m = Speed control notch 7 selected
n = Speed control notch 0 selected + emergency brake applied

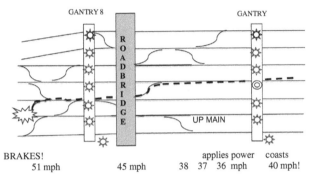

BRAKES!   51 mph   45 mph   applies power   coasts
                     38 37 36 mph   40 mph!

Record of Hodder's train control actions and speeds

Figure 1. Schematic of track layout near Ladbroke Grove, with signal settings and record of speeds reached by Driver Hodder. Not to Scale. Distance between signal gantries is 600 metres. Tracks weave left and right over this distance; so, visual alignment of signals with tracks changes frequently (see Figure 3). Signals are to the left of the track to which they apply.

Considerations of system infrastructure

Signal conventions

Signals are identified as SNn, where n is a number. The signal at which Hodder should have stopped is SN109,

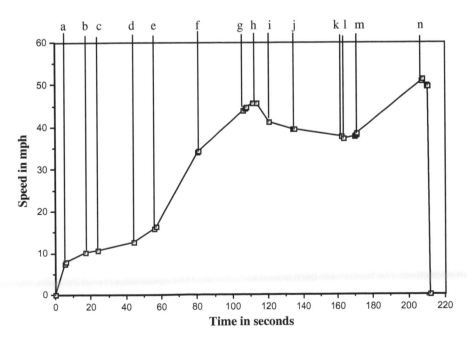

Figure 2. Speed profile of train with main events denoted.

displayed on Gantry 8. Signals on this section of track can show one of the four states (called 'aspects' in the industry): Green (G), Double Yellow (YY), Yellow (Y) or Red (R). The usual geometry of signals, as seen on all the signal gantries between Paddington Station and SN87, (the last signal prior to SN109 at which the SPAD occurred,) is shown in Figure 3. Four lamps are presented in a vertical column, just to the left of the left-hand rail, typically at a height of greater than three metres above the rail. A lamp subtends about 6' of arc at a distance of 275 yards (250 metres). An R aspect has a single red light at the bottom of the column, and requires the driver to stop the train before reaching the light. On a Single Y aspect, the driver may proceed, but must be prepared to stop, as the next signal is at that time R. On a Double Yellow aspect, YY, the driver may proceed without slowing, and should expect the next light to show G, YY or Y. On a G aspect, the driver may proceed at full speed. Speed limits are usually shown on trackside signs to the left of and about one metre above the rail. Where multiple tracks are present, most signals are displayed just to the left of his track as viewed by the approaching driver, on gantries that straddle the whole extent of the tracks. Some are displayed on vertical metal standards, cantilevers or lattice poles.

Figure 3. The tracks approaching SN109, showing the difficulty in identifying the signal aspects and their relation to tracks. The lower numbers identify the tracks. The upper row shows the signals that correspond to the tracks with the same numbers. These numbers have been added to the picture by the authors. The gantry on which they are mounted is *well beyond* the prominent white road bridge, which is why only SN105 shows all four lamps. (See also Figure 1.) Note the difficulty of identifying signal with track, that two of the signals are completely obscured by structures on the gantry, and that one or more of the lamps several signals are also obscured. The lamps have been drawn to make their position clearer. In reality, they are much less visually prominent at this point. White stars are red lamps. White circles are yellow and green lamps. The positions, not the aspects of the lamps, are shown. The rightmost tracks and signals are masked by the frame of the window.

Track layout

The track layout dates from the nineteenth century, at the time of steam engines and mechanical signalling systems. Main line electrification began in the late twentieth century, and the system was electrified using overhead catenaries to supply current. After leaving Paddington station, there are up to eight tracks, shared by up and down lines. There are many places where if the points are set appropriately, a train will pass from one track to an adjacent track. The points and signals are set and coordinated by the signalling system that functions independently of the driver of the train, who is required to observe and react to signal aspects as the train passes along and across the tracks. A detailed analysis of the signalling control system and its personnel relevant to Hodder's SPAD is given by Stanton and Baber (2008). On the section of track from Paddington to Ladbroke Grove, the lines curve sinuously left and right. (That is, they are not straight as shown in Figure 1) It is often difficult to tell which signal refers to which track because the line of sight straight ahead of the cab may be to the right or left of all signals, or towards some point along the gantry.

Supporting infrastructure

Several road bridges carry foot and automobile traffic across the railway. SN109, the signal that resulted in Hodder's SPAD, was one of the seven signals on Gantry 8 (numbered from the departure from London's Paddington station). Other structures carry insulators and support electrical power catenaries, and during the approach to signals, some or all signals are sometimes obscured by electrical equipment, bridges or gantries, other structures, foliage, etc. until a train is within one or two hundred metres of the signals. (See Figure 4.) To identify the signal relevant to

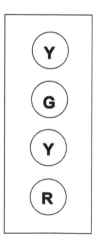
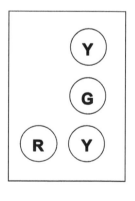
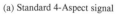

(a) Standard 4-Aspect signal

(b) Geometry of signals on the gantry carrying SN109

Figure 4. Signal geometry.

their track, drivers must often 'count across' the tracks and signals, pairing them off one against the other.

At the time of the accident, most trains were fitted with an Automatic Warning System (AWS) that sounds a horn in the driver's cab if a signal showing a non-G aspect is approached, or if a speed limit is approached. A bell sounds if a G aspect is approached. The AWS therefore distinguishes between a G aspect and any other aspect, but does not distinguish among Y, YY and R aspects. This auditory signal is driven by contact closure when a train passes over a magnet placed between the rails, some 185 metres (202 yards) before the signal.

The time line of the SPAD and collision

The following summary is based on evidence in Cullen and Lord (2000) and on reports from the black box recorder as analysed by the engineering consultants WS Atkins (1999, 2000a, 2000b, 2000c, 2000d, 2000e, 2000f) and cited in the Cullen report. It covers events from the moment Hodder passed signal SN87, the last signal relevant to his route prior to SN109, until the moment of the collision.

There were no 'trackside distractors' such as pedestrian intruders or maintenance operations, although as Figures 5–7 show, the signals are masked by a bridge on approach and a speed sign is displayed. The sun was low above the horizon, very bright and shining onto SN109 from *behind* Hodder.

Hodder's train passed SN87 at a speed of about 40 mph. No speed control notches on the driver's console were engaged. SN87 was showing a single yellow aspect (Y). Since Hodder was coasting, not accelerating or braking, it is probable that he saw the Y aspect on SN87 since a Y aspect means that the following signal may show a red aspect (R), and requires the driver to prepare to stop the train before that next signal. Hodder shortly afterwards engaged the speed control Notch 1, suggesting that he had decided to accelerate. At about 260 yards (239 m) before SN109, he

Figure 5. Approach to SN109, note that bridge is masking signal gantry.

Figure 6. Approach to SN109, note that signal gantry is revealed behind bridge.

Figure 7. Approach to SN109, note that signal gantry is completely revealed.

engaged Notch 5 and began to accelerate from a speed of 37 mph. An emergency brake application (a braking rate of 12% *g*) at this point would have stopped the train in approximately 137 m (150 yards), but slowing was unlikely to have been his intention. Hodder acknowledged the AWS horn associated with SN109 before reaching the latter, and SN109 was showing a R aspect. Approximately 4 s after the horn sounded (see Figure 2), at about 104 m (113 yards) before SN109, Hodder engaged Notch 7 and continued to accelerate. Stopping prior to the signal was now impossible. He passed SN109 with power still engaged at a speed of just over 40 mph. He applied the brakes at a speed of 51 mph (82kph) presumably on seeing the HST ahead, but by then the 250 m (273 yards) of clear track that would be required for the emergency braking to prevent the accident was no longer available.

Statement of problem

(1) Did Hodder correctly perceive the aspect of the previous signal SN87? Did he forget it? and if so, would this account for his SPAD at SN109?

(2) Did Hodder fail to see SN109 or did he misidentify the aspect of SN109 on first seeing it? Did he fail to check it again during his approach; and if so, why?

(3) Why did Hodder not interpret the AWS alert as indicating a danger aspect that required him to prepare to stop? Why did the signalling system not respond to his SPAD?

The most relevant cognitive and ergonomic factors are:

(1) The extent to which the layout of track, infrastructure and signals in the approach to SN109 impeded the visibility and interpretation of the signal state.

(2) The role of the driver's mental model of the route and his expectations of the upcoming signals.

(3) The high demands on dynamic visual attention.

(4) The fact that the AWS gives the driver ambiguous information.

The role of SN87

The fact that Hodder coasted past SN87 suggests that he had correctly identified the Y aspect of SN87, and was therefore expecting the next signal to be Y or R. There is a strong belief in the UK railway industry that drivers have almost perfect route knowledge of the routes they drive, and that these mental models (Moray 1990, 1997; Revell and Stanton 2012) are vital components of their skill as drivers. Their long-term mental models are thought to include the layout of the track, the times and places at which signals will appear, the location of speed limits and the probability of other rail traffic. Short-term mental models are thought to include the aspects of signals recently passed, and hence the probable state of upcoming signals, expected switches from track to track, expected occurrence of auditory AWS events and their implications, etc. Mental models are assumed to be used to predict imminent events and to plan appropriate responses. In theory, drivers could rely on strong mental models to reduce the need to scan the environment for information, thus reducing workload. Railway personnel commonly assert that experienced drivers will anticipate when a signal is due to appear and will be looking in the correct direction to see its aspect as soon as it appears. These claims support a strong belief in the high quality of drivers and imply the use of mental models. Reliance on mental models would also account for how drivers can be fixating the location of a signal before it appears. That drivers use mental models in this way is an assertion of faith since at the time of the Cullen enquiry, no empirical evidence existed about the distribution of train drivers' eye movements or any other measure of their dynamic attention or working memory. While the explanation for the SPAD in this paper emphasises overload in the driver's visual attention, other explanations have been proposed, such as that by Stanton and Walker (2011).

The dynamics of visual attention in train driving

We will estimate the cognitive workload on the driver by measuring eye movements. This is an uncommon technique, but very powerful. In particular, it allows us to investigate the moment-to-moment cognitive load in detail for a single driver. There has been other research into driver mental workload using interviews, subjective ratings, focus groups and similar methods (Zoer, Sluiter, and Frings-Dresen 2014; Naweed, Rainbird, and Chapman 2015). Such studies lead to qualitative models of mean performance at best, and resemble the classical research supported by NASA on pilot workload in the period 1970–1990 (see, e.g. Moray 2008). But those methods do not allow detailed real-time modelling of cognition. Balfe et al. (2012) believe that

> Ethnographic, naturalistic observation and structured interview methods can provide insights which may never be achieved with quantitative methods.

In this paper, we emphasise quantitative methods because they allow mathematical modelling since in real tasks, visual attention depends on the dynamics of eye movements, and it is well known that people have little or no idea of the patterns of their eye movements, and are always very surprised when shown eye movement recordings (Moray, Neil, and Brophy 1983). Quantitative design recommendations require quantitative analysis of behaviour and performance. We do not deny that qualitative methods can answer qualitative questions (Annett 2002).

Since the Cullen inquiry, there have been two extensive and detailed studies of train drivers' visual attention. Both analysed driver eye movements recorded in the real situation of driving commuter trains on real routes (Groeger et al. 2012; Luke et al. 2006). The quantitative data from these studies agree closely with each other. In this paper, we will use the quantitative data from the study by Groeger et al. (2004). The drivers, who took part in the experiment voluntarily and with the agreement of their Union, wore an eye camera that recorded the view in the direction of the driver's line of regard, and superimposed on the video recording a marker that indicated the point of fixation. They drove over a standard route on the Thameslink Bedford to Brighton line between St. Albans and East Croydon that included both rural commuter lines and a route through London, with many stations and signals. The journeys were part of the drivers' normal duty roster.

Groeger et al. (2004) summarise their qualitative analysis in Table 1 based on 470 approaches to signals, using 10 experienced drivers who were all familiar with the route. Their study did not include the section of track from Paddington to Ladbroke Grove, but enough signals were passed (470), and enough drivers (10) were observed to establish sound statistical estimates of typical behaviour. They were driving commuter trains similar to that driven by Hodder from Paddington. From their data, we can reasonably generalise to the Ladbroke Grove situation, although the fact that SN109 is a multi-SPAD signal, that is several drivers had committed a SPAD at the same location, means that there may be something special about it.

All the categories in Table 1 are needed to classify the distribution of driver visual attention. It is obvious from Table 1 that the dynamics of driver attention make it very unlikely that drivers will always be looking towards a signal at the moment it appears. Groeger et al. (2004, iii). state that:

The final 15 s of the approach to several hundred signals was analysed in detail. Despite the quantity of data collected, considerable variation in viewing behaviour (i.e. differences between drivers at the same signals, differences when the same driver viewed different signals) severely limits the generality of the findings, and the implications that can legitimately be drawn from them.

That acknowledged, the tentative indications are that drivers look at signals for the first time about 8 s before they are reached, thereafter signals are looked at on a number of separate occasions. About half of the 15 s before a signal is reached is spent 'fixating' on objects (i.e. glances lasting at least 0.12 s), 20% of the approach time is spent looking at signals. The timing of the first look at a signal and distribution of glances and fixation time across the approach depends on:

the signal aspect — cautionary signals are looked at earlier and for more time overall (i.e. adding duration of all fixations together);

the signal mounting — signals on gantries are looked at earlier and for longer;

the aspect of the signal-in-rear — having passed a SingleYellow or Red, the driver looks at the upcoming signal sooner."

During the last 15 s of a continuous approach to a signal, there are some 25 items of the visual environment to which attention may be paid (Table 1), and Groeger et al. reported that,

substantial numbers of first looks at signals are made after the AWS has sounded[1].

That means that the drivers' attention is often drawn by the Automatic Warning Signal to the appearance of the signal: drivers were not already looking towards the signal when it appeared, contrary to what is widely believed in the industry.

These data on visual attention to signals are relevant to the analysis of the crash in two ways. First, they allow us to estimate the probability that the signal SN109 was observed at least once, twice, etc. We can estimate the probability that Hodder looked at the signal at all, and that he checked its aspect in the time between its appearance and the arrival of his train at the signal. We will return to this calculation below.

Second, the data show that what is seen interacts with the driver's mental model to affect his or her expectations (Plant and Stanton 2012; Rafferty, Stanton, and Walker 2013; Revell and Stanton 2012; Salmon et al. 2013). Drivers are taught to drive 'defensively', so if a signal has a Y aspect, they prepare to stop at the next signal which will probably

Table 1. Distribution of train drivers' eye movements.

Category	Individual objects and object locations
Own signal	Own signal (separately coded for 'looks' within 1, 2 and 4 degrees of signal)
	Own direction indicator
Other signal	Signals to the left and right of driver's own signal (on the same gantry)
	Upcoming signal on the same line as driver's own signal Upcoming signal to the left or right of driver's own signal
	Direction indicators to the left or right of the driver's own signal
Own gantry	Any position on driver's own gantry, except the signal
Other gantry	Any signals not facing the driver
	Any gantries to the right or left of driver's own gantry
Signage	All speed restriction signs
	AWS magnet
	TPWS grid
	CDRA sign
	Station TV monitors
	Station clock
	Station electronic train information screen Platforms to the right and left
Moving objects	Travelling trains to the right and left of the driver Station or track crew
Off track	Fields to the right and left Verge between track and field Trees
	Oncoming bridges and city buildings Tunnel wall and ceiling
Track ahead	Driver's own track
	Track to the left or right of the driver
Sky	The sky
In cab	All objects and controls inside cab Driver's own window
	Side windows Instructor

have an R aspect when they reach it, while a G or YY aspect allows them to assume that they will not have to stop at the next signal (Stanton and Walker 2011), and can proceed at the current speed or accelerate to their desired running speed.

Long-term expectations become part of the 'route knowledge' contained in drivers' mental models. On this route, a driver will expect four aspect signals to have lights in a vertical column, reading Y–G–Y–R from top to bottom, as in Figure 2a since all signals between Paddington station and Gantry 8, which carries SN109, have this standard form, as do most other signals. The time of day will also produce expectancies about the density of train traffic en route, and hence the likelihood of conditions where cautionary aspects (Y and R) will predominate on certain sections. Short-term expectations will depend on perceptions made during the journey. If a driver has almost always driven over a stretch of track in which the signals show a green aspect, his long-term expectations will predict that to be the case again. If however he encounters a Y aspect instead of the usual green aspect, his expectation as to what he will see at the next signal should be altered, and that short-term expectation will be held in working memory (WM), and will bias what he expects to see and what he expects to happen next: the predictive nature of railway signalling encourages this. As is well known, the contents of WM are volatile and easily disrupted by the arrival of new perceptual information or recall from long-term memory (Reason 1990). So the association of features in the environment interacting with expectations can lead to strong-but-wrong assumptions about the state of the world (Moray 1990).

The W.S. Atkins report (1999) shows that Hodder's recent experience would not have led him to expect problems at SN109. A record of 21 recent journeys by Hodder when approaching Gantry 8 was analysed: 17 involved SN113, 2 involved SN111 and 2 involved SN109. Of these, SN109 once showed a G, (onto Down Main), and once showed a R aspect (when the SPAD occurred). All the other approaches showed a G aspect to proceed onto the Down Relief. Nothing in this experience that would have caused Hodder to develop a strong mental model for special attention to SN109, and the overwhelming evidence, with a probability of 19/20, $p = 0.95$, is to find a signal at Gantry 8 to be showing a G aspect. We may conclude that both Hodder's long- and short-term mental models would have primed him to expect a non-R aspect at Gantry 8. The AWS sounding on approaching SN109 would not change this expectation since that morning, all the signals after leaving Paddington had shown G, Y or YY aspect, and the AWS warning sound does not indicate R, but merely a non-G aspect. Hodder acknowledged the AWS within approximately 0.5 s, suggesting he expected

a Y or YY aspect at Gantry 8. Indeed, Stanton and Walker (2011) make a similar case when comparing alternative explanations for the behaviour of the driver. By comparing the driver's response time to other signals, they show that the driver's cancellation of the AWS warning at SN87 and SN109 was approximately half of that to SN63 (and earlier signals on the route).

Empirical data on visual attention

In real-world tasks, visual attention is allocated by eye movements. See Jones, Milton, and Fitts (1949) and Senders (1964) for classic examples. The train drivers' task is made difficult by the visual complexity of the environment, mechanical vibration in the cab, the rapid change in the visual array due to the movement of the train and the fine detail both in structure and colour of visual information. For example, on a signal sighted at a distance of 250 metres, the lamp subtends less than 0.1 degree, (about 6' of arc). For a driver to see accurately the colour and detail of such a signal, it must be foveated, that is appear within 2° of the centre of fixation, (Abramov, Gordon, and Chan 1991; van Esch et al. 1984). Moreover, there is some evidence that information is only read into the brain when the image on the retina is static, although eye movements are needed to prevent the fading of the image (Ishida and Ikeda 1989; Henderson 1993). In laboratory studies, eye movements may occur several times a second, but such values are extremely rare in real-world tasks where fixations of up to several seconds are common (Moray and Rotenberg 1989; Moray, Neil, and Brophy 1983). When a visual task involves fine discrimination or information of low probability, the duration of fixation increases and the frequency of eye movements decreases (Senders 1964; Senders et al. 1965).

Groeger et al. (2004) report several statistics of eye movements associated with the different categories of data in Table 1. Tables 2–4 are derived from their report. We have added a measure of the range of the data, by showing values associated with ± 1 standard deviation about the means. Does the design of the railway system impose a cognitive load on a driver that is so unreasonable that he may be led into making errors?

Table 2. Proportion of time spent fixating signals being approached.

Aspect being approached	Non-signals		Signals	
	Mean (SD)	Range ± 1SD	Mean (SD)	Range ± 1SD
Green	0.50 (0.28)	0.22–0.78	0.16 (0.14)	0.02–0.30
Double Yellow	0.43 (0.19)	0.24–0.62	0.22 (0.12)	0.10–0.34
Single Yellow	0.41 (0.21)	0.20–0.62	0.21 (0.15)	0.06–0.36
Red	0.44 (0.21)	0.43–0.63	0.22 (0.19)	0.03–0.42

Table 3. Fixation durations in seconds*.

Aspect being approached	Non-signals		Signals	
	Mean (SD)	Range +1SD	Mean (SD)	Range ±1SD
Green	0.91 (0.44)	0.47–1.35	0.40 (0.31)	0.09–0.71
Double Yellow	0.79 (0.35)	0.44–1.34	0.47 (0.26)	0.21–0.73
Single Yellow	0.76 (0.37)	0.39.–1.13	0.46 (0.29)	0.17–0.75
Red	0.75 (0.24)	0.51–0.99	0.43 (0.31)	0.12–0.74

*In the original paper by Groeger et al. 2004 this table is mistakenly labelled 'Fixations per second'. The present legend is correct.

Table 4. Mean proportion of time remaining until the signal is reached after the first fixation following the signal first becoming visible.

	Mean	SD	Range ±1SD
Green	0.59	0.33	0.26–0.92
Double Yellow	0.73	0.28	0.45–1.01
Yellow	0.76	0.28	0.48–1.04
Red	0.81	0.25	0.56–1.06

(A value of 0.81 means that when the first fixation on the signal occurs, 19% of the time between its first appearance and the moment it will be passed has elapsed. We can use the SD to define the range covered by ± 1 SD, but values greater than 1.0 have no meaning and will not occur in recorded data).

Examine Table 4. On an average, when approaching a single yellow signal, drivers first looked at that signal when about a quarter of the way to it from the point where it first appeared (mean proportion of distance remaining = 0.76, SD = 0.28). It would not therefore be unusual if a driver did not first look at the signal until he had covered half the distance from where it was first visible since 68% of the data fall within the range 0.48–1.00, that is ± 1.0 SD. The data suggest that about 16% of Groeger's drivers when approaching a Y aspect did not fixate it until they had traversed nearly 1/2 of the distance to it from where it first could have been seen had they looked at it.

From Tables 2–4 we should expect that as a signal is approached, drivers will spend more than 78% of the time looking at non-signals on about 16% of occasions, and hence only 22% of the time looking at signals.

When approaching a Red aspect, drivers tend to make their first fixation earlier, presumably because of their mental models, but the statistics suggest that on about 2% of approaches, drivers do not first fixate the signals until only about one-third of the distance to the signal remains (mean = 0.81. SD = 0.25; therefore 2 SD = 0.5, and 0.81–0.50 = 0.31.) The fact that 2% is a small probability is irrelevant: fortunately, SPADS are rare events.

SN109 is mounted on Gantry 8, which carries another six signals. (But note an important exception, SN105, to be discussed later.) On approaching Gantry 8, overhead equipment such as electric insulators, catenaries and bridges cause signals to appear in the following sequence. The R aspect of SN105 first becomes visible at a distance of 496 m (542 yards), that of SN107 at 349 m (381 yards), that

of SN115 at 327 m (357 yards), that of SN113 at 263 m (287 yards) and that of SN111 at 248 m (271 yards).

The red aspect of SN109 begins to appear from behind a large OLE insulator (See Figure 3) when the distance to the gantry is 219 metres. (Atkins, 1999).

In Figure 3, based on a photograph taken about 250 m (273 yards) from Gantry 8, the first three and the sixth signals are visible, but two are hidden by insulators carrying the catenary cables or by the road bridge. The later the first fixation, the less time there is for a second fixation to check the signal aspect before passing it. At the moment that SN109 became permanently visible at a distance of 168 m (183 yards), Hodder's train was travelling at 38 mph (61 kph). If on that occasion Hodder did not look at SN109 until half way to it, only 94 m (102 yards) remained, and that distance would be covered in about 5 s. If Hodder behaved as Groeger's drivers did on the average, and spent only 20% of the time looking at the signal, then he would only have about 1 s to examine SN109, (0.2 x 5.0 = 1.0). This would barely allow time even for one fixation on SN109, and would certainly afford no opportunity to recheck the signal, bearing in mind that signals are fixated for about 500msec, to which the time for an eye movement to that fixation would have to be added.

If the aspect of SN109 is so critical, why would not the driver keep his attention on the signal from the first time he fixates it until he passes it? Table 1 shows that he distributes his attention over more than 20 features of the driving environment when driving defensively, and to do so, he must switch his attention from one part of the visual environment to another. Furthermore, it is known that drivers 'count across' to determine which signal refers to which track. Even if counting across can be done without moving the eyes, Trick and Pylyshyn (1994) suggest that this may require a subitising time of more than 250 ms. The data in Tables 2 and 3 show that in the last 15 s, before passing a signal, drivers fixate the signal for only about 20% of the time. For another 44% of the time, they are fixating other parts of the environment (track side, rails, speed signs, oncoming trains, information and equipment inside the cab, etc.). The remaining 36% of the time they are not fixating anything: their eyes are moving, and therefore acuity will be much reduced.

Groeger et al. (2004, iii) report

"…20% of the approach time is spent looking at signals"

and,

On those approaches where the last 15 s were coded (a total of 154), an average of 17.05 fixations were made, lasting, on average, for 9.44 s. That is, there were 1.14 fixations per second of approach, and 63% of all approach time was spent fixating on various objects, with the remaining time spent scanning the visual scene. Perhaps not surprisingly, these figures vary with the length of the approach, with more fixations being made but the proportion of the approach when drivers were fixating remaining similar throughout. (Groeger et al. 14.)

The way in which signals are mounted also has a considerable effect. Groeger et al. report that signal fixation, averaged across all signals, irrespective of aspects, signal mounting, preceding signal, etc., is about 520 ms (SD = 400). Mean signal fixation is dramatically affected by the nature of the mounting. The mean times are approximately 380 ms for a lone signal on a post, 340 ms for one of a pair of signals on a cantilever and 450 ms for one signal among a group on a gantry. Total duration of fixations is similarly affected by signal mounting: post, 600 ms; cantilever, 660 ms; and gantry, 780 ms. This suggests that a situation such as SN109 where there are several signals on a gantry will adversely affect rapid scanning.

Table 2 shows that about 22% of time is spent looking at signals, and about 44% of time looking at non-signals. These are equivalent to about 3.3 s (out of 15 s) and 6.6 s (out of 15 s), respectively. At 38 mph, it takes only 11 s to cover the 198 metres to SN109 from when it first appears. Groeger's data suggest that there would be an average of fewer than two fixations on signals and about five fixations on non-signals during the last 15 s of an approach to a signal. Taking into account the values of the SDs in Table 2, and averaging over all four signal aspects during the last 15 s of the approach, it seems that on at least 16% of approaches, less than 10% of the time (that is less than 1.5 s) will be spent looking at signals. There will be as many as 16% of approaches where a driver will not have time for more than one look at the signal that he is approaching. If that look should, for whatever reason, lead to an incorrect identification of the signal aspect, there will not be time for a second look to correct that judgement, and the signal will be passed with the driver having an incorrect judgement of the aspect.

We emphasise that such behaviour would not be due to a driver's incompetence, lack of motivation or lack of training. Rather, it is a mark of well-developed driving skills and their associated mental models. It would be caused by the interaction of the design of the rail system (track layout, signals, train schedules, required speed, etc.) with the inherent properties of the human nervous system that limit the rate at which the environment can be visually attended. Even a well-trained, highly motivated, driver will suffer from these limitations (Atkins, 2000e, 2000f; 99817B).

We may conclude from the quantitative analysis by Groeger et al. (2004) that there is a very real possibility that Hodder had time only for one visual sample of the aspect of SN109 as he approached it. In addition, because of the electrical hardware, bridges, gantries etc., the signals become visible progressively, and given the curvature of

the tracks and the fact that Hodder's train was crossing from right to left, there would be considerable difficulty in identifying which signal was relevant to which track. (See Figure 3)

Why would Hodder fail to identify correctly the aspect of SN109?

The standard design of a four aspect signal is a vertical column of four lamps in order from top to bottom Y, G, Y and R. It is standard practice that the heights of the lamps above the rails should be identical for the several signals on a single gantry. If a driver scans laterally across the gantry, the R lamps will all be in a horizontal line, and similarly for the other kinds of lamps. Uniquely on this route at Gantry 8, both these expectations are violated. The leftmost signal is a four lamp vertical column, but this signal is not attached to Gantry 8. It is separate from the gantry, and the lamps are lower than any of those on Gantry 8, which are aligned horizontally as would be expected. Moreover, the sets of lamps on Gantry 8 have an abnormal geometry. They all have a unique 'Reversed-L' geometry, probably because of the spatial constraint on the location of the signals. During the approach, the structures of the bridge and of the electrical equipment obscured signals on the gantry from time to time. The R aspect light of SN109 was obscured partially or wholly until the driver was quite close to the signal, at about 168 m (183 yards).

Since the normal geometry of signals is a vertical array with red at the bottom, this is what the driver's mental model would predict and the driver consciously expect. If the red aspect, although illuminated, could not be seen because it was obscured by the bridge or by insulators, then the driver, if paying attention to the vertical YGY lamps, might assume that red was vertically below the visible lights but not illuminated, rather than that it was illuminated, to the left, and obscured. As he neared Gantry 8, the driver could have seen other signals on that gantry, and would have been able, in principle, to note their abnormal geometry, and hence could have realised that an absent red aspect of SN109 might be hidden rather than absent, but this would violate a strong expectation of the signal geometry. This possibility furthermore depends on his having enough time to look at several of the signals, which we have seen is unlikely. Furthermore, a R aspect to the left of a Y but at the same level might be interpreted as being on a different four vertical signal beyond that carrying the Y aspect. Hodder may have interpreted what he saw of SN109 during the approach as indicating that SN109 was in a state of 'not-R'. It is believed that the backscatter of the bright morning sun from the yellow lamp lenses probably did not 'shine' like a Y or YY aspect, (a so-called 'phantom' signal,) but in the absence of any visible R aspect, the appearance would have been ambiguous. According to official Sighting Standards, the signal controlling a driver's track should be adjusted such that it is the brightest signal on view. Although the backscatter of the sunlight from the lamp did not produce a 'phantom' light, the bright morning sun may have saturated the colour information available and may have caused the Y to be the most prominent if the R was not visible or was misinterpreted.

Interpreting a quick glimpse is made even more difficult by the free-standing SN105 vertical four aspect signal to the left of Gantry 8. This is the first signal to become visible as Gantry 8 is approached (Wilkins in Atkins 99817A1). If this were the first signal that was fixated by Hodder, it would confirm his expectation that as usual R lamps were located below YGY lamps, thus further biasing his expectations about any other signals he might fixate. Wilkins (Atkins 99817A1) found that SN105 becomes visible at 496 m (542 yards) from Gantry 8, and it is evident from the photographs in the Cullen reports and those we took during a ride in a train cab (Figure 3) that the red aspect lamp of SN105 is substantially lower than those on Gantry 8, including SN109. The height of the red lamp of SN105 is 3353 mm above rail level. SN105 will therefore have appeared earlier because it was not occluded by the structure of the bridge that hid the higher lights on the gantry. The heights of the other R aspects on Gantry 8, including SN109, are all over 5000 mm above rail level. SN109 is at 5085 mm, and the others differ only slightly from that value. Another question concerns the cognitive demands of 'counting across', which was mentioned earlier (Trick and Pylyshyn 1994). Potential confusion may occur about whether to include the off-gantry signal in any count. If drivers in a dynamic rather than static viewing context tend to rely on enumeration rather than estimation that will complicate the driver's task further.

As the train passed over the points from Line 4 to Line 3, it would have tended to point towards SN105 which would have been prominent. The default expectation of a driver on seeing one of a set of multiple signals at a single location will be that if one is visible, the others should be visible at the same height, because of the design doctrine of 'parallel' location of signals. If SN105 was seen by Hodder to be showing a R aspect, and if he were then to look rapidly across the lines at the same height above the ground as the red light of SN105 in order to see the other lights, no R aspects would be visible since the other R lamps were at a higher level and some at least were concealed by infrastructure. (Figure 3) If his default belief was that they were at the same height, and if moreover his mental model was of the standard signal geometry, this would generate a belief that they must be showing a non-R aspect. This belief could have been corrected as the other signals came into view, but the British Railways Sighting

Table 5. Probability of at least one fixation on Own Signal as a function of the number of fixations.

Number of fixations	2	4	6	8	10	12
P that at least one will be on Own Signal	0.39	0.63	0.78	0.87	0.92	0.95

Committee Report (1994) says that the R aspect of SN109 becomes unobscured only at 168 metres (183 yards).

Did Hodder ever look at SN109? The probability of at least one fixation

We know from Groeger et al. (2004) that as a driver approaches a signal, he spends a mean of about 22% of the final 15 s looking at the signal, and that the SD of that mean is about 15, if we change the proportions to percentages. This implies that the mean probability that the driver's first fixation is on the signal just after it first appears is about 0.22. Therefore, the probability that the first fixation will *not* be on the signal will be (1.00 - 0.22 = 0.78). Hence, the probability that none of the first n fixations after the signal comes into view will be on the relevant signal is given by $(1.00 - 0.78^n)$, and the probability of at least one of the n being on the signal is the complement of that number. Hence, we can construct Table 5. 'Own Signal' here means the signal that is relevant to the track on which the driver is travelling.

We have already seen that there would not have been time to examine every signal on Gantry 8. After reaching a point 168 m (183 yards) from Gantry 8, Hodder will have made very few fixations before passing Gantry 8. At 38 mph, the time to travel 168 metres would be approximately 9.5 s, which would allow for at most two fixations on SN109 if we take Groeger et al.'s mean value. If the train were travelling faster and if the driver were counting across to identify SN109, there would be even fewer fixations on the latter, the identification of SN109 might well be uncertain and there would be no time to check its aspect by a second fixation.

It may be thought that the driver's route knowledge would ensure that he would remember that the signals on Gantry 8 were of a 'reversed-L' shape, and that if he had kept alert as required by defensive driving, he would have been aware of this. But as Wilkins says in Atkins 99817B (20.25),

Perfect signalling design and perfect sighting arrangements would demand little or no route knowledge on the part of the train driver. Conversely, the ideal in respect of SPAD risk minimisation would require drivers to possess perfect photographic knowledge of the many hundreds of signals along the routes over which they drive. Neither ideal is capable of achievement in the practical world, . . . it would be wholly unreasonable to expect drivers to learn in photographic detail all the complexities of signal viewing in a complex layout affected by significant OLE obstruction.

The YGY column of lights on the vertical part of SN109, although probably not showing a phantom aspect, (Watt 2000; Wilkins, Atkins 99817A1, 12.0–12.1), was reflecting very bright sunlight, and may well have been interpreted as a washed out Y or YY, given that SN87 had shown a Y. The coloured photos available in Cullen and Lord (2000) do not show a phantom aspect, but nonetheless the surface reflection, in the absence of a clear R or G, might well be interpreted as a Y or YY, washed out by the sunlight, since they must be showing *some* aspect. If Hodder had decided that SN109 was 'not-Red', and did not see a bright G, this would further reinforce the memory carried from SN87 that the aspect of SN109 was Y or YY, and the reflected sunlight would reinforce this belief.

To summarise, we have two strong biases due to expectations from the past and current experience of the driver: because SN87 showed a Y aspect, Hodder expects a 'not-green' and probably a Y aspect ahead. Groeger et al.'s drivers saw Y followed by R on about 50% of occasions. Data in Atkins 99823A suggest that when approaching SN109 from SN87, the probability of finding SN109 showing a G aspect is 0.70, finding it showing a Y or YY aspect is 0.07, and showing an R aspect is 0.22. The data are not sufficient to decide what the probability of each aspect would be conditional on a Y at SN87, but if we combine those data with Groeger's, the probability of an R aspect at SN109, given a Y at SN87, would seem to be only about 0.11. A reasonable default expectation on approaching SN109 from SN87 showing Y would be that it would show not-Red with a probability of 0.89. If Hodder expected the red lamp to be below the YGY lamps, he would interpret his 'not-R' perception to be due to an unilluminated red lamp below the lower Y lamp, not to a hidden illuminated red lamp to the left of the lower Y lamp. Note also that the reverse-L format at Gantry 8 means that the red aspect of each signal is not as close as possible to the line of sight (as is required by regulations). In fact, the red aspect lamp of SN109 was 1505 mm to the left of the left rail running edge further violating default expectations (Atkins 99817A1, 10.5.10).

The Atkins report notes that,

> The fact that Hodder's eyes apparently did not revisit SN109 should not be regarded as unusual; unlike road traffic lights, railway signals are not in the habit of reverting to red as a driver approaches. There is therefore a tendency for a driver to mentally 'put a tick in the box' for a signal once it has been read and not to review the signal thereafter.

That comment supports the analysis made in the present paper.

After a point about 168 m (183 yards) before Gantry 8, the R aspect of SN109 remains permanently unobstructed and is clearly visible until the driver's cab passes the latter, but it was not fixated again because the dynamics of

eye movements and the tactics of visual attention in this task did not allow it. The fixation times in Table 3 equate to about 0.5 s per fixation on signals, and about 0.8 s per fixation for non-signal items. Therefore, *there would be an average of less than two fixations on signals and about five fixations on non-signals during the last 10 s of an approach to a signal*. From the values of the SDs in Table 2, and averaging over all four signal aspects, we predict that on at least 16% of approaches, less than 10% of the time (that is less than 1.0 s) will be spent looking at signals. That is, there is almost a 16% probability that not even one fixation will occur on the required signal during the final 168 m approach to the signal, if Groeger et al.'s measurements of driver eye movements are typical. It should be stressed that they and those of Luke et al. (2006) which agree with them, are the only objective data available.

Fixation rates might have been greater (i.e. more fixations per second) had Groeger et al.'s measurements been taken at Ladbroke Grove, but this is rather unlikely, given the complexity of the visual array on much of the Paddington–Ladbroke Grove route and that the duration of fixations increases with complexity. There is very little opportunity for a driver to take either several looks at the signal or one prolonged look, if he is also to pay attention to all the other aspects of the environment that are required to maintain situation awareness (Endsley 1995; Stanton et al. 2006) and drive defensively (Stanton and Walker 2011). Furthermore, the magnitude of the SD means that even if our estimate of the proportion of time spent looking at signals during approach were out by 100%, so that proportion was really about 0.4, there would still be more than 15% of approaches during which signals are fixated for less than 20% of the time. The approach to SN109 when made at a speed around 40 mph (64 kph) is very demanding and places a heavy attentional load on the driver. Because of the unusual geometry of the signal and the way in which the signal is obscured from time to time, there will seldom be time for the driver to correct a misperception of the aspect of the signal.

The role of the AWS

Hodder acknowledged the AWS within half a second and then applied maximum acceleration. Why did he not brake? There would have been time for him to switch his gaze in response to the horn at least once; and we know from Groeger et al. (2004) that drivers often direct their attention to signals in response to an auditory signal from the AWS. Any explanation is necessarily speculative, but can be based on ambiguities in the AWS, whose design has two major ergonomic flaws. If a train passes the AWS magnet when approaching a G aspect signal, a bell sounds in the cab. If the aspect of the signal is not G, a horn sounds.

But there is no difference between the sound when approaching an R aspect, a Y aspect or a YY aspect. The horn indicates 'Signal aspect is not-G'. It does not indicate 'Signal aspect is R', which indicates immediate danger and a need to stop the train. That is a fundamental design flaw.

There is a second ambiguity in the AWS. It is sometimes used as a Permissible Speed Warning Indication (PSWI) when a driver approaches a speed limit. Wilkins (Atkins 99817B) notes that if Hodder thought the AWS was a warning for the 80-mph limit, he would not have perceived it as a warning of the not-G aspect of SN109 and instead would have thought it appropriate to accelerate. The AWS is thus ambiguous in two senses, greatly reducing its effectiveness (Stanton and Walker 2011).

If Hodder continued to monitor the whole driving environment to maintain situation awareness in accordance with defensive driving, he did not have time to look back at SN109. The AWS horn sounded, indicating that SN109 showed a 'not-G' aspect confirming his mental model of YY at SN109. Groeger et al. (2004) found that 250 of the signals approached showed a G aspect (probability = 0.72), 15 a YY aspect (probability = 0.04), 52 a single Y aspect (probability = 0.15) and 29 a R aspect (probability = 0.08), (Groeger et al. 2004; 18). The probability that any signal will show a R aspect is less than half the probability that it will show Y or YY. Hodder's recent experience will have led him to expect 'not-R' at Gantry 8 with a probability of 0.95. The horn confirmed his belief in the presence of a YY or Y aspect, and his past experience of this route did not lead him to expect an R aspect. He therefore did not direct his attention to the signal, but merely cancelled the horn as a sign that an 80-mph speed limit was now applicable (Stanton and Walker 2011). Although the displays in the signalling control centre showed the occurrence of the SPAD, the work of Stanton and Baber (2008) shows that there was not enough time for the signallers to intervene effectively before the crash occurred.

Finally, on the morning of the crash, there would have been a powerful reflected glare from a yellow and black track identification panel at the top of the gantry (W. S. Atkins, 99817A1, Photograph 12). This would tend to make Hodder direct his eyes downwards to avoid being temporarily blinded by the glare, making it even less likely that he would look up at the signal at the last moment that it was visible. If he did not lower his gaze, he may have been temporarily blinded by the flash from the reflected sun, in which case his visual acuity would have been greatly reduced. Believing the signal to be showing Y or YY, having had this belief 'confirmed' by the AWS, or thinking the AWS referred to the speed limit at that point, and not having time to make another fixation, he took his train onto the track in the path of the express arriving at high speed.

Summary of factors affecting driver cognition and performance

A summary of the interaction of system characteristics and the way they affected Hodder's cognition and behaviour will lead to design recommendations.

(1) Previous experience of the sequence of signal aspects after leaving Paddington produced a strong expectation that signals ahead would not have an R aspect, so Hodder accelerated progressively from coasting as he approached SN109.
(2) The track curvature and the masking of signals on the approach to Gantry 8 created uncertainty as to which signal applied to which track.
(3) There was not sufficient time to examine the signals repeatedly due to the constraints on the physiology of eye movements and the need to foveate selected items.
(4) Nonetheless, SN109 was visually located, but the mental model of expected signal geometry, plus the fact that other signals had the normal geometry, made the driver expect that an R aspect would be vertically below the YGY of SN109. Moreover, an early sighting of the R aspect of SN105 at a visual angle nearer to the ground would increase the expectation that SN109 was not-R since no light was visible on SN109 or any other signal on Gantry 8 at the same height.
(5) No R was visible below the YGY of SN109, and its R lamp was masked by overhead structures until late in the approach.
(6) Hodder probably concluded that the state of SN109 was 'not-Red'.
(7) Since no bright G was visible, he concluded that the state of S N109 was Y or YY, probably the latter because the two Y lamps looked similar in the reflected sunlight.

If he remembered the aspect of SN87 as being Y, he would expect that SN109 would be not-R with a probability of almost 0.9 (see text above) and no R was visible when he fixated the signal.

From analysis to design recommendations

From the above analysis of cognitive ergonomics and applying quantitative modelling of visual attention, the following detailed recommendations for changes in systems design are evident.

(1) The AWS ambiguity as to whether it is signalling approach to a signal or approach to a speed limit should be removed.

(2) The AWS should provide an unambiguous indication of an approach to a R aspect instead of signalling 'non-G'.

(3) The geometry of the signals and gantries should be unambiguously standardised.

(4) There should be no ambiguity as to which signal applies to which track even when an approach is made over curving track[2].

(5) If it is impossible to relocate the gantries or rebuild the bridges, electrical equipment and other infrastructure, an appropriate speed limit must be set so that there will be adequate time for the drivers to pay attention to the signals and detect their aspects with a high probability of being correct.

(6) Appropriate quantitative models should be developed to allow these recommendations to be implemented as follows.

Conclusions for quantitative modelling

Given measures of dynamic visual attention such as that used by Groeger et al. (2001, 2012) and by Luke et al. (2006), the choice of speed limits can be made on the basis of a rational quantitative model (Moray 1999). We can proceed directly from the data of Groeger et al. (2004) and Luke et al. (2006). Use risk analysis to identify the minimum number of fixations that are required during the approach to any signal, and what probability is acceptable that a driver will make that number of fixations. Using the statistics derived from the data of Groeger et al. (2004) calculate the time required for the occurrence of that number of fixations and the probability of making them. Examine the track layout and determine at what distance the signal of interest becomes unambiguously and continuously visible during the approach. If we divide that distance by the time required, we obtain the required speed limit during the approach.

For example, suppose we decide we can accept a probability of $p = 0.9$ that at least one fixation will be on the relevant signal. Then, from Table 5, we find we need time for at least 10 fixations. Suppose that the signal is only visible without interruption for the final 100 metres. Then that in turn requires a speed of not more than 11 mph. If for operational reasons a higher speed is required by the train operating companies, say 22mph, on this section of track, then the signal *must* be made visible for 200 metres, and so on for other speeds and distances. Thus, we obtain an ergonomic design decision that satisfies the known constraints on driver cognition and behaviour. This approach could be used to supplement the guidelines on signal viewing time (Railway Group Guidance Note GE/

GN8537). Care would need to be taken when using these data in sighting signals, to take account of the context of the driving and signal sighting task and all of the issues that may lead to SPADs (including those identified in the analysis presented in this paper). These include different signal heights on the route, different signal configurations, the influence of the AWS and different mental models built up due to past experience on the route. More data are needed and should be collected on other trains, routes and with other drivers.

A second method is based on more elaborate modelling. Given enough data, we could represent the dynamics of visual attention as a frequency transition matrix among the set of fixated objects in Table 1. We can then collapse the data to make a transition matrix between fixations on Own Signal and any other selected subsets of Table 1. We could develop a Markov model, and derive statistics such as the Mean First Passage Time (Kemeny and Snell 1960). These in turn provide a more powerful tool for predicting how often a particular feature of the visual environment will be fixated within a given period (Moray 2007; Moray, Neil, and Brophy 1983). However, the first method will suffice to make a major improvement in safety, collecting a more extensive set of data on eye movements in relation to various patterns of signals, auditory warnings, infrastructure and cab design. This is particularly important because there are many changes occurring in the technology of train design, including the use of in-cab displays rather than track-side signalling, head-up displays, etc. System design in the face of such extensive evolution requires the use of quantitative predictive modelling where possible (Moray 1999).

Notes

1. Our italics.
2. Since the Public Inquiry, large boards carrying track numbers have been placed beside each set of lamps. Other changes are planned in the modernisation of the railway. This paper is primarily concerned with the state of the system at the time the Ladbroke Grove accident occurred.

Acknowledgements

This paper is based on an internal report originally prepared for the Health and Safety Executive of the UK. We thank Mr. S. Jones for permission to use that source and for the opportunity to ride in the cab of a train over the route on which the accident occurred in order to view the signal infrastructure and to obtain photographs for analysis. Except for the data on driver eye movements, the technical data are derived from the final report of the Public Enquiry (The Cullen Report) and evidence cited therein. The recording of in-service train drivers' eye movements was a challenging task, and relied on the goodwill and cooperation of drivers, train companies, trade unions and Railway Safety, who sponsored the empirical research. We thank the Railway Unions and their members for their cooperation, especially those drivers who participated in the eye movement research. Natasha Merat and David Field collected and coded the data. Finally, our then colleague and co-investigator on the Railway Safety project, Dr. Mark Bradshaw, died soon after that project was completed, and we dedicate this paper to him.

Disclosure statement

No potential conflict of interest was reported by the authors.

ORCID

Neville Stanton ⓘ http://orcid.org/ORCID=0000-0002-8562-3279

References

Abramov, I., J. Gordon, and H. Chan. 1991. "Color Appearance in the Peripheral Retina: Effects of Stimulus Size." *Journal of the Optical Society of America, a* 8: 404–413.

Annett, J. 2002. "Subjective Rating Scales: Science or Art?" *Ergonomics* 45 (14): 966–987.

Balfe, N., J. R. Wilson, S. Sharples, and T. Clarke. 2012. "Development of Design Principles for Automated Systems in Transport Control." *Ergonomics* 55 (1): 37–54. doi: http://dx.dol.org/10.1080/00140139.2011.636456.

Cullen, Rt. Hon, and P. C. Lord. 2000. *The Ladbroke Grove Rail Inquiry.* Norwich: Her Majesty's Stationery Office.

Endsley, M. R. 1995. "Toward a Theory of Situation Awareness in Dynamic Systems." *Human Factors: The Journal of the Human Factors and Ergonomics Society* 37 (1): 32–64.

Groeger, J. A., M. F. Bradshaw, J.E. Everatt, N. Merat. 2001. "Pilot Study of Train Drivers' Eye Movements: In-service Drivers. London: Railway Safety & Standard Directorate. 40 pages.

Groeger, J., M. Bradshaw, J. Everatt, and N. Merat. 2012. Allocation of visual attention among train drivers. In A.G. Gale, J. Bloomfield, G. Underwood, J. Wood (eds.), Vision in Vehicles IX, Proceedings of 9th International Conference on Vision in Vehicles, Loughborough University, pp. 72–79.

Henderson, J. M. 1993. "Visual Attention and Saccadic Eyemovements." In *Perception and Cognition. Advances in Eye Movement Research,* edited by G. d'Ydewslle and J. van Rensbergen, 37–50. Amsterdam: North-Holland.

Ishida, T., and M. Ikeda. 1989. "Temporal Properties of Information Extraction in Reading Studies by a Text-mask Replacement Technique." *Journal of the Optical Society of America a* 6: 1624–1632.

Jones, R. E., J. L. Milton, and P. M. Fitts. 1949. *Eye Fixations of Aircraft Pilots. 1. A Review of Prior Eye-movement Studies and a Description of a Technique for Recording the Frequency, Duration and Sequences of Eye-fixations during Instrument Flight.* USAF Technical Report No 5837. Dayton, OH: Wright-Patterson Air Force Base.

Kemeny, J. G., and J. L. Snell. 1960. *Finite Markov Chains.* New York: Van Nostrand.

Luke, T., N. Brook-Carter, A. M. Parkes, E. Grimes, and A. Mills. 2006. "An Investigation of Train Driver Visual Strategies." *Cognition, Technology and Work* 8 (1): 15–29.

Moray, N., 1990. "A Lattice Theory Approach to the Structure of Mental Models." *Philosophical Transactions of the Royal Society B: Biological Sciences,* 327, 577–583.

Moray, N. 1997. "Models of Models of ... Mental Models." In *Perspectives on the Human Controller,* edited by T. B. Sheridan and T. Van Lunteren, 271–285. Mahwah, NJ: Lawrence Erlbaum.

Moray, N. 1999. "The Psychodynamics of Human-Machine Interaction." In *Engineering Psychology and Cognitive Ergonomics: Vol. 4,* edited by D. Harris, 225–235. Aldershot: Ashgate.

Moray, N. 2007. "Real Prediction of Real Performance." In *People and Rail Systems,* edited by J. R. Wilson, B. Norris, and T. Clarke, and A. Mills, 9–22. Aldershot: Ashgate.

Moray, N. 2008. "Subjective Mental Workload." *Human Factors* 24: 481–496.

Moray, N., and I. Rotenberg. 1989. "Fault Management in Process Control: Eye Movements and Action." *Ergonomics* 32 (11): 1319–1342.

Moray, N., G. Neil, and C. M. Brophy. 1983. *Selection and Behaviour of Fighter Controllers.* London: Contract Report, Ministry of Defence.

Naweed, A., S. Rainbird, and J. Chapman. 2015. "Investigating the Formal Countermeasures and Informal Strategies Used to Mitigate SPAD Risk in Train Driving." *Ergonomics* 58 (6): 883–896. doi:http://dx.doi.org/10.1080/00140139.2014.1001448.

Plant, K. L., and N. A. Stanton. 2012. "Why Did the Pilots Shut down the Wrong Engine? Explaining Errors in Context Using Schema Theory and the Perceptual Cycle Model." *Safety Science,* 50 (2): 300–315.

Rafferty, L. A., N. A. Stanton, and G. H. Walker. 2013. "Great Expectations: A Thematic Analysis of Situation Awareness in Fratricide." *Safety Science* 56, 63–71.

Reason, J. 1990. *Human Error.* Cambridge: Cambridge University Press.

Revell, K. A., and N. A. Stanton. 2012. "Models of Models: Filtering and Bias Rings in Depiction of Knowledge Structures and Their Implications for Design." *Ergonomics* 55 (9): 1073–1092.

Salmon, P. M., G. J. M. Read, N. A. Stanton, and M. G. Lenné. 2013. "The Crash at Kerang: Investigating Systemic and Psychological Factors Leading to Unintentional Non-Compliance at Rail Level Crossings." *Accident Analysis and Prevention,* 50: 1278–1288.

Senders, J. W. 1964. "The Human Operator as a Monitor and Controller of Multidegree of Freedom Systems." *IEEE Transactions on Human Factors in Electronics,* 5 (1): 2–5.

Senders, J. W., J. I. Elkind, M. C. Grignetti, and R. Smallwood. 1965. *An Investigation of the Visual Sampling Behavior of Human Observers.* Technical Report NASA-3860. Cambridge, MA. Bolt, Beranek, and Newman.

Stanton, N. A., and C. Baber. 2008. "Modelling of Human Alarm Handling Response times: A Case Study of the Ladbroke Grove Rail Accident in the UK." *Ergonomics* 51 (4): 423–440.

Stanton, N. A. and G. H. Walker. 2011. "Exploring the Psychological Factors Involved in the Ladbroke Grove Rail Accident." *Accident Analysis & Prevention,* 43 (3): 1117–1127.

Stanton, N. A., R. Stewart, D. Harris, R. J. Houghton, C. Baber, R. McMaster, P. Salmon, G. Hoyle, G. Walker, M. S. Young, M. Linsell, R. Dymott, and D. Green. 2006 "Distributed Situation Awareness in Dynamic Systems: Theoretical Development and Application of an Ergonomics Methodology." *Ergonomics,* 49 (12–13), 1288–1311.

Trick, L. M., and Z. W. Pylyshyn. 1994. "Why Are Small and Large Numbers Enumerated Differently? A Limited-capacity Pre-attentive Stage in Vision." *Psychological Review* 101 (1): 80–102.

Van Esch, J. A., E. E. Koldenhof, A. J. van Doorn, and J. J. Koenderink. 1984. "Spectral Sensitivity and Wavelength Discrimination of the Human Peripheral Visual Field." *Journal of the Optical Society of America, A* 1: 443–450.

Watt, R. 2000. *Visibility Issues and Signal SN109 at Paddington.* Submission to Cullen Inquiry: Department of Psychology, University of Stirling.

WS Atkins Rail Investigative Services, 99817A1. McKenzie, D. H. 2000a. *Ladbroke Grove Collision 05/10/99: Analysis of SSI Logging Tapes* Submission to Cullen Inquiry

WS Atkins Rail Investigative Services, 99817A1. Walter, K. 2000b. *Professional Discussion of Findings.* Submission to Cullen Inquiry

WS Atkins Rail Investigative Services, 99817A1. Wilkins, S. J. 2000c. *Ladbroke Grove Collision 05/10/99: Signal Sighting.* Submission to Cullen Inquiry

WS Atkins Rail Investigative Services, 99817A2. Wilkins, S. J. 2000d. *Ladbroke Grove Collision 05/10/99: Signal Sighting.* Submission to Cullen Inquiry

WS Atkins Rail Investigative Services 99817B *Addendum.* 2000e. Submission to Cullen Inquiry

WS Atkins Rail Investigative Services 99817B. Wilkins, S. J. 2000f. *Ladbroke Grove Collision 05/10/99: Signal Sighting.* Submission to Cullen Inquiry

WS Atkins Rail Investigative Services, 99819A. McKenzie, D. H. 1999. *Ladbroke Grove Collision 1999 – Driver Aspect Sequences.* Submission to Cullen Inquiry.

Zoer, I., J. K. Sluiter, and M. H. W. Frings-Dresen. 2014. "Psychological Work Characteristics, Psychological Workload and Associated Psychological and Cognitive Requirements of Train Drivers." *Ergonomics* 57 (10): 1473–1487. doi: http://dx.doi.org/10.1080/00140139.2014.938130.

Beyond human error taxonomies in assessment of risk in sociotechnical systems: a new paradigm with the EAST 'broken-links' approach

Neville A. Stanton (iD) and Catherine Harvey

ABSTRACT

Risk assessments in Sociotechnical Systems (STS) tend to be based on error taxonomies, yet the term 'human error' does not sit easily with STS theories and concepts. A new break-link approach was proposed as an alternative risk assessment paradigm to reveal the effect of information communication failures between agents and tasks on the entire STS. A case study of the training of a Royal Navy crew detecting a low flying Hawk (simulating a sea-skimming missile) is presented using EAST to model the Hawk-Frigate STS in terms of social, information and task networks. By breaking 19 social links and 12 task links, 137 potential risks were identified. Discoveries included revealing the effect of risk moving around the system; reducing the risks to the Hawk increased the risks to the Frigate. Future research should examine the effects of compounded information communication failures on STS performance.

Practitioner Summary: The paper presents a step-by-step walk-through of EAST to show how it can be used for risk assessment in sociotechnical systems. The 'broken-links' method takes a systemic, rather than taxonomic, approach to identify information communication failures in social and task networks.

Introduction

The term 'Socio-Technical Systems' (STS) is used to refer to the interaction between humans and machines, from the small and simple to the large and highly complex (Walker et al. 2008, 2010; Read et al. 2015). These subsystems operate and are managed as independently functioning (autonomous) entities, with their own goals, but must collaborate with other subsystems to achieve the higher goals of the STS (Dul et al. 2012; Wilson 2012). A key characteristic is that these goals can only be achieved by the STS and not by individual subsystems functioning in isolation (Von Bertalanffy 1950; Rasmussen 1997). STS present unique challenges for safety management and risk assessment (Rasmussen 1997; Alexander and Kelly 2013; Flach et al. 2015; Waterson et al. 2015). Traditional approaches to risk assessment, such as THERP (Technique for Human Error Rate Prediction: Swain and Guttmann 1983), TRACEr (Technique for the Retrospective and Predictive Analysis of Cognitive Errors: Shorrock and Kirwan 2000) and SHERPA (Systematic Human Error Reduction and Prediction Approach: Embrey 1986), are typically reductionistic in

nature (Stanton et al. 2013), focusing on individual tasks and technologies rather than the system as a whole (Stanton and Stevenage 1998; Stanton 2006; Stanton et al. 2009; Waterson et al. 2015). These methods use error taxonomies to identify risk but recent research has suggested that the term 'human error' is obsolete (Dekker 2014). In its place the term 'human performance variability' has been proposed, which includes both normative and non-normative performance. This latter approach emphasises the broad spectrum of human behaviour, rather than a dichotomy, and therefore a need to build resilient systems (Hollnagel, Woods, and Leveson 2006). The Systemic Accident Analysis (SAA) approach treats systems as whole entities with complex, non-linear, networks (Underwood and Waterson 2013). A number of SAA methods were assessed for their potential for prospective risk analysis within STS in a previous study (Stanton, Rafferty, and Blane 2012). Some system methods incorporate error taxonomies, such as CREAM (Cognitive Reliability and Error Analysis Method: Hollnagel 1998), HFACS (Human Factors Analysis and Classification System: Shappell and Wiegmann 2001) and STPA (System-Theoretic Process

Analysis: Leveson 2012) which, given recent shift away from the term of 'human error', is something of a conundrum. Rather than considering risks in systems to be the result of error, the approach taken in this paper is to propose risks as the failure to communicate information via social and task networks. This type of failure may be seen in several major incidents. For example, in the MS Herald of Free Enterprise accident (1987), the state of the bow doors was not communicated to the ships bridge (Noyes and Stanton 1997). So the ship left harbour and subsequently capsized. In the Kegworth air disaster (1989), the aircraft failed to communicate which engine was on fire, leading to the pilots shutting down the wrong engine (Griffin, Young, and Stanton 2010; Plant and Stanton 2012). In the Ladbroke Grove Rail incident (1999), the signals failed to communicate to the train driver that the section of the rail network was protected (Moray, Groeger, and Stanton Forthcoming). Rather than stopping, the driver actually increased his speed as he passed the red signal leading to a collision with an oncoming high-speed intercity train (Stanton and Walker 2011). So, rather than conceive these behaviours as errors, we have reconceived them as the failure to communicate information in the system. This is a new paradigm for risk assessment that incorporates the value of a holistic perspective for appreciating the relationships between the various subsystems and the network diagrams for visualising important aspects of STS, such as the constraints on communications (see Flach et al. 2015). In order to analyse the information communications in systems, the Event Analysis of Systemic Teamwork (EAST) method was selected. EAST takes a different approach to the error taxonomic methods, by modelling and analysing STS-level interactions. In a previous study, Stanton (2014) analysed communications between various actors within a submarine control room: in contrast, this case study analyses a retrospective account of actions within a Royal Navy training activity and is conducted at the macro-level (Grote, Weyer, and Stanton 2014). The aim of this paper is to extend the EAST network-level analysis to include risk prediction by 'breaking' links within networks.

The EAST method was first proposed by Stanton et al. (2005, 2013) and further elaborated by Stanton, Baber, and Harris (2008) for modelling distributed cognition in STS. The method represents distributed cognition in networks, which enables both qualitative and quantitative investigations to be performed (Stanton 2014). One of the main advantages of EAST is its aim to capture the whole system, as opposed to reductionist methods which split a system into constituent parts for analysis (Walker et al. 2010). It is therefore considered in this study to be a suitable technique for representing a STS and potential non-normative behaviours. The analysis describes a system as three different types of network:

- Social; representing the agents (human, technical and organisational) within a system and communications between them,
- Task; representing the activities performed by the system and the relationships between them,
- Information; representing the information that is used and communicated within a system and links between different information types.

The social, task and information networks are developed individually and then combined to create a complete social–task–information network diagram, showing all the links and information flows (i.e. distributed cognition) within a network-of-networks. EAST has been applied in many domains, including aviation (Stewart et al. 2008; Walker et al. 2010; Stanton et al. 2016), military (Stanton et al. 2006; Stanton 2014), road (Salmon et al. 2014), rail (Walker et al. 2006) and the emergency services (Houghton et al. 2006; Baber et al. 2013). The aim of this work is to extend the EAST method to consider risk in systems via a case study and provide initial STS method evaluation criteria presented by Harvey and Stanton (2014). The premise of the risk assessment being that STS failures are predominately caused by the failure to communicate information between agents and tasks. This will be studied within the context of the following case study.

Case study of Hawk missile simulation training

Operation of the Hawk jet to simulate missile attacks against surface ships by the UK Royal Navy was selected as the case study. This activity is viewed as a STS because it comprises many interconnected subsystems, which are themselves complex. The context for this study is illustrated in an AcciMap (Rasmussen 1997; Jenkins et al. 2010) in Figure 1. The AcciMap places different subsystems within the STS at different levels and shows the links in communication and decision-making between the subsystems. Each node in the Accimap is labelled (a, b, c...) to correspond with the description in the text in the following paragraphs. The year in which each event occurred is also included where applicable, to give an indication of timescale.

The Royal Navy uses the Hawk Jet to simulate air attacks on ships during sea training of ships' gunners and radar operators (event 'a' in Figure 1). The Hawks are used to simulate enemy aircraft attacks and high-speed skimming missiles fired against ships (Royal Navy 2012). In order to perform these simulation activities, the Hawk must be flown at a low height above the sea level (b); however, the Hawk is not equipped with a Radar Altimeter (Rad-Alt), which provides a highly accurate measure of the altitude of the aircraft above the sea. This makes flying the Hawk accurately at very low levels extremely difficult, and requires

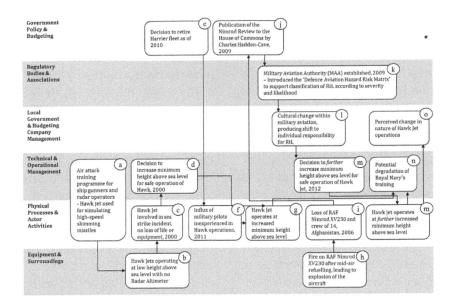

Figure 1. AcciMap showing subsystems within the Hawk Jet STS.
Note: Please note that the labels do not indicate a timeline, rather they are added for clarity of the description below.

a high level of expertise to perform safely. Prompted by events over a number of years, there have been some significant changes to the method for assessing the safety of the Hawk STS.

In 2000, a Hawk Jet was involved in a sea strike incident as a consequence of very low level flight (c). Although there was no resulting loss of life, this incident prompted a decision by the Royal Navy to increase the minimum allowable flying height above the sea level for the Hawk (d).

As part of its Strategic Defence and Security Review, H M Government (2010) took the decision to retire the Harrier jet from service in October 2010 (e). As a result of this, a number of Royal Navy pilots who would have flown the Harrier were diverted into the Hawk programme (f). Traditionally, the Hawk has been flown by Civilian pilots under contract to the Royal Navy (2012): these pilots have extensive military experience in fast jets, which includes low level flight supported by a Rad Alt. This experience provided mitigation against the RtL for the Hawk air attack simulation task; however, the cohort of military pilots did not have this same level of experience and the RtL had to be reassessed in the light of this (g).

In 2006, RAF Nimrod XV230 suffered a catastrophic explosion after a mid-air refuelling procedure (h): this caused the deaths of all 12 crew members plus two mission specialists and total loss of the aircraft (i). The Government requested a comprehensive review into the airworthiness and safe operation of the Nimrod (j), which was delivered by Haddon-Cave (2009). The report described the development of the safety case for the Nimrod as 'a story of incompetence, complacency, and cynicism' (161) and concluded

that it was undermined by the widespread assumption that the Nimrod was safe because it had been flown successfully for the preceding 30 years (Haddon-Cave 2009). The report also identified organisational changes in the years prior to the Nimrod accident as having significant influence; these included a shift in organisational culture towards business and financial targets 'at the expense of functional values such as safety and airworthiness' (Haddon-Cave 2009, 355). As a consequence of the findings, Haddon-Cave (2009) recommended the establishment of an *independent* Military Aviation Authority (MAA) to properly assess Risk to Life (RtL) and shape future safety culture (k). Further recommendations included the need for strong *leadership*, a greater focus on *people* to deliver 'high standards of safety and airworthiness' (Haddon-Cave 2009, 355) and increased *simplicity* of rules and regulations. The tragic consequences of the Nimrod accident, along with the recommendations of the Haddon-Cave report, effected a culture change within military aviation: this resulted in a decision to assign *individual* accountability for RtL assessments to 'Duty Holders' (DH), where previously responsibility for risk had been held at the organisation level (l). The newly established MAA produced guidelines for the assessment of RtL, in the form of the Defence Aviation Hazard Risk Matrix (MAA 2011) which supports the classification of single risks according to their estimated severity (catastrophic, critical, major, minor) and likelihood (frequent, occasional, remote, improbable). The resulting risk level determines at which level of DH the risk is held.

The organisational changes brought about by the events described above (i.e. influx of junior pilots) prompted reassessment of the RtL for the Hawk air attack simulation activity. The goal of safety management in the UK military is to reduce risk to a level which is As Low As Reasonably Practicable (ALARP): this is reached when 'the cost of further reduction is grossly disproportionate to the benefits of risk reduction' (Ministry of Defence 2007). The RtL for all Hawk operations is frequently reassessed and the shift in pilot experience levels, as described above, prompted changes to the RtL for the Hawk air attack simulation activity. In order to reduce this RtL to a level which was ALARP, a decision was taken by the Royal Navy DH with SME advice to further increase the minimum height above sea level (m). A potential consequence of this decision is the degradation of Royal Navy surface fleet training against very live low level targets, as the Hawk can no longer accurately simulate sea skimming missile attacks on surface ships (n). These events have changed the nature of Hawk operations within the UK MoD (o).

Potential risks to the safe operation of the missile simulation activity are, in part, assessed according to the Military Aviation Authority's (MAA) Regulatory Articles (MAA 2011). This assessment is based on the principle that risks can be tolerated provided they are reduced to ALARP. The MAA regulatory policy outlined its approach to the management of Risk to Life (RtL):

> Aviation DHs [Duty Holders] are bound to reduce the RtL within their AoR [Area of Responsibility] to at least tolerable and ALARP; the application of effective and coherent risk management processes will be fundamental to achieving this. (MAA 2011, 18)

Regulatory Article (RA) 1210 – Management of Operating Risk (Risk to Life) – defined risk as:

> a measure of exposure to possible loss [combining] severity of loss (how bad) and the likelihood of suffering that loss (how often). (MAA 2011, 1)

The MAA suggested that risks can be identified via a number of different methods including previous occurrences, checklists, HAZOPS, zonal hazard (safety) analyses and error trend monitoring. Previous work has showed that these techniques are likely to be inadequate for the analysis of STS (Stanton, Rafferty, and Blane 2012). RA 1210 specifically encourages the use of fault trees as accident models 'to assist understanding of the interrelationship between risks and to support the prioritisation of effort to maximise safety benefit' (MAA 2010, 6). This technique, along with other traditional error and risk prediction methods, does not account for the interactions of distributed actors within a STS (Salmon et al. 2011). Furthermore, there is also no clear method outlined by the MAA for structuring risk identification, for example, the recommendation is that a combination of these methods should be used

with the aim of identifying all credible risks, but there is no way of knowing when all possible credible risks have been defined and therefore how many methods to use and when to stop applying them.

The Hawk RtL case study was identified through interviews with a subject matter expert (SME) as part of this project. The analysts were provided with a high-level overview of the case study in an initial interview with the SME. This was followed up by a second, in-depth interview about the case study with the SME, conducted by two analysts. This resulted in a detailed account of the Hawk-Frigate STS, which was supplemented by extra information from official documentation including Military Aviation Authority (MAA 2010, 2011) guidelines, the official report into the Nimrod accident (Haddon-Cave 2009) and Royal Navy safety assessment guidance (Royal Navy 2012). The EAST method (Stanton 2014) was used to develop the three network diagrams, based upon the analysis of all case-study information, in an iterative process which involved the SME providing feedback during development.

Analysis of networks

Social Network Analysis (SNA) metrics provide quantitative measures that represent the structures and relations between nodes in the EAST networks (Driskell and Mullen 2005; Walker et al. 2009; Baber et al. 2013). The SNA metrics describe individual nodes (including reception, emission, eccentricity, sociometric status, centrality, closeness, farness and betweeness). The SNA metrics applied in the current study, along with their descriptions, are presented in Table 1. Analysis software, AGNA version 2.1 (Benta 2005), was used to calculate the SNA metrics. For each EAST network, key nodes were identified according to sociometric status. Sociometric status was selected to define key nodes because it identifies the prominence of an individual node's communications with the rest of the network, which influences the whole network's performance (Stanton 2014). In a STS, all of the nodes will have complex safety management rules and behaviours; however, as the 'key' nodes have the largest number of connections to the rest of the network, these nodes will have the highest degree of influence over the behaviour of the entire STS. Sociometric status key nodes are defined as nodes which have a higher sociometric status score than the sum of the mean sociometric status score plus the standard deviation sociometric status score for all nodes in the network. SNA metrics were calculated for the EAST networks created for this case study: key agents for sociometric status are indicated in the social, task and information network diagrams below.

Table 1. SNA Metrics, along with their descriptions.

	Safety constraint	Description
Node-level metrics	Emission	The number of edges (links to other nodes) originating at that node
	Reception	The number of edges incident to that node
	Sociometric status	Number of communications received and emitted relative to the number of nodes in the network
	Bavelas–Leavitt (B–L) centrality	Degree of connectivity to other nodes in the network
	Eccentricity	Length of the longest geodesic path originating in that node (a geodesic path is between two given nodes that has the shortest possible length)
	Closeness	Inverse of the sum of the geodesic distances from that node to all the other nodes, i.e. extent to which a node is close to all other nodes
	Standard closeness	Closeness multiplied by (g^{-1}), where g is the number of nodes in the network
	Farness	Sum of the geodesic distances from that node to all other nodes
	Betweeness	Frequency with which a node falls between pairs of other nodes in the network

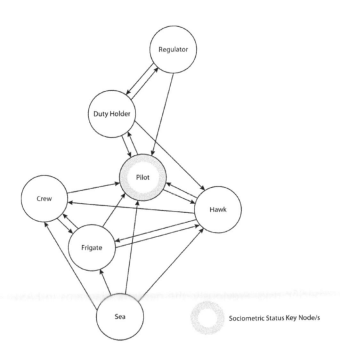

Figure 2. Social nodes and their links within the STS.

Results

The step-by-step application of the shortened version of EAST is described in detail in the following sections. This is accompanied by the outputs of the method along with interpretation of the results. The first stage in EAST was the identification of all social, task and information nodes within the Hawk missile simulation case study, based on the SME's account of activities, which informed the analysts' knowledge of the case study. The nodes were arranged in social, task and information networks and links drawn between related nodes. Related nodes were those between which some information was transferred. As well as providing a visual representation of a STS, the EAST network diagrams can be analysed to produce quantitative SNA metrics.

Social network

Seven social 'agents' and their connections were identified from the Hawk RtL case study with the SME: these are shown in Figure 2. The social network was constructed by first identifying the main agents that are in the system, then by examining the interdependencies between those agents. The SME agreed that the social network was a reasonable representation of the main agents and their relationships. The 'edges', or links, between the agents show where information is transferred and the direction of transfer. There are 19 edges in total; in some cases information transfer is reciprocal but in others it only goes in one direction between two agents. The 'pilot' node was identified as the key agent according to sociometric status (1.33): it has the highest number of links to and from other nodes in the network; in fact, the pilot receives and/or emits information from/to all of the other agents in the social network.

The pilot had the highest betweeness score (10.0) as it is located on the paths between a number of other agent pairs. The pilot also had the highest score for reception (6), highlighting a high degree of connectivity to other nodes in the network and indicating that the pilot's actions and communications are integral to the functioning of the STS. The high farness score for the regulator (10) indicates that this agent is located furthest from most other nodes and this is supported by the information in the case study, which showed that the regulator really only communicates with the DH and possibly the pilots but has no contact with the frigate or crew. This is because the regulator in this case is the MAA, which does not have direct control over the Navy's surface ship operations. The sea scored highest for emission (4) and lowest for reception (0) as it does not receive information from any other nodes but is used for feedback only. In this sense, the sea can be regarded as a 'passive' agent, as it cannot respond to feedback; the social agents can only respond to it.

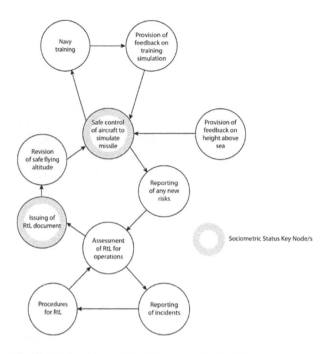

Figure 3. Task nodes and their links within the STS.

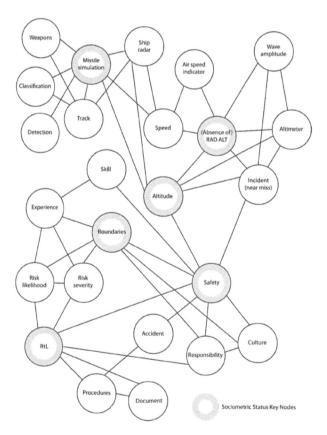

Figure 4. Information nodes and their links within the STS.

Task network

Ten tasks nodes and their connections were identified from the Hawk RtL case study with the SME: these are shown in Figure 3. The task network was constructed by first identifying the main tasks that are performed by the system, then by examining the interdependencies between the tasks. The SME agreed that the task network was a reasonable representation of the system. There are 12 edges in total and in all cases the transfer is unidirectional.

In the task network, key nodes were identified as 'safe control of aircraft to simulate missile' and 'issuing of RtL document', which had sociometric status scores of .56 and .44, respectively. The task network contains more nodes but fewer edges than the social network, indicating that there are fewer communications between tasks. Cohesion is zero because there are no mutual, or bidirectional, links between nodes. The highest score for betweeness was for 'safe control of aircraft to simulate missile' (45) demonstrating that this task is integral in the STS as it is located between a high number of other task nodes. This is unsurprising as this task can be considered to be the main objective of the STS configuration investigated in this case study. This task also scored highest on emission (2), reception (3) and B-L centrality (6.26), as well as sociometric status (.56), showing a high level of connectivity to other nodes.

Information network

EAST identified 25 information nodes and their connections were based on the Hawk RtL case study and further knowledge of the STS from the SME: these are shown in Figure 4. There are 50 edges in total; however, in this case, the links are not directional.

Six information nodes were identified as key nodes according to sociometric status: missile simulation, (absence of) Rad Alt, altitude, boundaries, safety and RtL. Safety had the highest betweeness (315.8), standard closeness (.53) and B-L centrality scores (18.51) and this reflects the importance of this in the case study: the aim within the STS is to achieve a safe solution for missile simulation. Density and cohesion were relatively low, i.e. compared with the social network, as the edges between nodes were single and non-directional.

Broken-links analysis

Studies of networks have discussed the effects of removing one or more nodes from a network on the resilience of that network to systemic failures and the resulting destabilisation (Houghton et al. 2008; Baber et al. 2013; Stanton 2014). This has been used to explore the resulting influence on network structure, rather than as a method for predicting specific risks. Previously, the network diagrams in EAST have been used to provide a visual representation of a system to further the users' understanding of distributed cognition (Stanton 2014). In this study however, the EAST network analysis was extended to identify and examine possible risks by 'breaking' the links between the various nodes, in a similar approach to the removal of nodes, to explore system effects. Broken-links represent failures in communication and information transfer between nodes in the networks and these failures can then be used to make predictions about the possible risks within the STS. Previously, 'broken-links' have only been investigated by EAST analysts when looking retrospectively at accidents to identify underlying causes. Griffin, Young, and Stanton 2010 demonstrated that the broken link between the Engine Vibration Indicator and the pilots in the cockpit was a causal factor in their failure to shut down the correct engine in the Kegworth accident. If this information had been communicated more effectively it could have helped to prevent the crash. Similarly, the EAST method has been adapted to analyse incidents of fratricide (Rafferty, Stanton, and Walker 2012) although this has been conducted as retrospective and concurrent, rather than predictive, analyses. The broken-links analysis was performed for the Hawk missile simulation case study, on the social and task networks as shown in Figures 2 and 3, respectively. The information network was not subject to the broken-links analysis because broken-links between information nodes were not considered to represent risks, as they are caused by a failure in either the social or task networks. In other words, information does not fail in isolation; it is the failure to use or communicate the information correctly and in all cases this can be attributed to social nodes, task nodes or both. For the social and task networks, each link was identified and documented in a table. The combined EAST networks diagram (see Figure 5) shows the information network tagged with the social networks nodes (to show who owns each information node in the network) and grouped by the task network nodes (to show which task each information node belongs to). Details on construction of the combined network have been reported by Stanton (2014). This combined network was used to identify what information (from the information network) should be communicated from the origin node to the destination node in the task and social networks and therefore what information would not be communicated if the links between the nodes in the task and social networks were removed.

Figure 5 also shows the combined information–task–social network as a single depiction of the entire STS. This shows the overlaps between the three networks, in other words, which information is being communicated by which agents in which tasks, and how these nodes are interlinked.

In order to conduct the broken-links analysis, the social and task networks were compared to the combined information–task–social network in turn. For example, there is a reciprocal relationship between the duty holder and the pilot (in the social network shown in Figure 2) and the duty holder and the pilot share the nodes of boundaries, RtL, risk likelihood, risk severity, procedures, document, responsibility and safety (in the combined information–task–social network shown in Figure 5). The risk assessment procedure requires that the relationship between the duty holder and pilot be interrogated to see what would happen if each information element was not transmitted, as shown in Table 2. The pilot was identified as having the highest Sociometric Status in the analysis presented in Figure 2, so was chosen for the illustration of the broken-links analysis in Table 2. Although the pilot is linked to all other agents in the social network, for the purpose of illustration just their reciprocal relationship with the duty holder is presented in Table 2.

Table 2 shows the risks resulting from the failure to pass relevant information between duty holder and pilot and vice versa. Anecdotal evidence from our SME suggests that there is variability in what individual pilots will chose to report back to the duty holder, as they have different interpretations of what they consider to be a risk and near miss. This shows that there is at least some face validity for the approach we have proposed.

In the similar manner to the social-information broken-links analysis shown in Table 2, there is a task-information broken-links analysis in Table 3. From the task network, there is a uni-directional relationship between the 'Issuing of RtL document' and the 'Revision of safe flying altitude' (in the task network shown in Figure 3) and they overlap in the combined information–task–social network (shown in Figure 5). The risk assessment procedure requires that the relationship between the 'Issuing of RtL document' and the 'Revision of safe flying altitude' be interrogated to see what would happen if each information element was not transmitted, as shown in Table 3. The 'Issuing of RtL document' was chosen as it has the highest Sociometric Status in the analysis presented in Figure 3. The 'Safe control of aircraft to simulate missile' was chosen for the same reason and is paired with 'Navy training' for the purposes of offering an illustrative example of the method in Figure 3.

Examination of the analysis in Tables 2 and 3 offers a systematic approach for examining a system of operation in a holistic manner. For example, increasing the safe flying attitude (see Figure 1) has led to the altitude profile of the Hawk not matching that of the low flying missile (see Table 3). This has meant that reducing the risk for the Hawk pilots could have a negative effect on training of the crew on the Frigate, ultimately increasing their risk. So whilst the top of Table 3 is about improving the safety of the pilot, by increasing altitude, for example, the bottom of table three shows that this could reduce the safety of the Navy frigate crew as they do not receive realistic training.

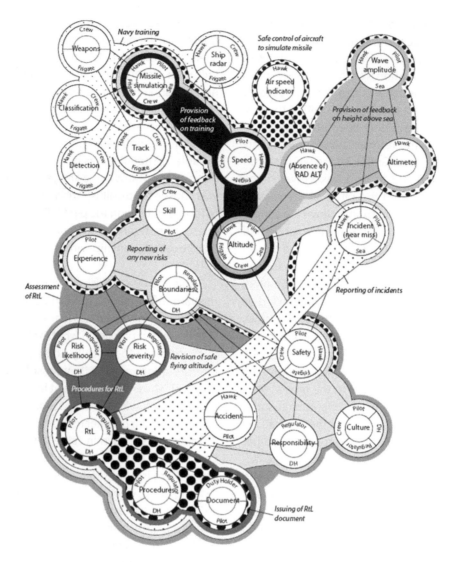

Figure 5. Combined information–task–social network for the Hawk case study.
Note: Shading represents the different tasks being undertaken.

The benefit of systems approaches is that the emergent properties become more readily apparent.

Discussion

This work aimed to explore the use of a modified version of EAST (network modelling and broken-links analysis, see Stanton 2014) in a case study of a Royal Navy training activity. First, the findings of this study are discussed in terms of criteria which were identified as essential for methods designed to analyse the human component of STS (Stanton, Rafferty, and Blane 2012; Harvey and Stanton 2014). This enabled comparisons to be made between the method and the current RtL procedure used in the Hawk missile simulation case study. Second, the modifications and extensions to EAST are discussed with reference to use of the method as an assessment of potential risks within a STS.

Aviation accidents, as with most accidents in STS, usually occur due to a conjunction of factors (Jenkins et al. 2010; Hodgson, Siemieniuch, and Hubbard 2013) and it is therefore essential that analysis methods are able to explore all of these factors by taking an integrated and holistic approach (Salmon et al. 2011; Ramos, Ferreira, and Barceló 2012). EAST specifically enables the exploration of the social, task and information components of the STS, allowing a high-level model of the STS to be created (visual diagrams) and analysed (social network metrics). This visual component is likely to help analysts and other stakeholders to understand the interactions within networks (Flach et al. 2015): this is an advantage over many other methods such as HAZOP, Fault Tree analysis, as well as the MAA's RtL/HRM approach. Baber et al. (2013) argued that it is sensible to speak of a 'useful' (rather than 'complete') network, as there will always be a possibility that some connections have been left out due to not being observed,

Table 2. Extract from broken-links analysis for EAST social network.

From (agent)	To (agent)	Information not communicated	Resulting risk	Mitigation strategy
Duty Holder	Pilot	*Boundaries*	Pilots are not aware of the boundaries for flight operations and for the identification and reporting of risks within this	Boundaries for risk reporting must be made clear to pilots as part of the RtL process
Duty Holder	Pilot	*RtL*	Pilots are not made aware of the results and consequences of the RtL assessment process after it is conducted at DH level	Results and consequences of the RtL assessment process must be effectively communicated to pilots
Duty Holder	Pilot	*Risk likelihood*	Pilots are not made aware of risks assessed that their likelihood of occurrence	Risks identified as having a high likelihood of occurrence must be reported to pilots
Duty Holder	Pilot	*Risk severity*	Pilots are not made aware of risks assessed and their severity of impact	Risks deemed as having a high severity of impact must be reported to pilots
Duty Holder	Pilot	*Procedures*	Pilots are not aware of how the RtL process is conducted at DH level and of procedures for reporting incidents to the DH	Pilots must be provided with clear procedures describing the assessment of RtL at DH level and the reporting of risks to DH
Duty Holder	Pilot	*Document*	Pilots are not provided with documentation covering the RtL process and its results	Pilots must be provided with documentation covering the RtL process and its results
Duty Holder	Pilot	*Responsibility*	Pilots are not aware of the DH's nor their own responsibilities for safety	The responsibilities of both the pilot and DH for safety must be clearly defined and understood by pilots
Duty Holder	Pilot	*Safety*	Pilots do not receive information about the safety of operations, based on the RtL assessment process	The safety of operations, as assessed during the RtL process, must be reported to the pilots
Pilot	Duty Holder	*RtL*	The DH does not receive information about new risks identified by the pilots	Pilots must clearly report all relevant risks to the DH
Pilot	Duty Holder	*Risk likelihood*	The DH does not receive information about the likelihood of new risks identified by the pilots	Pilots must report their estimate of the likelihood of occurrence of all relevant risks
Pilot	Duty Holder	*Risk severity*	The DH does not receive information about the severity of new risks identified by the pilots	Pilots must report their estimate of the severity of impact of all relevant risks
Pilot	Duty Holder	*Incident (near miss)*	The DH does not receive information about incidents (or near misses) which occur during Hawk operations	Pilots must clearly report all relevant incidents which occur during Hawk operations to the DH
Pilot	Duty Holder	*Experience*	The DH cannot learn from the pilots' experience of Hawk operations and the risks encountered	Pilots must clearly report their experience levels to the DH. Pilots must report their assessment of risks and any consequent assumptions, based on this experience
Pilot	Duty Holder	*Skill*	The DH cannot learn from the pilots' skill in Hawk operations	Pilots must clearly report their skill levels to the DH. Pilots must report their assessment of risks and any consequent assumptions, based on this skill level
Pilot	Duty Holder	*Safety*	The DH does not receive information about the safety of Hawk operations	Pilots must report their estimates of the safety impact of any risks identified in Hawk operations to the DH
Pilot	Duty Holder	*Culture*	The DH is not aware of the culture of safety amongst the Hawk pilots	Pilots must consider and report the estimated influence of safety culture on the risks to Hawk operations

reported and/or documented. This is certainly applicable to the networks generated by EAST, as it is impossible to know whether an analysis has been exhaustive and it is therefore safest to assume that it has not. It is also particularly true in this case as the analysis was performed on an SME's reports of activities within the STS rather than communications between STS actors (as in Stanton 2014). A consequence of this approach is a lack of richness of information, although if the main contribution of EAST lies in its ability to visually represent an STS then this may not be a significant issue. EAST includes the calculation of SNA metrics, which provide the analyst with quantitative values to represent various characteristics of the networks. In this way, the analysis encompasses all elements of a STS and provides the analyst with an understanding of the structure of a system as a whole and the relationships

between individual system components. These metrics can provide potential insight into the resilience of the networks (Stanton, Harris, and Starr 2016).

The inclusion of particular agents in an accident model is dependent on the information put into the analysis and therefore on the analysts and SMEs involved. This also is true for more traditional HAZOP and error identification methods and the current RtL assessment process, as well as EAST. However, because HAZOP and RtL assessment essentially focus on a list of potential errors, there is no formal procedure for identifying the decision-makers involved. In contrast, EAST enabled a visual representation of the decision-making agents (Flach et al. 2015) and their relationships with other nodes in a STS to be constructed, thereby encompassing the identification of decision-makers into the analysis process. This can allow analysts to understand

Table 3. Extract from broken-links analysis for EAST task network.

From (task)	To (task)	Information not communicated	Resulting risk	Mitigation strategy
Issuing of RtL document	Revision of safe flying altitude	*Document*	The information contained in the RtL document does not trigger a revision of safe flying altitude	The RtL document must be used by regulators to inform changes to regulations and safety guidance where appropriate
Issuing of RtL document	Revision of safe flying altitude	*RtL*	The outcome of the RtL process outlined in the RtL document does not trigger a revision of safe flying altitude	The outcomes of RtL assessment must be used by regulators to inform changes to regulations and safety guidelines where appropriate
Issuing of RtL document	Revision of safe flying altitude	*Risk likelihood*	The outcome of the Risk likelihood assessment, conducted as part of the RtL process and outlined in the RtL document, does not trigger a revision of safe flying altitude	The outcome of the Risk likelihood assessment, conducted as part of the RtL process and outlined in the RtL document, must be used to inform changes to regulations and safety guidelines where appropriate
Issuing of RtL document	Revision of safe flying altitude	*Risk severity*	The outcome of the Risk severity assessment, conducted as part of the RtL process and outlined in the RtL document, does not trigger a revision of safe flying altitude	The outcome of the Risk severity assessment, conducted as part of the RtL process and outlined in the RtL document, must be used to inform changes to regulations and safety guidelines where appropriate
Issuing of RtL document	Revision of safe flying altitude	*Safety*	The safety implications of the RtL process outlined in the Rtl document do not trigger a revision of safe flying altitude	The safety implications of RtL assessment must be used by regulators to inform changes to regulations and safety guidelines where appropriate
Issuing of RtL document	Revision of safe flying altitude	*Responsibility*	Responsibility for the revision of safe flying altitude is not outlined in the RtL document	Responsibility for changes to regulations and safety guidelines based on RtL assessment must be clearly assigned and accepted
Safe control of aircraft to simulate missile	Navy training	*Missile simulation*	The overall control of the Hawk does not adequately simulate missile attack on the frigate to aid with training	The operation of the Hawk must aid Navy training for missile attack situations
Safe control of aircraft to simulate missile	Navy training	*Speed*	The speed profile of the Hawk does not adequately simulate missile attack on the frigate to aid with training	The speed of the Hawk during missile simulation must be sufficiently realistic to aid Navy training for missile attack situations
Safe control of aircraft to simulate missile	Navy training	*Altitude*	The altitude profile of the Hawk does not adequately simulate missile attack on the frigate to aid with training	The altitude of the Hawk during missile simulation must be sufficiently realistic to aid Navy training for missile attack situations
Safe control of aircraft to simulate missile	Navy training	*Track*	The track of the Hawk does not adequately simulate missile attack on the frigate to aid with training	The track of the Hawk during missile simulation must be sufficiently realistic to aid Navy training for missile attack situations

where responsibility for risks resides within the STS and so target mitigation strategies appropriately (Lundberg, Rollenhagen, and Hollnagel 2010). This case study showed that EAST provided a useful visual representation of relationships between the various components of the STS. EAST examines the links between nodes and so is focused on communications, and therefore on the consequences of an action at a node, rather than its causes (Walker et al. 2010; Rafferty, Stanton, and Walker 2012).

In this case study, the analysts used a modified version of EAST, concentrating on the social, information and task networks (Stanton 2014). Guidance is provided on structuring a model of the STS under investigation and there are numerous examples of previous EAST models (e.g. Griffin, Young, and Stanton 2010; Walker et al. 2010; Rafferty, Stanton, and Walker 2012; Stanton 2014) in the literature. The 'break-link' process is very straight forward indeed and would be a useful addition to the current RtL assessment process (Haddon-Cave 2009). The current guidance from the MAA states that risks should be identified from a number of sources including HAZOP, error data

and experience of previous events; however, there are no explicit instructions on how many of these methods to use and when to stop this analysis. This means that the RtL assessment may proceed without a comprehensive list of potential risks. It appears that EAST could be a useful model for ensuring that this does not happen; however, it is important to note that provision of guidance may not be sufficient for successful application of STS methods. The training requirements of these methods can often be high for practitioners, with many citing a lack of time and difficulty accessing new information as barriers to STS analysis (Underwood and Waterson 2013).

Stanton, Rafferty, and Blane (2012), Stanton et al. (2014) previously suggested that EAST could be suitable for prospective analysis of STS risks; however, these studies only demonstrated the utility of methods for retrospective analysis (Salmon, Cornelissen, and Trotter 2011; Waterson et al. 2015). In this study, EAST has been applied to a STS that is currently in operation in order to investigate the ability of methods to model the future state of a STS. The Hawk missile simulation STS has already experienced and been impacted by incidents (e.g. Hawk sea strike) and accidents (e.g. Nimrod) but this analysis focused on the prediction of a future state given the changes in the STS, such as the alteration in safe flying altitude for the Hawk and the effects of this on missile simulation for the frigate and crew. Having said this, the emphasis with EAST is not on predicting accidents per se; rather, it is about creating a comprehensive model of the links and information flows within the STS and by doing so making the analysts aware of potential breakdowns and failures that may occur in the future. This means that the success of EAST for prospective analysis is dependent on the participation in the assessment process of those who will be impacted by these failures and those that can apply the appropriate mitigation strategies.

In summary, this study used a modified version of EAST, following the examples in Stanton (2014). In this case, only the network analysis phases of EAST were applied (followed by the new paradigm of the broken-links analysis which has not been previously reported) because the preliminary stages of EAST were negated by having already collected and represented the data via interviews with an SME. Furthermore, some of the EAST methods require communications data, which was lacking in this particular case study as the information came from a SME's account of the STS. Compared to Stanton's (2014) analysis of the operations within a submarine control room, the current study analysed activity at a macro level (Grote, Weyer, and Stanton 2014), using an SME's account of activities within the STS rather than a transcript of direct communications between STS actors. Recording and transcribing communications within a working system in real time is difficult, time-consuming and potentially disruptive to the STS under investigation. The approach presented in the current paper would be easier for personnel within the STS itself to apply, to support their own safety management and risk prediction activities, as it relies on a macro-level account of actions and relationships with a STS. This also allows these personnel to create a systems view of the STS of interest, which, as previously discussed, offers benefits over traditional taxonomic techniques. The absence of communications data obviously means that the analysis lacks detail and a richness of information which comes from speech data. This also meant that frequency of communications could not be represented in the same way as Stanton (2014). So in effect, the network diagrams in this case study offer a basic visual representation of a STS which could aid understanding of the relationships between agents, tasks and information as well as their combination. Of greater importance is the extension to EAST presented in this paper: the broken-links analysis.

In order to identify potential risk in the STS, these links between nodes in the networks had to be examined in more detail; this was accomplished in the broken-links analysis. In this phase, each link between the task and social nodes was 'broken' to illustrate the effect of a communication breakdown between nodes. In this case study, 19 social links and 12 task links were broken and assessed against numerous information nodes, resulting in the identification of 137 risks in total. These breakdowns would result in a failure in information transfer, so each broken link was analysed against the information nodes to identify potential risks. This extension to the EAST method provides a structured method for identifying all of the risks within a STS. The broken-links can be listed in table form, along with 'to' and 'from' information detailing the origin and destination nodes between which information is transferred. The broken-links analysis is concluded by developing mitigation strategies for each of the identified risks, in a similar way to traditional human error analysis methods.

Conclusions

This paper presented a case study of the extension to the EAST method applied in the analysis of a STS, specifically Hawk missile simulation to aid with training of Navy crew. The approach models the STS in two phases. In the first phase, the system is modelled as three connected networks (social, information and task) and the second phase the social and task network links are systematically broken to reveal what risks are introduced by the failure to communicate information. The broken-links approach is a substantial and novel innovation over traditional human error taxonomic approaches to assessing risks in systems. The approach is based on the premise that most, if not all, accidents and near misses are caused, at least in part, by the failure to communicate information between agents and tasks. By enabling the generation of a system-model, EAST ensures that all of the components of interest within a STS have at been identified and this should lead to a more comprehensive analysis of potential risks. The extension to EAST offers a holistic, structured and systematic approach to the identification of information communication failures in task and social networks. The EAST network broken-links approach is a new paradigm for risk assessment in systems. The approach can be applied to any STS in any domain where an EAST model has been constructed. Future work should explore the risks associated with multiple communication failures occurring simultaneously as well as considering the degree of resilience offered by different system network models.

Acknowledgements

The authors would like to thank Wing Commander Neil Bing (Bingo) of Air Cap SO1 Lightning, RAF High Wycombe, for his account of the Hawk Risk-to-Life case study and his very valuable insights into the challenges faced by this complex Socio-Technical System.

Disclosure statement

No potential conflict of interest was reported by the authors.

Funding

This work was part-funded by the Defence Human Capability Science and Technology Centre (DHCSTC) grant reference [TIN 2.002].

ORCiD

Neville A. Stanton (iD) http://orcid.org/0000-0002-8562-3279

References

Alexander, R., and T. Kelly. 2013. "Supporting Systems of Systems Hazard Analysis Using Multi-Agent Simulation." *Safety Science* 51: 302–318.

Baber, C., N. A. Stanton, J. Atkinson, R. McMaster, and R. J. Houghton. 2013. "Using Social Network Analysis and Agent-Based Modelling to Explore Information Flow Using Common Operational Pictures for Maritime Search and Rescue Operations." *Ergonomics* 56: 889–905.

Benta, M. I. 2005. "Studying Communication Networks with AGNA 2.1." *Cognition, Brain, Behaviour* 9: 567–574.

Dekker, S. 2014. *The Field Guide to Understanding 'Human Error'*. Aldershot: Ashgate.

Driskell, J. E., and B. Mullen. 2005. "Social Network Analysis." In *Handbook of Human Factors and Ergonomics Methods*, edited by N. A. Stanton, A. Hedge, K. Brookhuis, E. Salas, and H. Hendrick. Boca Raton, FL: CRC Press.

Dul, J., R. Bruder, P. Buckle, P. Carayon, P. Falzon, W. S. Marras, J. R. Wilson, and B. van der Doelen. 2012. "A Strategy for Human Factors/Ergonomics: Developing the Discipline and Profession." *Ergonomics* 55: 377–395.

Embrey, D. E. 1986. "SHERPA: A Systematic Human Error Reduction and Prediction Approach." Paper presented at the International meeting on Advances in Nuclear Power Systems, Knoxville, TN.

Flach, J. M., J. S. Carroll, M. J. Dainoff, and W. I. Hamilton. 2015. "Striving for Safety: Communicating and Deciding in Sociotechnical Systems." *Ergonomics* 58 (4): 615–634.

Griffin, T. G. C., M. S. Young, and N. A. Stanton. 2010. "Investigating Accident Causation through Information Network Modelling." *Ergonomics* 53: 198–210.

Grote, G., J. Weyer, and N. A. Stanton. 2014. "Beyond Human-centred Automation – Concepts for Human–Machine Interaction in Multi-layered Networks." *Ergonomics* 57 (3): 289–294.

Haddon-Cave, C. 2009. *The Nimrod Review. An Independent Review into Broader Issues Surrounding the Loss of the RAF Nimrod MR2 Aircraft XV230 in Afghanistan in 2006*. London: The Stationery Office.

Harvey, C., and N. A. Stanton. 2014. "Safety in System-of-Systems: Ten Key Challenges." *Safety Science* 70: 358–366.

H M Government. 2010. *Securing Britain in an age of uncertainty: the strategic defence and security review*. London: The Stationery Office.

Hodgson, A., C. E. Siemieniuch, and E.-M. Hubbard. 2013. "Culture and the Safety of Complex Automated Sociotechnical Systems." *IEEE Transactions on Human-machine Systems* 43: 608–619.

Hollnagel, E. 1998. *Cognitive Reliability and Error Analysis Method*. Oxford: Elsevier Science.

Hollnagel, E., D. D. Woods, and N. C. Leveson, eds. 2006. *Resilience Engineering: Concepts and Precepts*. Aldershot: Ashgate.

Houghton, R. J., C. Baber, M. Cowton, G. H. Walker, and N. A. Stanton. 2008. "WESTT (Workload, Error, Situational Awareness, Time and Teamwork): An Analytical Prototyping System for Command and Control." *Cognition, Technology and Work* 10: 199–207.

Houghton, R. J., C. Baber, R. McMaster, N. A. Stanton, P. M. Salmon, R. Stewart, and G. H. Walker. 2006. "Command and Control in Emergency Services Operations: A Social Network Analysis." *Ergonomics* 49: 1204–1225.

Jenkins, D. P., P. M. Salmon, N. A. Stanton, and G. H. Walker. 2010. "A Systemic Approach to Accident Analysis: A Case Study of the Stockwell Shooting." *Ergonomics* 53: 1–17.

Leveson, N. G. 2012. *Engineering a Safer World: Systems Thinking Applied to Safety*. Cambridge: MIT Press.

Lundberg, J., C. Rollenhagen, and E. Hollnagel. 2010. "What You Find is Not Always What You Fix – How Other Aspects than Causes of Accidents Decide Recommendations for Remedial Actions." *Accident Analysis and Prevention* 42: 2132–2139.

MAA (Military Aviation Authority). 2010. "RA 1210: Management of Operating Risk (Risk to Life)." In *RA 1210*, edited by Military Aviation Authority. London: MAA.

MAA (Military Aviation Authority). 2011. *Regulatory Instruction MAA RI/02/11 (DG) – Air Safety: Risk Management*. London: MAA.

Ministry of Defence (MoD). 2007. *Defence standard 00-56, issue 4, part 2*. London: MoD.

Moray, N., J. Groeger, and N. A. Stanton. Forthcoming. Quantitative Modelling in Cognitive Ergonomics: Predicting Signals Passed at Danger. *Ergonomics*. http://dx.doi.org/10.1080/00140139.2016.1159735

Noyes, J., and N. A. Stanton. 1997. "Engineering Psychology: Contribution to System Safety." *Computing & Control Engineering Journal* 8 (3): 107–112.

Plant, K. L., and N. A. Stanton. 2012. "Why Did the Pilots Shut down the Wrong Engine? Explaining Errors in Context Using Schema Theory and the Perceptual Cycle Model." *Safety Science* 50 (2): 300–315.

Rafferty, L., N. A. Stanton, and G. H. Walker. 2012. *The Human Factors of Fratricide*. Surrey: Ashgate.

Ramos, A. L., J. V. Ferreira, and J. Barceló. 2012. "Model-Based Systems Engineering: An Emerging Approach for Modern Systems." *IEEE Transactions on Systems, Man, and Cybernetics, Part C (Applications and Reviews)* 42: 101–111.

Rasmussen, J. 1997. "Risk Management in a Dynamic Society: A Modelling Problem." *Safety Science* 27: 183–213.

Read, G. J. M., P. M. Salmon, M. G. Lenné, and N. A. Stanton. 2015. "Designing Sociotechnical Systems with Cognitive Work Analysis: Putting Theory Back into Practice." *Ergonomics* 58: 822–851.

Royal Navy. 2012. *Fleet Requirements Air Direction Unit (FRADU) [Online]*. Royal Navy. Accessed January 11, 2013. http://www.royalnavy.mod.uk/sitecore/content/home/the-fleet/air-stations/rnas-culdrose/fleet-requirements-air-direction-unit-fradu

Salmon, P. M., M. Cornelissen, and M. J. Trotter. 2011. "Systems-based Accident Analysis Methods: A Comparison of Accimap, HFACS, and STAMP." *Safety Science* 50: 1158–1170.

Salmon, P. M., M. G. Lenne, G. H. Walker, N. A. Stanton, and A. Filtness. 2014. "Using the Event Analysis of Systemic Teamwork (EAST) to Explore Conflicts Between Different Road User Groups When Making Right Hand Turns at Urban Intersections." *Ergonomics* 57: 1628–1642.

Salmon, P. M., N. A. Stanton, M. Lenné, D. P. Jenkins, L. Rafferty, and G. H. Walker. 2011. *Human Factors Methods and Accident Analysis*. Surrey: Ashgate.

Shappell, S., and D. Wiegmann. 2001. "Applying Reason: The Human Factors Analysis and Classification System." *Human Factors and Aerospace Safety* 1: 59–86.

Shorrock, S. T., and B. Kirwan. 2000. "Development and Application of a Human Error Identification Tool for Air Traffic Control." *Applied Ergonomics* 33: 319–336.

Stanton, N. A. 2006. "Hierarchical Task Analysis: Developments, Applications, and Extensions." *Applied Ergonomics* 37: 55–79.

Stanton, N. A. 2014. "Representing Distribution Cognition in Complex Systems: How a Submarine Returns to Periscope Depth." *Ergonomics* 57 (3): 403–418.

Stanton, N. A., C. Baber, and D. Harris. 2008. *Modelling Command and Control: Event Analysis of Systematic Teamwork*. Surrey: Ashgate.

Stanton, N. A., D. Harris, and A. Starr. 2016. "The Future Flight Deck: Modelling Dual, Single and Distributed Crewing Options." *Applied Ergonomics* 53: 331–342.

Stanton, N. A., L. A. Rafferty, and A. Blane. 2012. "Human Factors Analysis of Accidents in Systems of Systems." *Journal of Battlefield Technology* 15: 23–30.

Stanton, N. A., P. Salmon, D. Harris, A. Marshall, J. Demagalski, M. S. Young, T. Waldmann, and S. W. A. Dekker. 2009. Predicting Pilot Error: Testing a New Methodology and a Multi-Methods and Analysts Approach. Applied Ergonomics 40 (3): 464–471.

Stanton, N. A., P. M. Salmon, L. A. Rafferty, G. H. Walker, C. Baber, and D. P. Jenkins. 2013. *Human Factors Methods: A Practical Guide for Engineering and Design*. 2nd ed. Aldershot: Ashgate.

Stanton, N. A., P. M. Salmon, G. H. Walker, C. Baber, and D. P. Jenkins. 2005. *Human Factors Methods: A Practical Guide for Engineering and Design*. 1st ed. Aldershot: Ashgate.

Stanton, N. A., and S. Stevenage. 1998. "Learning to Predict Human Error: Issues of Reliability, Validity and Acceptability." *Ergonomics* 41 (11): 1737–1756.

Stanton, N. A., R. Stewart, D. Harris, R. J. Houghton, C. Baber, R. McMaster, P. Salmon, G. Hoyle, G. H. Walker, M. S. Young, M. Linsell, R. Dymott, and D. Green. 2006. "Distributed Situation Awareness in Dynamic Systems: Theoretical Development and Application of an Ergonomics Methodology." *Ergonomics* 49: 1288–1311.

Stewart, R., N. A. Stanton, D. Harris, C. Baber, P. Salmon, M. Mock, K. Tatlock, L. Wells, and A. Kay. 2008. "Distributed Situation Awareness in an Airborne Warning and Control System: Application of Novel Ergonomics Methodology." *Cognition, Technology and Work* 10: 221–229.

Stanton, N. A., and G. H. Walker. 2011. "Exploring the Psychological Factors Involved in the Ladbroke Grove Rail Accident." *Accident Analysis & Prevention* 43 (3): 1117–1127.

Swain, A. D., and H. E. Guttmann. 1983. *A Handbook of Human Reliability Analysis with Emphasis on Nuclear Power Plant Applications, NUREG CR-1278*. Washington, DC: USNRC.

Underwood, P., and P. Waterson. 2013. "System Accident Analysis: Examining the Gap between Research and Practice." *Accident Analysis and Prevention* 55: 154–164.

Von Bertalanffy, L. 1950. "An Outline of General System Theory." *The British Journal for the Philosophy of Science* 1: 134–165.

Walker, G. H., H. Gibson, N. A. Stanton, C. Baber, P. Salmon, and D. Green. 2006. "Event Analysis of Systematic Teamwork (EAST): A Novel Integration of Ergonomics Methods to Analyse C4i Activity." *Ergonomics* 49: 1345–1369.

Walker, G. H., N. A. Stanton, C. Baber, L. Wells, H. Gibson, P. M. Salmon, and D. Jenkins. 2010. "From Ethnography to the EAST Method: A Tractable Approach for Representing Distributed Cognition in Air Traffic Control." *Ergonomics* 53: 184–197.

Walker, G. H., N. A. Stanton, P. M. Salmon, and D. P. Jenkins. 2008. "A Review of Sociotechnical Systems Theory: A Classic Concept for Command and Control Paradigms." *Theoretical Issues in Ergonomics Science* 9: 479–499.

Walker, G. H., Stanton, N. A., Salmon, P. M., and Jenkins, D. P. 2009. *Command and Control: The Sociotechnical Perspective*, Aldershot: Ashgate.

Walker, G. H., N. A. Stanton, P. M. Salmon, D. P. Jenkins, and L. A. Rafferty. 2010. "Translating Concepts of Complexity to the Field of Ergonomics." *Ergonomics* 53: 1175–1186.

Waterson, P., M. M. Robertson, N. J. Cooke, L. Militello, E. Roth, and N. A. Stanton. 2015. "Defining the Methodological Challenges and Opportunities for an Effective Science of Sociotechnical Systems and Safety." *Ergonomics* 58 (4): 565–599.

Wilson, J. R. 2012. "Fundamentals of Systems Ergonomics." *Work* 14: 3861–3868.

Detection of error-related negativity in complex visual stimuli: a new neuroergonomic arrow in the practitioner's quiver

Ben D. Sawyer, Waldemar Karwowski, Petros Xanthopoulos and P. A. Hancock

ABSTRACT

Brain processes responsible for the error-related negativity (ERN) evoked response potential (ERP) have historically been studied in highly controlled laboratory experiments through presentation of simple visual stimuli. The present work describes the first time the ERN has been evoked and successfully detected in visual search of complex stimuli. A letter flanker task and a motorcycle conspicuity task were presented to participants during electroencephalographic (EEG) recording. Direct visual inspection and subsequent statistical analysis of the resultant time-locked ERP data clearly indicated that the ERN was detectable in both groups. Further, the ERN pattern did not differ between groups. Such results show that the ERN can be successfully elicited and detected in visual search of complex static images, opening the door to applied neuroergonomic use. Harnessing the brain's error detection system presents significant opportunities and complex challenges, and implication of such are discussed in the context of human-machine systems.

Practitioner Summary: For the first time, error-related negativity (ERN) has been successfully elicited and detected in a visually complex applied search task. Brain-process-based error detection in human-machine systems presents unique challenges, but promises broad neuroergonomic applications.

1. Introduction

Error related negativity (ERN) is one of the brain's evoked response potentials (ERP) that occurs when a human actor becomes aware of their own error. Relative to the time of such an erroneous response, for averaged data, the ERN appears within a latency around 100 ms in the form of a pronounced negative deflection for errors as compared to non-errors (see Gehring et al. 2012; for a more detailed overview). ERNs are generally best detected over the frontal scalp, closest to the 'Cz', electrode of the 10–20 system (Homan, Herman, and Purdy 1987), although they can also be measured at the Fz and Pz electrodes (Luck and Kappenman 2011). The anterior cingulate cortex (ACC) has been identified as the most likely primary causal neural structure for the ERN (Gehring et al. 1993; Debener et al. 2005).

The ERN is a well-studied ERP, but very little existing work explores the applied potential of this phenomenon. Clinical psychopathology has sought to diagnostically use individual differences in ERN magnitude, for example, as a marker for potential disturbances (for an overview see Olvet and Hajcak 2008). Similar efforts exist outside of pathology; for example, the ERN's magnitude correlates positively with academic performance (Hirsh and Inzlicht 2010). An ERP marker for error detection would be of great value to the neuroergonomic practitioner (see Parasuraman 2003), but would also need to be robust to the more complex environmental stimuli that occur beyond controlled laboratory environments. This challenge presents distinct difficulties; there has been little exploration of the ERN in complex applied contexts, with associated complex visual stimuli, and no evidence that it could be either elicited or detected in these latter circumstances.

ERN is essentially a subjective response, the elicitation of which is informed by the cognitive model of what is 'correct'. Such models are unique by individual and situation (Hester, Fassbender, and Garavan 2004; Luck and Kappenman 2011). A participant in an ERN experiment may therefore detect errors beyond the context of the experiment at hand (Luck and Kappenman 2011). These include, for example, social errors, past errors recalled in mid-task, the suppression of behaviours (e.g. checking a watch that has been removed for the experiment) or the error of inattention to the task itself; each might produce a similar ERN response. The experimenter therefore faces problems disambiguating these from 'correct' errors expected from the experimental protocol itself. Such potential for confounds has understandably led past researchers to adopt conservative choices in their manipulated stimuli. Unambiguous choices such as letters thus prove to be a common choice for ERN experiments (Riesel et al. 2013), and even the most complex visual stimuli used to date have been limited to icon-like images of tools and guns (Amodio et al. 2004; see Figure 1).

Unambiguous methods are likewise preferential for ERN experiments. Simple, binary forced-choice tasks were a regular feature of early ERN research (Renault, Ragot, and Lesèvre 1979) and continue to be used to the present. In a classic example, Gehring et al. (1993) elicited ERN with a speeded letter-based flanker task (Eriksen and Eriksen 1974; Eriksen 1995; see Figure 1A). This task proved an effective way to consistently generate errors even with well-trained participants. Flanker tasks and letter stimuli continue to be the most common choice for ERN experiments, alongside similar error-prone, automaticity-resistant options such as Stroop and go/no go tasks (see Riesel et al. 2013 for a discussion).

To date, the most complex stimuli used in eliciting ERN has been icon-like images of tools (Amodio et al. 2004; Fleming, Bandy, and Kimble 2010). Such photographs involve a level of visual complexity previously unseen in the ERN literature. However, they still fall short of revealing whether ERN is robust enough to be detected in applied tasks. Operations in complex environments, where stimuli are complicated by numerous distractive elements, might provoke cortical reactions that effectively mask or suppress the ERN pattern. In order to test whether detection of the ERN was feasible in the face of such challenges, an applied task in a visually complex environment with parity to a known, replicable ERN task needed to be identified. Ideally, the task would be naturalistic, occurring in a context familiar to the participant. Here, we chose a motorcycle conspicuity task in the context of driving as best fulfilling these requirements.

1.2. Motorcycle conspicuity

The relative inability of motorcycles to attract the attention of other drivers, as compared to other vehicles in the traffic stream, has been a subject of study for an extended period of time (Engel 1971, 1977; Thomson 1980; Hancock et al. 1990; Caird and Hancock 1994; Ledbetter et al. 2012). A disproportionate number of motorcycle collisions involve the colliding party's report of not having seen the motorcycle, an effect which can be reproduced in the laboratory (Hurt, Ouellet, & Thom, 1981; Wulf, Hancock, and Rahimi 1989). For the present study, this difficult detection problem was identified as an effective way to consistently generate errors among highly trained participants in an unambiguous applied context. Arguably, the flanker letter task has some parity to this motorcycle detection task. In the former, a series of static images is observed to produce a choice between two letters. This decision is complicated by distractors, while speeded binary responses are collected. In the latter motorcycle task, a series of static images are observed to elicit a binary decision, motorcycle or no motorcycle. This decision is also complicated by distractors in the form of environmental variation and other roadway vehicles, and again, speeded binary responses are collected. In this study, the visually complex motorcycle detection problem becomes a source of applied errors that may be detected through ERP analysis.

Formally, then, the present work sought to determine whether ERN could be elicited in a motorcycle detection task. For comparison purposes, a replication of the flanker letter task used by Gehring et al. (1993) was collected. The result was a within-participant 2(task: flanker, motorcycle) x 2(response: correct, incorrect) design. We predicted that, for both tasks, EEG voltage level at Cz would vary significantly between response types so that incorrect responses would result in a negative deflection. It was further hypothesised that this pattern would not be significantly affected by task type, i.e., no significant interaction of task type on ERN status was predicted. There was a concern that differences in baseline error rate between the motorcycle and

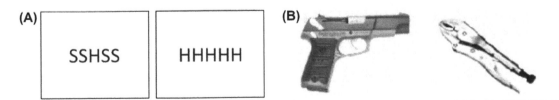

Figure 1. Stimuli previously used in evoking the ERN have been simple and unambiguous. (A) The flanker task used in Gehring et al. (1993) elicited errors by asking for a binary decision regarding the centre letter in an array. (B) Tools and guns from Amodio and colleagues' 2004 shoot/do not shoot racial bias in decision-making study represent the most complex visual stimuli yet used to evoke the ERN.

(A)

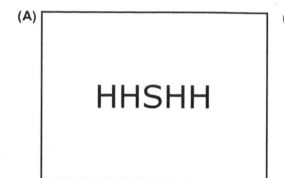

(B)

Figure 2. The present study elicited errors through discrimination tasks in simple and complex stimuli. (A) Simple stimuli used in the letter flanker task. The target in this trial, an 'S', is present in the central position and should be reported. Participants made errors in 7.7% of such letter flanker trials. (B) Complex stimuli used in the motorcycle conspicuity task. The target in this trial, a motorcycle, is present, and should be reported. Participants made errors in 6.2% of such motorcycle conspicuity trials.

letter flanker tasks could unbalance the design, as could a substantial difference in cognitive workload between tasks. Therefore, as a manipulation check, overall error rates and subjective workload via the NASA TLX (Hart and Staveland 1988) were collected for each task.

2. Method

2.1. Participants

Twenty-five participants, who were undergraduates at a major southeastern university, provided three hours of participation in return for class credit. Two participants were removed from the study due to hairstyles that would not allow the attachment of EEG leads. An additional participant was removed due to the occurrence of seasonal allergies so severe as to prohibit effective EEG recording. As a result, our final sample included 22 participants, 12 males and 10 females, who ranged in age from 18 to 59 years (mean = 20.00, SD = 10.36).

All participants were required to have 20/20 or corrected to 20/20 vision, a valid driver's license and no self-reported history of neurological disorders. All were additionally right-handed. No participant had a motorcycle endorsement on their license or reported any history of motorcycle or scooter use. The sample contained a mix of novice and experienced drivers; omitting one very experienced driver (with 19 years' experience) the average experience reported was 4.5 years.

2.2. Apparatus

An Advanced Brain Monitoring (ABM) X-10 nine channel, wireless EEG collected data at 256hz from sensors over prefrontal, ventral, parietal and occipital regions (sites F3/F4, C3/C4, Cz/PO, F3-Cz, Fz-C3 and Fz-PO). This unit applied a hardware 0.1 Hz high bandpass filter and a 5th order low bandpass 100 Hz filter to all data. An ABM External Sync Unit (ESU) connected wirelessly to the EEG providing time-stamping of data packets and response signals.

Informed consent was administered and collection of demographic data was performed using Qualtrics (2013) survey software. Experimental data collection and stimuli were handled by a single i7 Windows 8 laptop with 8 GB of RAM and a 512 GB SSD. To minimise interference, this machine was positioned over two meters from the EEG collection area. Visual stimuli were presented on a Dell LCD monitor at 1024×768 resolution (Hancock, Sawyer and Stafford 2015). Responses were recorded on a Dell QWERTY keyboard with all keys removed except 'a' and 'apostrophe', both of which were blacked out, resulting in an input device with symmetrically placed left and right keys. Tasks were built in ePrime (Schneider, Eschman, and Zuccolotto 2002), which presented stimuli, recorded user responses, and transmitted response signals via USB-to-serial adapter to the ESU for time-stamping.

2.3. Stimuli

Two categories of stimuli were built. Letter flanker stimuli (Figure 2A) displayed an array of five letters, each an 'S' or an 'H'. The centre letter either matched or did not match the four outer letters, which were the same. As such, the possible letter arrays were HHHHH, HHSHH, SSSSS and SSHSS. For motorcycle stimuli (Figure 2B), various images of the same intersection were presented. In these images, one or more of the following elements could appear: pedestrian, traffic cone, car, SUV, mail truck and motorcycle. The same motorcycle and rider were present in half of all images. These images were drawn from a stimuli set previously used in the motorcycle conspicuity work of Ledbetter and colleagues (2012) and Al-Awar Smither and Torrez (2010).

2.4. Task

Motorcycle and letter flanker stimuli shared a similar presentation format. Before every trial an asterisk appeared in the centre of the screen for 1 s. Participants were instructed to orient on this symbol and wait for the stimuli. Flanker letter stimuli consisted of an array of five letters and participants were trained to press the left key if the centre letter was an 'S', and the right key if the letter was an 'H'. In the motorcycle conspicuity, task participants were presented with a photo of a traffic scene. Participants were trained to press the left key if a motorcycle was present, and the right key if no motorcycle was present.

2.5. Procedure

After informed consent, participants were asked to surrender all electronics, which were held outside the experimental area. Each participant sat in a chair facing the LCD screen while a research assistant fitted the EEG cap and checked impedance at all electrode sites. During both training and actual trials, no experimenter was in the room, although participants could summon one upon request. Participants were instructed to follow on-screen instructions and not to speak unless necessary. The training portion of the experiment guided participants through 16 practice trials of each set of stimuli. This was followed by an opportunity to ask questions and repeat the training if desired. The experimental portion consisted of four blocks of 64 trials of each task, for a total of 512 trials; 256 on each task type. Between blocks, the experimenter entered the room and asked the participant to get up and move around, a request facilitated by the wireless nature of the X-10 EEG device. Upon participant's return to the seat, the researcher again checked impedance levels and made any necessary corrections. Participants were then verbally advised that they should try to beat their previous speed. On-screen instructions before each block further instructed participants to go as fast as they were able. Upon completing all trials, participants completed a demographic survey while the EEG headset was removed by the researcher. After disclosure, participants were thanked for their time and released.

2.6. Post-processing and Analysis

All EEG data was analysed in MATLAB. (2012) using EEGLAB 12.0.1.0b (Delorme and Makeig 2004) and ERPLAB (Lopez-Calderon and Luck 2014). Recordings were hand-trimmed to include only the experimental tasks. The result was submitted to the EEGLAB runica function, an implementation of Bell and Sejnowski's (1995) infomax ICA (for a more detailed discussion see Delorme and Makeig 2004).

Identified components primarily containing eyeblinks, saccades and/or EMG were removed, although in the case of any doubt components were retained. Data were then imported into ERPLAB, and individual time-stamped event markers were assigned to four bins based upon task (flanker letter or motorcycle conspicuity) and response (correct or incorrect). All data were then segmented into 400 ms epochs, from −100 ms before the time-stamped event to 300 ms after. In each epoch, the mean amplitude between 0 and 100 ms was calculated at the Cz electrode, relative to a baseline of −50–0 ms. These mean values were averaged by bin within each participant and then transferred to R 3.1.0 (R Development Core Team, 2008) for statistical analysis in a 2(response type: correct, incorrect) x 2(task type: flanker, motorcycle) within-participant repeated measures ANOVA. Visualisations were also produced using the ERPLAB pop_gaverager function run against all participant data sets and results were submitted to the ERPLAB pop_ploterps function. This output was saved as an .eps file, and final adjustments to fonts and the legend were made with Adobe Illustrator.

3. Results

3.1. Manipulation check results

As a manipulation check, error rates and subjective workload were collected for each task. Across participants, the error rate for the flanker letter task (Figure 2A) was 7.7%, as compared to 6.2% for the motorcycle detection task (Figure 2B). Likewise, across participants the NASA TLX composite workload score for the flanker letter task was 49, and 48 for the motorcycle detection task. The two tasks elicited comparable rates of error and subjective workload.

3.2. Graphical results

Visula inspection of aggregate waveform data (Figure 3) reveals that error state, but not task type, is clearly discriminable. The negative deflection seen in error trials conforms to previously described error related negativity (ERN) evoked response potential (ERP) patterns (for example, Gehring et al. 2012).

3.3. Statistical results

A significant main effect of response type was shown between error and no error trials Wilk's Lamda = .426, $F(3, 11) = 4.94$, $p = .02$, $\eta^2_p = 0.57$. No significant main effect of task type was present, Wilk's Lamda = .76, $F(3, 11) = 1.14$, $p = .38$, $\eta^2_p = 0.24$. Likewise, no significant interaction between task type and response type was evident, Wilk's Lamda = .86, $F(3, 11) = .62$, $p = .62$, $\eta^2_p = 0.14$. These results

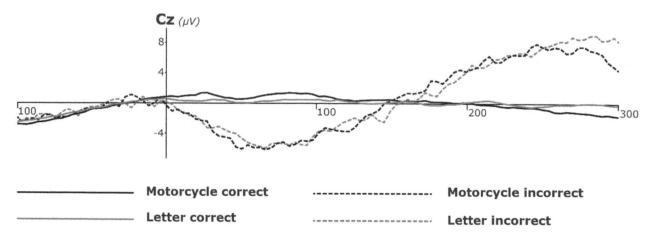

Figure 3. Inspection reveals a pronounced negative deflection for error trials, as is typical for the error related negativity (ERN) evoked response potential (ERP) (as in Gehring et al. 2012).
Notes: This pattern is similar for both letter and motorcycle stimuli, suggesting that the visual complexity of the latter is not a barrier to detection of the ERN ERP. Waveform data are plotted negative-down relative to a 50 ms baseline, while a full 100 ms of preresponse activity is shown for evaluative purposes (see Luck 2005).

indicate that the presence or absence of errors was discriminable in ERN data of both tasks, and furthermore that the task type did not significantly impact ability to discriminate ERN.

4. Discussion

As hypothesised, the ERN was detectable in the flanker letter task (for a visual representation, see Figure 3), which represents a successful evocation of this ERP using the methods of Gehring et al. (1993). These results represent a replication of extant findings using the traditional method to elicit ERN. Aggregated trials in which an error was committed revealed a pronounced negative deflection in activity measured at the Cz electrode, as compared to trials in which no error was committed. The similarity between data from this successful replication and that collected in the motorcycle detection task is striking. Despite the muscular and cognitive 'noise' associated with searching a more complex image, a clear ERN pattern emerged in the latter motorcycle detection (Figure 2). ERN thus joins the company of other ERPs shown to be robust in applied settings, such as the N2pc, a marker for selective attention, and the P300, a marker for evaluation of stimuli as novel (Woodman and Luck 1999). This important new evidence reveals the ERP as a practical neuroergonimic (Parasuraman 2003) tool that can be immediately applied to a wider spectrum of experimental procedures and applied concepts.

Some appropriate caution is well advised. This is because our motorcycle detection task was ultimately a controlled stimuli set and binary response a controlled method. Applied contexts are likely to require more latitude of stimuli and response. However, there remain many cognitive phenomena that require binary response to complex images that have found utility in the applied domain. For example, the implicit association test (IAT, Greenwald, McGhee, and Schwartz 1998) is used as a part of bias training in police departments (Rudman, Ashmore, and Gary 2001). We anticipate that other realms will now become available for exploration via the ERN assessment (e.g. nuclear power control design; Reinerman et al. 2015).

4.1. Next steps

From detecting ERN in a static scene, it appears to be a feasible step to attempt detection in more dynamic environments. There are, however, some theoretical complications. ERN is a subjective, time-locked ERP, and out-of-context ERNs that fall outside of experimenters' expectations seem more likely to arise from tasks where the stimuli are constantly changing. In the face of the flow of time, modality matters as well, as auditory or tactile stimuli onset can be far more easily temporally defined (Hancock et al. 2013). Further, in immersive visual stimuli, mere presentation does not mean a target is perceptually available; participants must look before they can see. This visiomanual requirement of a foveal fixation means that, at the least, eye tracking will be a bridging requirement for accurate time-locked response data in dynamic environments. It is however well documented that visual-manual requirements, even when met, may not result in perception of a target, especially if the observer is engaged in multitasking (Strayer, Drews, and Johnston 2003). This 'looking without seeing' (O'Regan et al. 2000;) represents both a challenge and a potential opportunity, as applied use of the ERN might provide valuable evidence in applied differentiation of errors of structural versus cognitive origin (a question posed in Sawyer et al. 2014).

The present experiment detected ERN in averaged data, but applied systems will likely need to detect in real time. Real-time ERP classification has been available for some time (Vidal 1977), and efforts to classify ERN in simple stimuli have since approached sensitivity and specificity of over 85% in under 150 ms (Ventouras et al. 2011; Vi and Subramanian 2012). There is a question as to whether these present findings will generalise to such real-time applications. Future efforts to develop a real-time ability to detect ERN in complex visual stimuli seem a logical progression from the present effort. Such steps are vitally necessary to understand the viability of many potential applications.

It is in application of the ERN where the most interesting future questions lie. Consider, for example, the ERN in the context of training. As novices move toward proficiency, they pass through various skill acquisition stages in which they may be well aware what is correct and yet still provide an incorrect answer (for a detailed discussion of such skill acquisition, see Anderson 1983). Expert teachers often use their own intuition and experience to determine whether a pupil has made a misstep of which they are conscious or, in contrast, an unwitting error. While the former might warrant a mild rebuke (itself optional, a witting error often needs no feedback), the latter necessitates time to educate the student as to what connotes the correct action. In machine-tutoring systems, such subtleties are lost, and students are punitively drilled on material they know, but have not yet completely mastered. ERN distinction could provide an automated system using a context to approach training in a much more intelligent fashion, or provide a human teacher augmented insight. Both would save time and resource while facilitating the learning process.

The crux of the aforementioned issue is that when humans and machines work together, at present, the machine has relatively minimal information regarding the state of the human (Parasuraman and Riley 1997). Humans can communicate their errors through manual interfaces, such as keyboards or mouse-driven 'undo' commands. While in word processing this may still be considered an acceptable interface, it is woefully inadequate for tasks where the time in which an error must be reported falls within the threshold of human response time. The idea of an undo button in tasks such as piloting, combat, and surgery is, at present, somewhat grimly humorous, and precisely so because of the severe consequences that stopping the task to push such a button would entail. ERNs hold the possibility of making such interface a reality, recruiting the operator's own error detection capacity to inform automation in much closer to real-time (150 ms in Vi and Subramanian 2012). Latencies of this order would allow human-machine systems to be informed quickly of perceived errors, and to weight possible ameliorative actions accordingly.

Such advanced neuroergonomic applications are only hinted at by the present step forward. It is tempting to relegate notions of brain-activity-mediated human–machine interaction to the realm of science fiction, but this future is approaching. To effectively harness the applied potential of ERN, the detection of human error must be moved away from artificially simple laboratory conditions in favour of the complexity of real-world environments. It is through present exploration and iterative implementation of such applied ERN detection that this tool may be understood and intelligently applied to future critical systems.

Disclosure statement

No potential conflict of interest was reported by the authors.

References

Amodio, D. M., E. Harmon-Jones, P. G. Devine, J. J. Curtin, S. L. Hartley, and A. E. Covert. 2004. "Neural Signals for the Detection of Unintentional Race Bias." *Psychological Science* 15 (2): 88–93.

Anderson, J. R. 1983. *The Architecture of Cognition*. Cambridge, MA: Harvard University Press.

Bell, A. J., and T. J. Sejnowski. 1995. "An Information-Maximization Approach to Blind Separation and Blind Deconvolution." *Neural Computation* 7 (6): 1129–1159.

Caird, J. K., and P. A. Hancock. 1994. "The Perception of Arrival Time for Different Oncoming Vehicles at an Intersection." *Ecological Psychology* 6 (2): 83–109.

Debener, S., M. Ullsperger, M. Siegel, K. Fiehler, D. Y. Von Cramon, and A. K. Engel. 2005. "Trial-by-trial Coupling of Concurrent Electroencephalogram and Functional Magnetic Resonance Imaging Identifies the Dynamics of Performance Monitoring." *Journal of Neuroscience* 25 (50): 11730–11737.

Delorme, A., and S. Makeig. 2004. "EEGLAB: An Open Source Toolbox for Analysis of Single-Trial EEG Dynamics including Independent Component Analysis." *Journal of Neuroscience Methods* 134: 9–21.

Engel, F. L. 1971. "Visual Conspicuity, Directed Attention and Retinal Locus." *Vision Research* 11: 261–270.

Engel, F. L. 1977. "Visual Conspicuity, Visual Search and Fixation Tendencies of the Eye." *Vision Research* 17: 95–108.

Eriksen, B. A., and C. W. Eriksen. 1974. "Effects of Noise Letters upon the Identification of a Target Letter in a Non-Search Task." *Perception & Psychophysics* 16: 143–149.

Eriksen, C. W. 1995. "The Flankers Task and Response Competition: A Useful Tool for Investigating a Variety of Cognitive Problems." *Visual Cognition* 2 (2–3): 101–118.

Fleming, K. K., C. L. Bandy, and M. O. Kimble. 2010. "Decisions to Shoot in a Weapon Identification Task: The Influence of Cultural Stereotypes and Perceived Threat on False Positive Errors." *Social Neuroscience* 5 (2): 201–220.

Gehring, W. J., B. Goss, M. G. Coles, D. E. Meyer, and E. Donchin. 1993. "A Neural System For Error Detection and Compensation." *Psychological Science* 4 (6): 385–390.

Gehring, W. J., Y. Liu, J. M. Orr, and J. Carp. 2012. "The Error-Related Negativity (ERN/Ne)". *Oxford Handbook of Event-Related Potential Components*, (pp. 231–291). XXX

Greenwald, A. G., D. E. McGhee, and J. L. Schwartz. 1998. "Measuring Individual Differences in Implicit Cognition: The Implicit Association Test." *Journal of Personality and Social Psychology* 74 (6): 1464–1480.

Hancock, P. A., J. E. Mercado, J. Merlo, and J. B. Van Erp. 2013. "Improving Target Detection in Visual Search through the Augmenting Multi-Sensory Cues." *Ergonomics* 56 (5): 729–738.

Hancock, P. A., B. D. Sawyer, and S. Stafford. 2015. "The Effects of Display Size on Performance." *Ergonomics* 58 (3): 337–354.

Hancock, P. A., G. Wulf, D. Thom, and P. Fassnacht. 1990. "Driver Workload during Differing Driving Maneuvers." *Accident Analysis & Prevention* 22 (3): 281–290.

Hart, S. G., and L. E. Staveland. 1988. "Development of NASA TLX (Task Load Index): Results of Empirical and Theoretical Research." In *Human Mental Workload*, edited by P. A. Hancock and N. Meshkati, 139–183. Amsterdam: North-Holland.

Hester, R., C. Fassbender, and H. Garavan. 2004. "Individual Differences in Error Processing: A Review and Reanalysis of Three Event-Related FMRI Studies Using the GO/NOGO Task." *Cerebral Cortex* 14 (9): 986–994.

Hirsh, J. B., and M. Inzlicht. 2010. "Error-Related Negativity Predicts Academic Performance." *Psychophysiology* 47 (1): 192–196.

Homan, R. W., J. Herman, and P. Purdy. 1987. "Cerebral Location of International 10–20 System Electrode Placement." *Electroencephalography and Clinical Neurophysiology* 66 (4): 376–382.

Hurt, H. H., J. V. Ouellet, and D. R. Thom. 1981. *Motorcycle Accident Cause Factors and Identification of Countermeasures*, Volume 1: Technical Report No. DOT HS-5-01160. Los Angeles, California: University of Southern California, Traffic Safety Center.

Ledbetter, J. L., M. W. Boyce, D. K. Fekety, B. Sawyer, and J. A. Smither. 2012. "Examining the Impact of Age and Multitasking on Motorcycle Conspicuity." *Work: A Journal of Prevention Assessment and Rehabilitation* 41: 5384–5385.

Lopez-Calderon, J., and S. J. Luck. 2014. "ERPLAB: An Open-Source Toolbox for the Analysis of Event-Related Potentials." *Frontiers in Human Neuroscience* 8: 213.

Luck, S. J. 2005. *An Introduction to the Event-Related Potential Technique*. Cambridge, MA: MIT press.

Luck, S. J., and E. S. Kappenman, eds. 2011. *The Oxford Handbook of Event-Related Potential Components*. Oxford: Oxford University Press.

MATLAB. 2012. [Computer Software] Version 7.10.0. Natick, MA: The MathWorks Inc.

Olvet, D. M., and G. Hajcak. 2008. "The Error-Related Negativity (ERN) and Psychopathology: Toward an Endophenotype." *Clinical Psychology Review* 28 (8): 1343–1354.

O'Regan, J., H. Deubel, J. J. Clark, and R. A. Rensink. 2000. "Picture Changes during Blinks: Looking without Seeing and Seeing without Looking." *Visual Cognition* 7 (13): 191–211.

Parasuraman, R. 2003. "Neuroergonomics: Research and Practice." *Theoretical Issues in Ergonomics Science* 4 (1-2): 5–20.

Parasuraman, R., and V. Riley. 1997. "Humans and Automation: Use, Misuse, Disuse, Abuse." *Human Factors: The Journal of the Human Factors and Ergonomics Society* 39 (2): 230–253.

Qualtrics. 2013. *Version 37892*. Provo, UT.

R Development Core Team. 2008. *R: A Language and Environment for Statistical Computing*. R Foundation for Statistical Computing, Vienna, Austria. ISBN 3-900051-07-0, URL http://www.R-project.org.

Renault, B., R. Ragot, and N. Lesèvre. 1979. "Correct and Incorrect Responses in a Choice Reaction Time Task and the Endogenous Components of the Evoked Potential." *Progress in Brain Research* 54: 547–554.

Reinerman, L., J. Mercado, R. Leis, J. L. Szalma, and P. A. Hancock. 2015. "Assessing Operator Response on Nuclear Power Plant Control Tasks through Objective Measures, Subjective Report, and Physiological Reactions." *Submitted*.

Riesel, A., A. Weinberg, T. Endrass, A. Meyer, and G. Hajcak. 2013. "The ERN is the ERN is the ERN? Convergent Validity of Error-Related Brain Activity across Different Tasks." *Biological Psychology* 93 (3): 377–385.

Rudman, L. A., R. D. Ashmore, and M. L. Gary. 2001. "'Unlearning' Automatic Biases: The Malleability of Implicit Prejudice and Stereotypes." *Journal of Personality and Social Psychology* 81 (5): 856–868.

Sawyer, B. D., V. S. Finomore, A. A. Calvo, and P. A. Hancock. 2014. "Google Glass: A Driver Distraction Cause or Cure?" *Human Factors: The Journal of the Human Factors and Ergonomics Society* 56 (7): 1307–1321.

Schneider, W., A. Eschman, and A. Zuccolotto. 2002. *E-Prime: User's Guide*. Psychology Software Incorporated.

Al-Awar Smither, J. A. A., and L. I. Torrez. 2010. "Motorcycle Conspicuity: Effects of Age and Daytime Running Lights." *Human Factors: The Journal of the Human Factors and Ergonomics Society* 52 (3): 355–369.

Strayer, D. L., F. A. Drews, and W. A. Johnston. 2003. "Cell Phone-Induced Failures of Visual Attention during Simulated Driving." *Journal of Experimental Psychology: Applied* 9 (1): 23.

Thomson, G. A. 1980. "The Role Frontal Motorcycle Conspicuity Has in Road Accidents." *Accident Analysis & Prevention* 12: 165–178.

Ventouras, E. M., P. Asvestas, I. Karanasiou, and G. K. Matsopoulos. 2011. "Classification of Error-Related Negativity (ERN) and Positivity (Pe) Potentials Using KNN and Support Vector Machines." *Computers in Biology and Medicine* 41 (2): 98–109.

Vi, C., and S. Subramanian. 2012. "Detecting Error-related Negativity for Interaction Design." *Proceedings of the SIGCHI Conference on Human Factors in Computing Systems* (pp. 493–502). ACM. http://dl.acm.org/citation.cfm?id=2207744

Vidal, J. J. 1977. "Real-Time Detection of Brain Events in EEG." *Proceedings of the IEEE* 65 (5): 633–641.

Woodman, G. F., and S. J. Luck. 1999. "Electrophysiological Measurement of Rapid Shifts of Attention during Visual Search." *Nature* 400 (6747): 867–869.

Wulf, G., P. A. Hancock, and M. Rahimi. 1989. "Motorcycle Conspicuity: An Evaluation and Synthesis of Influential Factors." *Journal of Safety Research* 20: 153–176.

Towards continuous and real-time attention monitoring at work: reaction time versus brain response

Pavle Mijović, Vanja Ković, Maarten De Vos, Ivan Mačužić, Petar Todorović, Branislav Jeremić and Ivan Gligorijević

ABSTRACT

Continuous and objective measurement of the user attention state still represents a major challenge in the ergonomics research. Recently available wearable electroencephalography (EEG) opens new opportunities for objective and continuous evaluation of operators' attention, which may provide a new paradigm in ergonomics. In this study, wearable EEG was recorded during simulated assembly operation, with the aim to analyse P300 event-related potential component, which provides reliable information on attention processing. In parallel, reaction times (RTs) were recorded and the correlation between these two attention-related modalities was investigated. Negative correlation between P300 amplitudes and RTs has been observed on the group level ($p < .001$). However, on the individual level, the obtained correlations were not consistent. As a result, we propose the P300 amplitude for accurate attention monitoring in ergonomics research. On the other hand, no significant correlation between RTs and P300 latency was found on group, neither on individual level.

Practitioner Summary: Ergonomic studies of assembly operations mainly investigated physical aspects, while mental states of the assemblers were not sufficiently addressed. Presented study aims at attention tracking, using realistic workplace replica. It is shown that drops in attention could be successfully traced only by direct brainwave observation, using wireless electroencephalographic measurements.

Introduction

Studies in the Human factors and ergonomics (HFE) regarding mental, cognitive and emotional functions are perceived through theoretical constructs and are still dependent on behavioural indicators (Fafrowicz and Marek 2007), subjective questionnaires and measurements of operators' overall performance (Parasuraman 2003). However, these methods are often unreliable (Lehto and Landry 2012; Parasuraman 2003; Parasuraman and Rizzo 2008; Simpson et al. 2005). Additionally, they are unable to provide real-time and continuous performance and attention measurement at work places (Jagannath and Balasubramanian 2014), where the continuous focus is essential (Jung et al. 1997). On the other hand, wearable electroencephalography (EEG) can provide the possibility to continuously and objectively assess the attention level of the operators, which may provide a new paradigm in ergonomics research for human performance monitoring.

In the early years of industrialisation, accidents were reported mainly in terms of technological malfunctions, ignoring the human element as the cause (Gordon 1998). However, as technology became increasingly reliable, failures related to it have been dramatically reduced, attributing majority of the remaining accidents to human elements in the system (Stanton et al. 2009). Regardless of all the technological advancements, resulting in the increase in the process automation, majority of the manufacturing processes still rely on human participation and intelligence (Hamrol, Kowalik, and Kujawińsk 2011). This is especially notable in manual assembly tasks, which are still unavoidable in variety of modern industries (Hamrol, Kowalik, and Kujawińsk 2011; Michalos et al. 2010; Tang et al. 2003).

Throughout the industrial history, studies of human performance in assembly tasks were mainly concerned with postures of the operators (Fish, Drury, and Helander 1997; Li and Haslegrave 1999; Rasmussen, Pejtersen, and Goodstein 1994), which are still one of the main causes

ⓘ Supplemental data for this article can be accessed at http://dx.doi.org/10.1080/00140139.2016.1142121

for work-related musculoskeletal disorders (Leider et al. 2015). However, far less attention has been dedicated to the cognitive and perceptual factors that can cause errors in operating (Fish, Drury, and Helander 1997). For example, the decrease in attention often precedes human error (Arthur, Barret, and Alexander 1991; Kletz 2001; Reason 1990; Wallace and Vodanovich 2003), and therefore, its timely detection could help avoidance of dangerous situations including workers injuries, material damage and even accidents with casualties.

In order to provide more objective parameters of workers cognitive state, Parasuraman (2003) proposed a novel path in ergonomics research, which was tentatively named neuroergonomics (Parasuraman 2003). The main objective of neuroergonomics is to objectively assess how the brain carries out everyday and complex tasks in naturalistic work environments (Mehta and Parasuraman 2013; Parasuraman 2003). In its essence, the neuroergonomics is able to provide precise analytical parameters depending on the work efficiency of individuals, by directly investigating relationship between neural and behavioural activity (Fafrowicz and Marek 2007). In this way, unreliable user state evaluation based on theoretical constructs, which are mostly describing cognitive states of the workers related to the task execution, can be avoided (Fafrowicz and Marek 2007).

Widely used technique for neuroergonomics studies was functional near-infrared spectroscopy (fNIRS), mainly due to its high mobility and low cost. However, fNIRS provides indirect metabolic indicators of neural activity and it has low temporal resolution (Mehta and Parasuraman 2013). On the other hand, techniques for direct measurement of neural activity that provide high temporal resolution, EEG and event-related potentials (ERPs) were moderately mobile and the most of the research was confined in the laboratory space or simulators, thus limiting the usefulness of such a measurements in neuroergonomics research (Fu and Parasuraman 2008; Mehta and Parasuraman 2013). However, as technology advanced EEG became increasingly mobile and eventually wearable, providing possibility to directly observe neural activity in applied environments (Mijović et al. 2014; Wascher, Heppner, and Hoffmann 2014).

EEG provides the possibility to both timely and objectively detect the critical behaviour of humans (e.g. drops in attention, error, etc.) and it has been confirmed as a reliable tool in estimating ones' cognitive state (Klimesch et al. 1998; Luck, Woodman, and Vogel 2000; Murata, Uetake, and Takasawa 2005; Yamada 1998). Analysis of the ERPs, extracted from continuous EEG recording, represents commonly employed method in evaluating ones' neural activity (Hohnsbein, Falkenstein, and Hoormann 1998). Picton et al. (2000) defined ERPs as 'voltage fluctuations that are

associated in time with certain physical or mental occurrence'. ERP components are usually defined in terms of polarity and latency with respect to a discrete stimulus, and have been found to reflect a number of specific perceptual, cognitive and motor processes (Brookhuis and De Waard 2010). In that sense, so-called P300 (also called P3) component is represented by the positive deflection in terms of voltage, appearing around 300 ms after the stimulus presentation (Gray et al. 2004; Polich and Kok 1995). Further, the P300 component is often used to identify the depth of cognitive information processing, being strongly related to the attention level (De Vos, Gandras, and Debener 2014; Johnson 1988; Polich 2007). It is usually considered that P300 component is not influenced by the physical attributes of the stimuli (Gray et al. 2004; Murata, Uetake, and Takasawa 2005). However, the recent study demonstrated that if P300 is indeed equivalent to centro-parietal positivity (CPP) in the gradual target detection task, physical attributes could influence the P300 component (O'Connell, Dockree, and Kelly 2012). Another modality which can provide a continuous-like assessment of human attention level is a behavioural measure of the reaction times (RTs, Larue, Rakotonirainy, and Pettitt 2010; Sternberg 1969). RT represents a time interval from the indicated start of operation (stimulation), until the moment of the action initiation and the main reason for wide usage of RT measurements is that they are easy to obtain and simple to interpret (Salthouse and Hedden 2002). However, the major drawback of experiments involving RT is that they usually consist of a stimulus followed by the response, without direct possibility to observe the mental processing that occurs between stimuli (Luck, Woodman, and Vogel 2000; Young and Stanton 2007).

Although Parasuraman (1990) proposed the idea of applying ERP recording in operational environments, in order to address various HFE problem areas, only very recent studies provided possibility of recording ERPs in applied environments by utilising available wireless connections (De Vos, Gandras, and Debener 2014; Debener et al. 2012; Wascher, Heppner, and Hoffmann 2014). This finally allowed merging EEG with the guiding principle of neuroergonomics, and examination of how the brain carries out complex everyday work tasks in realistic environments (Parasuraman and Rizzo 2008). Present study proposes a 'new paradigm' in ergonomics research through utilisation of ERP measurement in naturalistic workplace environment, where manual assembly operation was simulated. This is the first study (up to our knowledge), which utilises a wireless 24-channel EEG recordings for ERP extraction in naturalistic environment for purpose of studying the operators attention. The main aim of this study is proposal of novel methodology for attention monitoring of an assembly worker, which is based on real-time EEG

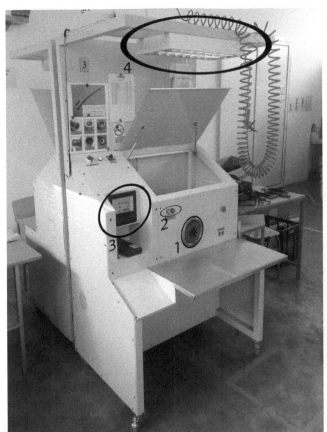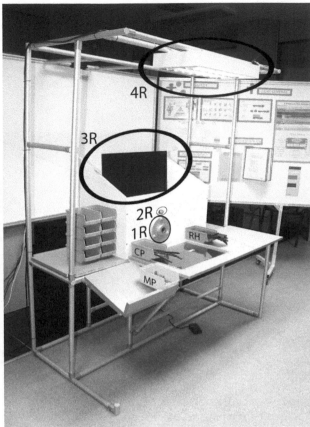

Figure 1. Left image – Real workplace (replicated from our industrial partner); Right image – Replicated workplace.
Notes: The numbered elements from the left image (1, 2, 3, 4) represent the machine operational parts that were replicated in laboratory settings (1R, 2R, 3R, 4R). 1 – the machine opening for the crimping operation; 2 – industrial lamps for identification of correct placement of the uncompleted parts (rubber hose and metal part); 3 – display for the information presentation to the worker; 4 – industrial lightning for the workplace. The right image also shows boxes for placing of the rubber hoses (RH), metal parts (MP) and completed parts (CP).

signal acquisition. As the main disadvantage of the EEG measurement, its immobility, is now overcome we strongly believe that its utilisation in the real workplace environments will be ubiquitous in the years to come.

In present study, we investigated the propagation of the P300 ERP component peak amplitude and latency in order to assess the operators' level of attention, utilising recently available mobile EEG equipment that did not alter the working process and enabled a 'truly unobtrusive' paradigm. In parallel, the propagation of behavioural component (RT) was examined. We tested the hypothesis that the decreased level of attention, reflected in the reduced P300 amplitude, would also be followed by the longer duration of RT, as the operator will need more time to complete the operation, and vice versa. We further examined the relationship between the RTs and P300 peak latency, in order to investigate whether the RT duration would influence the latency of the P300 peak. To address the problem of realistic work environment, an authentic replica of an existing assembly work position from a car sub-component manufacturer was created.

Materials and methods

Participants

Fourteen healthy subjects, all right-handed and white skin colour males, of age between 19 and 21 years volunteered as participants in the study. Two participants were excluded from further analysis due to abnormalities during the recording. Participants had no past or present neurological or psychiatric conditions and were free of medication and psychoactive substances. They have agreed to participation and signed informed consent after reading the experiment summary. The Ethical committee of the University of Kragujevac approved the study and procedures for the participants.

Experimental task

Our laboratory simulation replicates the production of rubber hoses used in the hydraulic brake systems in automotive industry. Full-scale replica of the specific workplace from car sub-component manufacturing company has been created in the laboratory of the

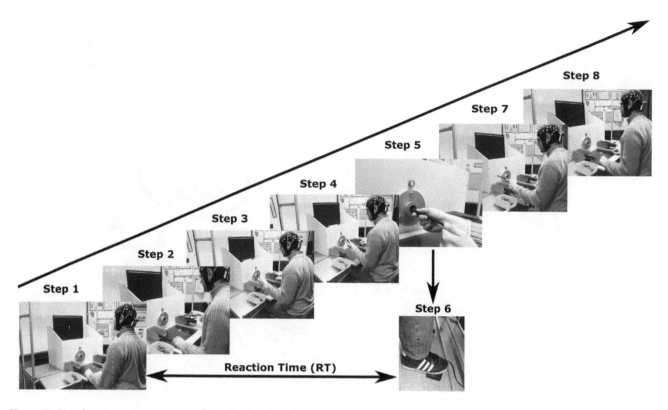

Figure 2. Step-by-step representation of the simulated working process.
Notes: Step 1 – Stimulus presentation; step 2 – taking the rubber hose; step 3 – taking the metal part; step 4 – placing metal part on the rubber hose; step 5 – insertion of the uncompleted part inside the improvised machine opening; step 6 – pressing the pedal in order to initiate the simulated crimping operation; step 7 – placing the completed into the box with completed parts; step 8 – waiting for the successive stimulus presentation.

Department for Production Engineering, University of Kragujevac (Figure 1). All major elements have been included, replicating microclimate conditions from manufacturing company (including ambient temperature, air humidity, noise and luminance), while preserving respective spatial ratios. An important notion is that in order to access the P300 ERP component a slight functional modification for this specific workplace was introduced, without significantly altering the work routine. Instead of the information which the workers would receive in the real workplace, here the participants were receiving information regarding the initiation of their assembly operation as they were presented with the 'go/no-go' psychological test, simultaneously with the simulated assembly operation (explained in detail in the subsequent section).

In the production process, an operator is carrying out the crimping operation in order to assemble the metal extension to the rubber hose. This single operation consists of eight simple steps (actions). Step-by-step simulated operation, carried out by participants in replicated working environment, is graphically presented in Figure 2 and explained in details further in the text.

Major production steps can be summarised as follows: firstly, the information in the form of visual stimulus is presented to the participant (step 1), upon which he is instructed to instantly initiate the operation by taking the metal part (step 2) and the rubber hose (step 3). Following this, participants should place the metal part on the hose (step 4), which is followed by placement of the incomplete element inside the crimping machine (step 5). Participant then proceeds by promptly pressing the pedal, upon which the improvised machine replicates the real machines' crimping sound in the duration of the 3500 ms (step 6). Upon completion of the simulated crimping process, the participant removes the component and places it in the box with completed parts (step 7). Finally, following these steps, the participant sits still, waiting for the subsequent stimulus (step 8) indicating the next-in-line operation.

Although the process is comprised of eight sub-actions, the whole operation lasts less than ten seconds and a single operator completes between 2500 and 3000 elements during a work shift. Therefore, this workplace represents a typical example of a repetitive, monotonous operational task in industrial assembly settings.

Preparation and experimental procedure

Each of the participants arrived in the laboratory at 9:00 am. Upon carefully reading the experiment summary and signing the informed consent for participation in the study, participants started the 15-min training session, in order to become familiar with the task, following which they confirmed the readiness to start the experiment. Finally, EEG cap and amplifier were mounted on the participants' head and the recording started around 9:30 am.

Participants were seated in the comfortable chair in front of an improvised workplace, while the modified version of Sustained Attention to Response Task (SART) was presented on the 24" screen from a distance of approximately 100 cm. The screen was height adjustable and the centre of the screen was set to be in level with participants' eyes.

In short, SART paradigm proposed by Robertson et al. (1997) consists of sequentially presenting the digits from '1' to '9'. Participants are required to respond to each digit by the single button press upon its presentation, with the exception of the digit '3', which is marked as a 'no-go' stimulus and participants are instructed to withhold the response. However, since in this study the original paradigm would impede the simulation of assembly operation, it was not possible to require speeded responses in (literal) sense of discrete button presses. Instead, participants were to remain still until the stimulus appeared on the screen and only then initiate the described operation. Further on, the digits were presented randomly to the participants, so that participants could not predict the appearance of the 'no-go' stimuli. Given all these changes, in further text we will refer to this paradigm as Numbers. In this way, in the Numbers paradigm participants were instructed to pay attention at all 'go', which are regarded here as the target stimuli, and to withhold their action otherwise.

All stimuli were presented for 1000 ms in a white font on a black screen background. The total experiment per subject duration was around one and a half hour during which 500 stimuli were presented in total (450 'go' trials and 50 'no-go' trials). Sequence of stimuli was randomised with the condition that forbade two consecutive appearances of the 'no-go' stimuli. A mean inter-stimulus interval (ISI) on 'go' trials was 11318 ms (STD = 529 ms), while the ISI between 'no-go' and the next 'go' trial was 2970 ms (STD = 48 ms), including a jitter between the end of operation and presentation of following stimuli that was set to be in the 1000–2000 ms range. Further, similarly to Dockree et al. (2007), five randomly allocated digit sizes (60, 80, 100, 120 and 140 points in Arial text font) were presented to increase the demands for processing the numerical value and to minimise the possibility of setting a search template for some perceptual feature of the 'no-go' trial.

The task specifications were programmed in Simulation and Neuroscience Application Platform (SNAP, https://github.com/sccn/SNAP). As explained in Bigdely-Shamlo et al. (2013), SNAP is a python-based experiment control framework that is able to send markers as strings to Lab Streaming Layer (LSL, https://code.google.com/p/labstreaminglayer/). LSL is a real-time data collection and distribution system that allows multiple continuous data streams as well as discrete marker timestamps to be acquired simultaneously.

EEG recording

EEG data acquisition was performed using state-of-the-art wireless and wearable EEG system 'SMARTING' (mBrainTrain, Serbia), with the sampling frequency of 500 Hz. The small in size and lightweight EEG amplifier ($80 \times 50 \times 12$ mm, 55gr) is tightly connected to a 24-channel electrode cap (Easycap, Germany), at the occipital site of the participants' head using an elastic band. The connection between the EEG amplifier and recording computer was obtained using Bluetooth connection, and the data were streamed to the mentioned LSL recorder. The design of the cap-amplifier unit ensured minimal isolated movement of individual electrodes, cables, or the amplifier, which strongly reduced electromagnetic interference and movement artefacts. Further, small dimensions of the recording system provided full mobility and comfort to the participants, as movement constraints were not imposed. The electrode cap contained sintered Ag/AgCl electrodes that are placed based on the international 10–20 System. The electrodes were referenced to the FCz and the ground electrode was AFz. During the recording, the electrode impedances were kept below 5 kΩ, which was confirmed by the device acquisition software.

Data analysis

The RTs were calculated as the difference between timestamps from the operation initiation and actual beginning of the crimping process. In other words, RTs are here regarded as the time elapsed between the stimulus presentation (step 1) and the moment when participant presses the pedal (step 6), as indicated in Figure 2.

EEG analysis was performed offline using EEGLAB (Delorme and Makeig 2004) and MATLAB (Mathworks Inc., Natick, MA). EEG data were first bandpass filtered in the 1–35 Hz range. The EEG signals were then re-referenced to the average of Tp9 and Tp10 electrodes. Further, an extended Infomax Independent Component Analysis (ICA) was used to semi-automatically attenuate contributions from eye blink and (sometimes) muscle artefacts (as explained in De Vos, De Lathauwer, and Van Huffel [2011];

De Vos et al. [2010]; Viola et al. [2009]). ERP epochs were extracted from continuous EEG signal in the time range −200 to 800 ms with respect to timestamp values of stimuli. Baseline values were corrected by subtracting mean values for the period from −200 to 0 ms from the stimuli. The identified electrode sites of interest for the ERP analysis in this study were Fz, Cz, CPz and Pz, as the P300 component is usually distributed and is most prominent over the central and parieto-central scalp locations (Picton 1992).

ERP processing – P300 amplitudes and latencies

In the ERP analysis, we have firstly calculated the mean grand average (GA) values of the ERPs for the 'go' and 'no-go' conditions. The GA methodology provides only the single value for the whole measurement period, thus the continuous evaluation of the ERP components was impossible. On the other hand, single-trial ERPs could be used for the continuous evaluation of ERP components, but they would have low signal-to-noise (SNR) ratio. However, it has been reported that good quality ERPs could be obtained with as few as 11-repeated stimulus trials (Humphrey and Kramer 1994; Prinzel et al. 2003). Therefore, in order to create a trade-off between reliability and temporal resolution we decided to employ a moving window on single-trial ERPs elicited by 'go' condition, averaging the last 15 trials for selected electrodes. The usage of this one-trial-step overlapping window left the total of 435 averaged ERPs for further analysis.

The P300 component obtained in this study was bifurcated containing two sub-components, P3a and P3b. Whilst the P3a is more frontally distributed, the P3b is more prominent in the centro-parietal region (Polich 2007). However, their latency varies depending on the stimulus events which elicit them, nature of task, population of participants included in the study, etc. In order to quantify and examine the propagation of P3a and P3b component amplitude and latency for 435 averaged ERPs, the following strategy was used: for the P3a and P3b sub-components, the latency of the maximum peak on the grand averaged ERPs for each subject was found and the 100 ms interval window surrounding the peak was chosen for the calculation of the amplitude, utilising mean peak amplitude method proposed by Luck (2005). Similarly, the latency value on each of the 435 averaged ERPs was calculated using peak latency measures (Luck 2005).

Comparison of ERP and RT

Similar to the ERP analysis, the data for RTs were also averaged using a 15-trial moving window, thus allowing examination of the RTs propagation during the task. This provided continuous-like time series of RTs, together with

the P3a and P3b amplitude and latency values, further enabling the observation of common trends between these two modalities of attention monitoring. In this way, it was possible to examine the correlation between the values of the P3a and P3b amplitudes and RTs.

Statistical analysis

In order to examine the difference of the GA ERPs between 'go' and 'no-go' condition, a paired t-test was performed. The ERPs used for 'go/no-go' comparison included all ERPs related to the 'no-go' condition and 50 ERPs related to 'go' stimuli preceding the 'no-go' condition. To identify latencies with significant difference of go and no-go stimuli, mean amplitude values of GA ERPs across subjects were extracted over fixed 20 ms time windows. 'Windows of interest' were defined as follows: where successive bins achieved statistical significance, one after first, and one before last bin in this significant 'run' respectfully marked its beginning and ending. That is to say, times were treated as the windows of interest only if neighbouring 20 ms bins were also significant ($p < .05$). After identification of these windows, mean amplitudes across the window were computed and further analysis was conducted. Due to multiple comparisons, Bonferroni corrections were applied where necessary and the reported pattern of data did not change.

The correlation between the values of the RTs and P3a and P3b peak amplitudes and latencies were statistically analysed: vectors of P3a and P3b mean amplitude/latency values, calculated from the 435 values of the averaged 15 ERPs, and analogous values of the RTs were fed to SPSS and Pearson correlation coefficients were extracted.

Results

EEG results

ERPs were successfully extracted confirming the validity of the set-up and accurate synchronisation of the stimuli-inferred marking of EEG stream. Figure 3 depicts GA ERPs for the go (full line) and no-go (dotted line) tasks for Fz, Cz, CPz and Pz electrode sites. The P3a and P3b values in the 'go' condition were significantly higher than in 'no-go' condition ($p < .05$), while the more prominent N2 component was elicited over 'no-go' trials ($p < .05$), as marked on the upper left image of Figure 3. Further, the P300 peak elicited in our task was bifurcated, containing its both sub-components (P3a and P3b), as shown on the upper-left image of Figure 3.

The P3a and P3b components were consistent throughout the trials, which is represented in the colour maps, on the upper trace of Figure 4. (a, c, d and f), that represents an example of data obtained from subject 11 (Table 1). The lower traces of Figure 4 (a, c, d and f) represent the average

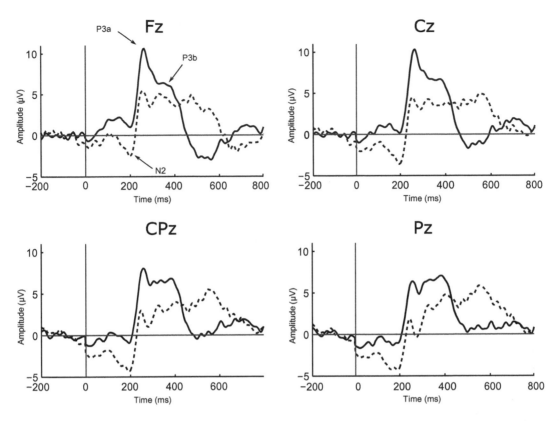

Figure 3. Grand average ERP waveform for 'go' (full line) and 'no-go' (dotted line) conditions across electrode sites under study. The N2, P3a and P3b ERP components are indicated on the upper left image.

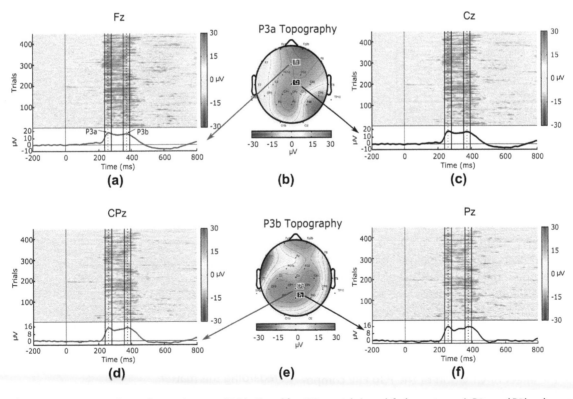

Figure 4. The average ERP waveforms, from subject 11 (Table 1) and for 450 go trials (a, c, d, f – lower traces); P3a and P3b sub-components of bifurcated P300 peak are indicated in the lower trace of image (a); the amplitudes were calculated for the window between the full lines for both P3a and P3b (as marked on images a, b, d, f). Further, the topography of P3a and P3b components are represented on images (b) and (e).

Table 1. Pearson's correlation values between the RTs and P3a and P3b mean amplitudes on the group level (upper part) and on the individual level (lower part of the table).

	Pearson's Correlation Values							
Component	P3a				P3b			
Electrode site	Fz	Cz	CPz	Pz	Fz	Cz	CPz	Pz
Group level	−.23	−.16	−.15	−.03	−.24	−.25	−.27	−.18
Individual subjects								
1	−.04	−.01	.03	.07	−.27	−.26	−.23	−.18
2	−.16	−.13	−.05	−.05	−.14	−.18	−.19	−.20
3	−.14	.01	.09	.09	.12	.23	.18	.08
4	−.33	−.35	−.36	−.36	−.10	−.14	−.20	−.27
5	−.03	.02	.02	.03	−.19	−.15	−.11	−.06
6	−.05	−.03	−.03	−.02	−.15	−.10	−.07	−.04
7	.22	.22	.16	.14	.15	.23	.22	.19
8	−.18	−.07	−.03	−.01	−.18	−.07	−.05	−.08
9	.03	.19	.13	.10	.17	.17	.02	−.05
10	−.07	.13	.16	.16	−.01	−.14	.02	.06
11	−.53	−.60	−.61	−.52	−.46	−.46	−.46	−.40
12	.36	.44	.41	.36	.15	.12	.02	.19

Negative correlations ($p < .05$);
Positive Correlations ($p < .05$);
Non significant values ($p > .05$).

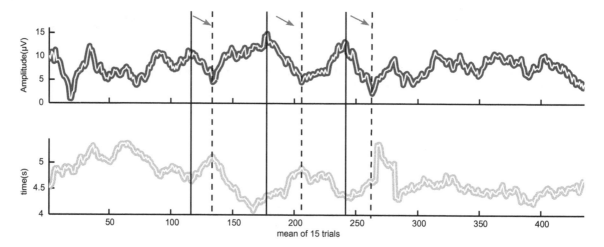

Figure 5. Visual representation of the time series of the 435-averaged P3b mean amplitude values (upper trace) vs. 435-averaged RT values (lower trace).

ERP waveform on the single subject level, which confirmed that our task paradigm was suitable for electing the P3a and P3b ERP waveforms for 'go' conditions in simulated workplace environment. Additionally, Figure 4(b) and (e) represent the topographic maps and the distribution of the P3a and P3b sub-components across the scalp locations.

Finally, the time series of the 435-averaged P3b components' mean amplitudes (upper panel of the Figure 5) and the corresponding averaged time series of the RTs (lower panel of the Figure 5) are presented for the visualisation of the effect of variation of the P3b ERP component and RTs. Vertical full lines indicate moments when P3b mean amplitude starts dropping, eventually reaching its lower peak (depicted with dashed lines). Red arrows on the top of the Figure 5 represent the direction of the decrease in P300 amplitude. It is notable that when the P3b amplitude is decreasing, opposite trend in RT can be observed.

Errors of commission

There was only one participant who executed errors on the 'no-go' trials (six errors of commission, approximately 10% of all 'no-go' trials). Additionally, none of the participants committed errors of omission. Given that there were very few errors in total, we did not carry out further analysis regarding this matter.

Go-No-go comparison

Paired sample *t*-test for the N2 ERP component at all four electrode sites revealed statistically significant difference between 'go' and 'no-go' trials (Fz: $t(1, 11) = 3.42, p < .01$; Cz: $t(1, 11) = 3.26, p < .01$; CPz: $t(1, 11) = 3.40, p < .01$; Pz: $t(1, 11) = 3.31, p < .01$). Similarly, we observed statistically significant differences across 'go/no-go' trials at all four channels for P3a (Fz: $t(1, 11) = 3.30, p < .01$; Cz: $t(1, 11) = 3.80, p < .01$; CPz: $t(1, 11) = 4.55, p < .001$; Pz: $t(1, 11) = 4.64, p < .001$) as well as for P3b (Fz: $t(1, 11) = 2.54, p < .05$; Cz: $t(1, 11) = 3.40, p < .01$; CPz: $t(1, 11) = 6.11, p < .001$; Pz: $t(1, 11) = 8.72, p < .001$) ERP components.

Pearson's correlation results

In order to evaluate the correlation between ERPs and RTs we used Pearson correlation. To further examine the strength of obtained correlation results, we also applied the Bootstrapping and Fisher-Z method on our data, verifying the consistence of the obtained results. The results of correlation between the RTs and P3a and P3b mean amplitudes are presented in the Table 1. These revealed that, on the group level, the correlation was negative on all electrode sites under study, with the high statistical significance ($p < .001$, Table 1).

However, compared to the group level, the overall significance of Pearson correlation varied substantially between individual participants at all four sites and in both P3a and P3b ERP windows. The results were less variable in the P3b compared to P3a window (values of correlation are presented in lower part of Table 1). Moreover, even in the P3b window, as obvious from the Table 1, only 4 out of 12 participants followed the general trend of negative correlation between ERPs and RTs at all four sites. Another four participants had significant negative correlations at 3, 2 or only 1 electrode site. Finally, one participant even had positive correlation over all sites, while the remaining three participants had positive correlations at 2 or 3 electrode sites under study.

Unlike the mean P3a and P3b amplitudes, the correlation between RTs and P3a and P3b latencies was inconsistent. Moreover, the distribution of latencies at all four sites of interest (Fz, Cz, CPz and Pz), across both P3a and P3b windows significantly differed from normal distribution. For that reason, the log instead of raw values was used, which approximated normal distribution somewhat better. At the group level, the P3b sub-component showed only two marginally significant negative correlations (at CPz and Pz electrode sites). On the other hand, P3a sub-component latencies showed positive correlation at all electrode sites ($p < .05$) at the group level. However, when analysed for the individual subjects, the pattern of results was inconclusive.

Based on the results reported beforehand, we identified two groups of participants, five participants who showed negative correlation between RTs and P3b amplitude in one group, and four who showed positive correlation

in the other. Regarding RTs, participants with negative correlation between RTs and P3b were faster ($t(RT) = 2.2$, $p < .05$), with higher P3b amplitudes ($t(Fz) = 35.21, p < .001$; $t(Cz) = 38.91, p < .001$; $t(CPz) = 39.68, p < .001$; $t(Pz) = 28.36$, $p < .001$) and shorter P3b latencies ($t(Fz) = 36.31, p < .001$; $t(Cz) = 30.74, p < .001$; $t(CPz) = 30.43, p < .001$; $t(Pz) = 34.61$, $p < .001$). On the other hand, the positively correlated participants showed slower RTs, lower P3b amplitudes and longer latencies.

Similarly, with regard to P3a component, two groups of participants (four in each) demonstrated the same pattern of results. Negatively correlated had higher amplitude ($t(Fz) = 22.2, p < .001$; $t(Cz) = 26.5, p < .001$; $t(CPz) = 27.14$, $p < .001$; $t(Pz) = 16.84, p < .001$) and shorter latencies ($t(Fz) = 18.77, p < .001$; $t(Cz) = 11.05, p < .001$; $t(CPz) = 7.51$, $p < .001$; $t(Pz) = 9.89, p < .001$), and vice versa for positively correlated. However, there were no significant group differences regarding RTs.

Discussion

The grand average comparison between ERPs extracted for 'go' and 'no-go' stimuli revealed that the higher P300 amplitude values are elicited for frequent 'go' condition. This is in contrast to most of the other findings, where participants were required to respond to deviant (infrequent) stimuli. However, this manipulation (with responding to frequent stimuli) was necessary, given that the study was conducted in simulated working environment, whereby the continuity of operation is essential. Therefore, the lower amplitude value of the 'no-go' P300 component is not surprising (Figure 3), since the passive stimulus processing generally produces reduced P300 amplitudes, as non-task events engage attention resources to reduce the amplitude (Polich 2007).

The Pearson's correlation between the RTs and P3a and P3b amplitudes, on the group level at all four sites of interest, showed significant negative correlation (Table 1). This confirms our main hypothesis, proving that the higher P300 amplitude values, which reflect the higher level of attention allocated to the task (Hohnsbein, Falkenstein, and Hoormann 1998; Murata, Uetake, and Takasawa 2005) correspond to the shorter RTs needed to complete the action. Additionally, higher values of negative correlation were obtained for the P3b, compared to P3a sub-component. However, the correlations between these modalities on the individual level were not consistent as within the group (Table 1), which constitutes one of the main finding of this study. This inconsistency could be attributed to the inter-individual differences, as the P300 component is influenced with the various factors, e.g. intelligence, introversion/extraversion, etc. (Picton 1992), but there can be also individual differences that are not functional but anatomical, such as scull thickness (Hagemann et al. 2008). Furthermore, the RT variability is also known to be subjected to inter-individual differences (MacDonald, Nyberg, and Bäckman 2006). Therefore, we support the notion of Hockey et al. (2009), where the importance of studying individual level data when performing psychophysiological measurements in ergonomics studies was emphasised.

Based on the Pearson's correlations between RT and P3b we identified two groups of participants, one of which was negatively correlated and the other one positively correlated. Negatively correlated group was faster with higher P3b amplitudes and shorter P3b latencies, whereby the positively correlated group showed slower RTs, lower amplitudes and longer latencies. Similar pattern of the results was observed for the P3a component (except for the RT comparisons, which were not significant). Therefore, it may be concluded that participants who showed negative correlation between P3b component and RTs were more focused on the task (given that they had higher P3b amplitude values) and were more efficient (given shorter RTs) than the positively correlated group. However, this finding should be examined in future studies and the consistency of the correlation results on individual basis needs to be confirmed through repeated measures on a single subject basis.

Another interesting comparison would be between ERPs on 'go' trials preceding correctly withhold 'no-go' trials and on 'go' trials preceding commission error on 'no-go' trials, as this could be an useful information on alerting the attention system (Robertson et al. 1997). However, the fact is that there was only one participant who executed actions on 'no-go' trials (6 errors in total, app. 10%). Interestingly enough, this was the participant (No. 12, from Table 1) who showed a positive correlation between RTs and P3 amplitudes, in contrast to the generally observed trend (negative correlation between RTs and P3 amplitudes). It is noteworthy that it was hard to set an objective criterion as to what action to mark as an error, given that participants would sometimes demonstrate slight movements without executing the action. Therefore, we chose the stricter criterion based on which the errors of commission were defined as completion of the action on 'no-go' trials (including the pedal press).

Although the P300 component is generally related to attention processing, the mechanisms that generate P3a and P3b sub-components differ significantly. P3a component is more related to novelty preference, processing of exogenous aspects of stimuli, i.e. low-level attention processes (Daffner et al. 2000; Polich 2007). This component usually follows the N2 component, which was also found to be increased in response to novel or deviant stimuli processing (Daffner et al. 2000), as also shown on Figure 3. On the other hand, P3b component was found to be more

related to high-level attention processing, processing of endogenous aspects of stimuli, context-updating information (working memory) and memory storage (Polich 2007). The P3b component is also related to decision processes (O'Connell, Dockree, and Kelly 2012), in which it mediates function between stimulus processing and required response (Verleger, Jaśkowski, and Wascher 2005). This is in line with our findings, since the P3b was more prominent in response to go-stimuli, which required action, particularly in central and centro-parietal sites.

We further examined the continuous-like time series of the RTs and P3a and P3b amplitudes and we noted the visible trends of fluctuation of these two modalities over time (Figure 5). Existing literature suggests that both RTs (Flehmig et al. 2007) and P300 component (especially P3b, Polich [2007]) are closely related to the attention, thus we can infer that fluctuation of these modalities correspond to the attention fluctuation on the neural as well as on the behavioural level. However, it is apparent from our results that not all the participants showed negative correlation between RTs and P3a and P3b components, which arises an obvious question: Which data are more closely related to the attention and should ERP or RT measures be used for evaluation of the assembler attention? Bishu and Drury (1988) pointed out that in assembly tasks translational stage from input information into output action is more complex than in conventional RT tasks and therefore, the structure of the response may influence the performance. Moreover, in RT experiments there are many possible processes that contribute to the RT and therefore it is difficult to isolate and address specific feature of interest, such as attention (Salthouse and Hedden 2002). On the other hand, the P3b component is found to be the direct correlate of the higher level attention processing (Verleger, Jaśkowski, and Wascher 2005). Following this logic, we speculate that findings in this study demonstrate that ERP correlates of attention offer a more detailed and sophisticated understanding of the nature of attention decline compared to robust, but rough RT measures. Not only that we are able to achieve precision of measurement with ERPs (which is recognised as 'reaction time of the twenty-first century', Luck, Woodman, and Vogel [2000]), but also gain more insightful understanding of the nature of the process as demonstrated through the analysis of P3a and P3b sub-components. However, further studies are desirable to confirm the generality of this finding.

The analysis of the relationship between RTs and P3a and P3b peak latencies revealed no statistically significant correlations between these components. Although Murata, Uetake, and Takasawa (2005) proposed that the P300 peak latency corresponds to the stimulus evaluation time and that it can be also directly correlated to the level of attention, this was not observed in our study. This finding is consistent with the recent work of Ramchurn et al. (2014) and it confirms that only the P300 component amplitude variation, but not its latency, correlates with the variation of the RTs. The P300 amplitude, on the other hand, was recognised as an index of the attention allocated to the task in numerous studies (De Vos, Gandras, and Debener 2014; Murata, Uetake, and Takasawa 2005; Polich 1988, 2007 and Ramchurn et al. 2014).

It was reported that the sudden drops in the attention, during a monotonous task, could be attributed to the e.g. daydreaming and mind wandering (Fisher 1998). However, the neural correlates of these phenomena are still not fully understood (Hasenkamp et al. 2012). For instance, potential benefit of real-time attention monitoring would be to provide the feedback to the operator once the attention level starts decreasing, thereby attempting to keep the attention level high and prevent possible human errors. The presented study indicates that 'periods of attention oscillation' are sufficiently long to make such a feedback system meaningful. However, one of the limitations of the present study is that the results were obtained in an off-line analysis. Therefore, one of the directions of future studies will be utilisation of one of the existing Brain–Computer Interface (BCI) software packages for real-time data processing in the desired time window and to provide proper visual, auditory or mechanical (e.g. vibration) feedback. The process could be automated in sense that once the amplitude values of the P3b component start decreasing with an obvious trend, as indicated by red arrows on Figure 5. (e.g. between 180th and 200th averaged trial), the feedback could be provided. It is important to investigate the effects of such a feedback also in relation with its content, all the while taking care of workers privacy and mental well-being.

Although, the authors believe that the measurement of covert attention-related modality (P3b) offers better understanding of attention processes than the overt performance measure of RTs, one of the limitations of present study is that EEG is still uncomfortable for everyday use and on-site recordings in naturalistic industrial environments. The main reason for this is that the reliable EEG recordings still depend on the wet gel-based electrodes (Mihajlovic et al. 2015) and an ethical question of EEG recording arises, in sense that the supervisor could have information about the physiological signals obtained from employees, raising privacy concerns (Fairclough 2014). Nevertheless, if the positive/negative correlation between P3b component's amplitude and RTs is held on a single-subject basis, then proposed methodology can be applied as that a primary (entry) test for workers. The benefits of such a testing can be twofold: firstly, the company management could be able to early detect whether the worker, for particular work position, is focused on the task (based on which group he belongs – positively/negatively correlated); secondly, the

reliable, comfortable and low-cost attention-monitoring system could be created based solely on non-invasive RTs recordings. Thus, the future studies should be directed towards investigation of the reliability of correlation between P3b and RTs on single-subject basis, upon which the proposed methodology could be applied in industrial settings.

The presented methodology was applied on a manual assembly work, where a single functional modification of the real workplace was needed, in the sense of on-screen stimulus presentation for the aim of eliciting the anticipated P300 ERP component. This modification was necessary, since the covert cognitive context is usually encrypted in complex brain dynamics and in naturalistic settings it is hard to isolate the specific cognitive processes, since they should firstly be evoked (Bulling and Zander 2014). Therefore, at current stage this methodology cannot be directly applied for the on-site recording in realistic industrial settings and other workplaces, as we would have had to modify the work routine. For that reason, either a more general approach needs to be developed for further application to this work position, or another work position has to be identified, where such attention-monitoring systems can be readily applied. These represent an additional direction for future research in this area.

Conclusions

In this study, we extended existing psychophysiological approaches in ergonomics by providing novel methodology for workers' continuous attention monitoring during the course of a monotonous assembly task and in the realistic workplace environment. We observed that, while on the group level P3a and P3b attention-related ERP component amplitudes and the RTs correlated in the negative fashion, that did not hold on individual subjects' level. This constitutes one of our major findings: overt performance measure of RTs alone is not reliable attention level measure per se, and covert physiological data also need to be employed for this task. Oscillating attention justifies the use of future feedback systems that would serve both to increase the attentiveness of workers and to prevent work-related errors. In that way, the potential accidents, which could lead to workers injuries and material damage, could be prevented, consequently increasing the workers overall well-being. Future studies are still needed to confirm the applicability of proposed methods, as well as to tune and sufficiently generalise them.

Acknowledgements

This research is financed under EU project 'Innovations Through Human Factors in Risk Analysis and Management', Marie Curie

Actions FP7-PEOPLE-2011-ITN-InnHF-289837. We would further like to acknowledge company 'Gomma Line' (Serbia), for their assistance and advisory during the experimental set-up phase.

Disclosure statement

No potential conflict of interest was reported by the authors.

Funding

This work was supported by the Seventh Framework Programme [FP7-InnHF-289837].

References

Arthur, W., G. V. Barret, and R. A. Alexander. 1991. "Prediction of Vehicular Accident Involvement: A Meta-Analysis." *Human Performance* 4 (2): 89–105. doi:10.1207/s15327043hup0402_1.

Bigdely-Shamlo, N., K. Kreutz-Delgado, K. Robbins, M. Miyakoshi, M. Westerfield, T. Bel-Bahar, C. Kothe, J. Hsi, and S. Makeig. 2013. "Hierarchical Event Descriptor (HED) Tags for Analysis of Event-Related EEG Studies." Paper presented In Global Conference on Signal and Information Processing for the IEEE society, Austin, TX, December 1–4. doi:10.1109/GlobalSIP.2013.6736796.

Bishu, R. R., and C. G. Drury. 1988. "Information Processing in Assembly Tasks — A Case Study." *Applied Ergonomics* 19 (2): 90–98. doi:10.1016/00036870(88)90001-4.

Brookhuis, K. A., and D. De Waard. 2010. "Monitoring Drivers' Mental Workload in Driving Simulators Using Physiological Measures." *Accident Analysis & Prevention* 42 (3): 898–903. doi:10.1016/j.aap.2009.06.001.

Bulling, A., and T. O. Zander. 2014. "Cognition-Aware Computing." *IEEE Pervasive Computing* 13 (3): 80–83. doi:10.1109/MPRV.2014.42.

Daffner, K. R., M. Mesulam, L. F. M. Scinto, V. Calvo, R. Faust, and P. J. Holcomb. 2000. "An Electrophysiological Index of Stimulus Unfamiliarity." *Psychophysiology* 37 (6): 737–747. doi:10.1111/1469-8986.3760737.

De Vos, M., S. Riès, K. Vanderperren, B. Vanrumste, F. X. Alario, S. Van Huffel, and B. Burle. 2010. "Removal of Muscle Artifacts from EEG Recordings of Spoken Language Production." *Neuroinformatics* 8 (2): 135–150. doi:10.1007/s12021-010-9076-8.

De Vos, M., L. De Lathauwer, and S. Van Huffel. 2011. "Spatially Constrained ICA Algorithm with an Application in EEG Processing." *Signal Processing* 91 (8): 1963–1972. doi:10.1016/j.sigpro.2011.02.019.

De Vos, M., K. Gandras, and S. Debener. 2014. "Towards a Truly Mobile Auditory Brain–Computer Interface: Exploring the P300 to Take Away." *International Journal of Psychophysiology* 91 (1): 46–53. doi:10.1016/j.ijpsycho.2013.08.010.

Debener, S., F. Minow, R. Emkes, K. Gandras, and M. De Vos. 2012. "How about Taking a Low-Cost, Small, and Wireless EEG for a Walk?" *Psychophysiology* 49 (11): 1617–1621. doi:10.1111/j.1469-8986.2012.01471.x.

Delorme, A., and S. Makeig. 2004. "EEGLAB: An Open Source Toolbox for Analysis of Single-Trial EEG Dynamics including Independent Component Analysis." *Journal of Neuroscience Methods* 134 (1): 9–21. doi:10.1016/j.jneumeth.2003.10.009.

Dockree, P. M., S. P. Kelly, J. J. Foxe, R. B. Reilly, and I. H. Robertson. 2007. "Optimal Sustained Attention is Linked to the Spectral Content of Background EEG Activity: Greater Ongoing Tonic Alpha (~ 10 Hz) Power Supports Successful Phasic Goal Activation." *European Journal of Neuroscience* 25 (3): 900–907. http://www.ncbi.nlm.nih.gov/pubmed/17328783.

Fafrowicz, M., and T. Marek. 2007. "Quo Vadis, Neuroergonomics?" *Ergonomics* 50 (11): 1941–1949. doi:10.1080/00140130701676096.

Fairclough, S. 2014. "Physiological Data Must Remain Confidential." *Nature* 505 (7483): 263–263. doi:10.1038/505263a.

Fish, L. A., C. G. Drury, and M. G. Helander. 1997. "Operator-Specific Model: An Assembly Time Prediction Model." *Human Factors and Ergonomics in Manufacturing & Service Industries* 7 (3): 211–235. doi:10.1002/(SICI)1520-6564(199722)7:3<211:AID-HFM4>3.0.CO;2-6.

Fisher, C. D. 1998. "Effects of External and Internal Interruptions on Boredom at Work: Two Studies." *Journal of Organizational Behavior* 19 (5): 503–522. doi:10.1002/(SICI)1099-1379(199809)19:5<503:AID-JOB854>3.0.CO;2-9.

Flehmig, H. C., M. Steinborn, R. Langner, A. Scholz, and K. Westhoff. 2007. "Assessing Intraindividual Variability in Sustained Attention: Reliability, Relation to Speed and Accuracy, and Practice Effects." *Psychology Science* 49 (2): 132–149.

Fu, S., and R. Parasuraman. 2008. "Event-Related Potentials (ERPs) in Neuroergonomics." In *Neuroergonomics: The Brain at Work*, edited by R. Parasuraman, and M. Rizzo, 32–50. New York: Oxford University Press.

Gordon, R. P. 1998. "The Contribution of Human Factors to Accidents in the Offshore Oil Industry." *Reliability Engineering & System Safety* 61 (1): 95–108. doi:10.1016/S0951-8320(98)80003-3.

Gray, H. M., N. Ambady, W. T. Lowenthal, and P. Deldin. 2004. "P300 as an Index of Attention to Self-Relevant Stimuli." *Journal of Experimental Social Psychology* 40 (2): 216–224. doi:10.1016/S0022-1031(03)00092-1.

Hagemann, D., J. Hewig, C. Walter, and E. Naumann. 2008. "Skull Thickness and Magnitude of EEG Alpha Activity." *Clinical Neurophysiology* 119 (6): 1271–1280. doi:10.1016/j.clinph.2008.02.010.

Hamrol, A., D. Kowalik, and A. Kujawińsk. 2011. "Impact of Selected Work Condition Factors on Quality of Manual Assembly Process." *Human Factors and Ergonomics in Manufacturing & Service Industries* 21 (2): 156–163. doi:10.1002/hfm.20233.

Hasenkamp, W., C. D. Wilson-Mendenhall, E. Duncan, and L. W. Barsalou. 2012. "Mind Wandering and Attention during Focused Meditation: A Fine-Grained Temporal Analysis of Fluctuating Cognitive States." *NeuroImage* 59 (1): 750–760. doi:10.1016/j.neuroimage.2011.07.008

Hockey, G. R. J., P. Nickel, A. C. Roberts, and M. H. Roberts. 2009. "Sensitivity of Candidate Markers of Psychophysiological Strain to Cyclical Changes in Manual Control Load during Simulated Process Control." *Applied Ergonomics* 40 (6): 1011–1018. doi:10.1016/j.apergo.2009.04.008.

Hohnsbein, J., M. Falkenstein, and J. Hoormann. 1998. "Performance Differences in Reaction Tasks Are Reflected in Event-Related Brain Potentials (ERPs)." *Ergonomics* 41 (5): 622–633. doi:10.1080/001401398186793.

Humphrey, D. G., and A. F. Kramer. 1994. "Toward a Psychophysiological Assessment of Dynamic Changes in Mental Workload." *Human Factors* 36 (1): 3–26. doi:10.1177/001872089403600101.

Jagannath, M., and V. Balasubramanian. 2014. "Assessment of Early Onset of Driver Fatigue Using Multimodal Fatigue Measures in a Static Simulator." *Applied Ergonomics* 45 (4): 1140–1147. doi:10.1016/j.apergo.2014.02.001.

Johnson, R. 1988. "The Amplitude of the P300 Component of the Event-Related Potential: Review and Synthesis." In *Advances in Psychophysiology: A Research Annual*, edited by P. Ackles, J. R. Jennings, and M. G. H Coles, 3 vols, 69–137. Greenwich, CT: JAI press.

Jung, T. P., S. Makeig, M. Stensmo, and T. J. Sejnowski. 1997. "Estimating Alertness from the EEG Power Spectrum." *IEEE Transactions on Biomedical Engineering* 44 (1): 60–69. doi:10.1109/10.553713.

Kletz, T. A. 2001. *An Engineer's View of Human Error.* Rugby: IChemE.

Klimesch, W., M. Doppelmayr, H. Russegger, T. Pachinger, and J. Schwaiger. 1998. "Induced Alpha Band Power Changes in the Human EEG and Attention." *Neuroscience Letters* 244 (2): 73–76. doi:10.1016/S0304-3940(98)00122-0.

Larue, G. S., A. Rakotonirainy, and A. N. Pettitt. 2010. "Real-Time Performance Modelling of a Sustained Attention to Response Task." *Ergonomics* 53 (10): 1205–1216. doi:10.1080/00140139.2010.512984.

Lehto, M. R., and S. J. Landry. 2012. *Introduction to Human Factors and Ergonomics for Engineers.* Boca Raton: Crc Press.

Leider, P. C., J. S. Boschman, M. H. W. Frings-Dresen, and H. F. Van der Molen. 2015. "Effects of Job Rotation on Musculoskeletal Complaints and Related Work Exposures: A Systematic Literature Review." *Ergonomics* 58 (1): 1–15. doi:10.1080/00140139.2014.961566.

Li, G., and C. M. Haslegrave. 1999. "Seated Work Postures for Manual, Visual and Combined Tasks." *Ergonomics* 42 (8): 1060–1086. doi:10.1080/001401399185144.

Luck, S. J. 2005. *An Introduction to the Event-Related Potential Technique.* Cambridge, MA: MIT press.

Luck, S. J., G. F. Woodman, and E. K. Vogel. 2000. "Event-Related Potential Studies of Attention." *Trends in Cognitive Sciences* 4 (11): 432–440. doi:10.1016/S1364-6613(00)01545-X.

MacDonald, S. W. S., L. Nyberg, and L. Bäckman. 2006. "Intra-Individual Variability in Behavior: Links to Brain Structure, Neurotransmission and Neuronal Activity." *Trends in Neurosciences* 29 (8): 474–480. doi:10.1016/j.tins.2006.06.011.

Mehta, R. K., and R. Parasuraman. 2013. "Neuroergonomics: A Review of Applications to Physical and Cognitive Work." *Frontiers in Human Neuroscience* 7 (889): 1–10. doi:10.3389/fnhum.2013.00889.

Michalos, G., S. Makris, N. Papakostas, D. Mourtzis, and G. Chryssolouris. 2010. "Automotive Assembly Technologies Review: Challenges and Outlook for a Flexible and Adaptive Approach." *CIRP Journal of Manufacturing Science and Technology* 2 (2): 81–91. doi:10.1016/j.cirpj.2009.12.001.

Mihajlovic, V., B. Grundlehner, R. Vullers, and J. Penders. 2015. "Wearable, Wireless EEG Solutions in Daily Life Applications: What Are We Missing?" *IEEE Journal of Biomedical and Health Informatics*, 19 (1): 6–21. doi: 10.1109/JBHI.2014.2328317.

Mijović, P., E. Giagloglou, P. Todorović, I. Mačužić, B. Jeremić, and I. Gligorijević. 2014. "A Tool for Neuroergonomic Study of Repetitive Operational Tasks." In Proceedings of the 2014 European Conference on Cognitive Ergonomics, 1–2, ACM: New York. doi:10.1145/2637248.2637280.

Murata, A., A. Uetake, and Y. Takasawa. 2005. "Evaluation of Mental Fatigue Using Feature Parameter Extracted from Event-Related Potential." *International Journal of Industrial Ergonomics* 35 (8): 761–770. doi:10.1016/j.ergon.2004.12.003.

O'Connell, R. G., P. M. Dockree, and S. P. Kelly. 2012. "A Supramodal Accumulation-to-Bound Signal That Determines Perceptual Decisions in Humans." *Nature Neuroscience* 15 (12): 1729–1735. doi:10.1038/nn.3248.

Parasuraman, R. 1990. "Event-Related Brain Potentials and Human Factors Research". In *Event-Related Brain Potentials: Basic Issues and Applications*, edited by J. W. Rohrbaugh, R. Parasuraman, and Jr. R. Johnson, 279–300. New York: Oxford University Press.

Parasuraman, R. 2003. "Neuroergonomics: Research and Practice." *Theoretical Issues in Ergonomics Science* 4 (1–2): 5–20. doi:10.1080/14639220210199753.

Parasuraman, R., and M. Rizzo. 2008. *Neuroergonomics: The Brain at Work.* New York: Oxford University Press.

Picton, T. W. 1992. "The P300 Wave of the Human Event-Related Potential." *Journal of Clinical Neurophysiology* 9 (4): 456–479. http://journals.lww.com/clinicalneurophys/Abstract/1992/10000/The_P300_Wave_of_the_Human_Event_Related.2.aspx.

Picton, T. W., S. Bentin, P. Berg, E. Donchin, S. A. Hillyard, R. Johnson, G. A. Miller, W. Ritter, D. S. Ruchkin, M. D. Rugg, and M. J. Taylor. 2000. "Guidelines for Using Human Event-Related Potentials to Study Cognition: Recording Standards and Publication Criteria." *Psychophysiology* 37 (2): 127–152. http://www.ncbi.nlm.nih.gov/pubmed/10731765

Polich, J. 1988. "Bifurcated P300 Peaks" *Journal of Clinical Neurophysiology* 5 (3): 287–294. http://www.ncbi.nlm.nih.gov/pubmed/3170723

Polich, J. 2007. "Updating P300: An Integrative Theory of P3a and P3b." *Clinical Neurophysiology* 118 (10): 2128–2148. doi:10.1016/j.clinph.2007.04.019.

Polich, J., and A. Kok. 1995. "Cognitive and Biological Determinants of P300: An Integrative Review." *Biological Psychology* 41 (2): 103–146. doi:10.1016/0301-0511(95)05130-9.

Prinzel, L. J., F. G. Freeman, M. W. Scerbo, P. J. Mikulka, and A. T. Pope. 2003. "Effects of a Psychophysiological System for Adaptive Automation on Performance, Workload, and the Event-Related Potential P300 Component." *Human Factors: The Journal of the Human Factors and Ergonomics Society* 45 (4): 601–613. doi:10.1518/hfes.45.4.601.27092.

Ramchurn, A., J. W. De Fockert, L. Mason, S. Darling, and D. Bunce. 2014. "Intraindividual Reaction Time Variability Affects P300 Amplitude rather than Latency." *Frontiers in Human Neuroscience* 8: 557. doi:10.3389/fnhum.2014.00557.

Rasmussen, J., A. M. Pejtersen, and L. P. Goodstein. 1994. *Cognitive Systems Engineering.* New York: Wiley.

Reason, J. 1990. *Human Error.* Cambridge: Cambridge University Press.

Robertson, I. H., T. Manly, J. Andrade, B. T. Baddeley, and J. Yiend. 1997. "'Oops!': Performance Correlates of Everyday Attentional Failures in Traumatic Brain Injured and Normal Subjects." *Neuropsychologia* 35 (6): 747–758. doi:10.1016/S0028-3932(97)00015-8.

Salthouse, T. A., and T. Hedden. 2002. "Interpreting Reaction Time Measures in between-Group Comparisons." *Journal of Clinical and Experimental Neuropsychology* 24 (7): 858–872. doi:10.1076/jcen.24.7.858.8392.

Simpson, S. A., E. J. K. Wadsworth, S. C. Moss, and A. P. Smith. 2005. "Minor Injuries, Cognitive Failures and Accidents at

Work: Incidence and Associated Features." *Occupational Medicine* 55 (2): 99–108. doi:10.1093/occmed/kqi035.

Stanton, N. A., P. Salmon, D. Harris, A. Marshall, J. Demagalski, M. S. Young, T. Waldmann, and S. Dekker. 2009. "Predicting Pilot Error: Testing a New Methodology and a Multi-Methods and Analysts Approach." *Applied Ergonomics* 40 (3): 464–471. doi:10.1016/j.apergo.2008.10.005.

Sternberg, S. 1969. "Memory-Scanning: Mental Processes Revealed by Reaction-Time Experiments." *American Scientist* 57 (4): 421–457. http://www.jstor.org/stable/27828738

Tang, A., C. Owen, F. Biocca, and W. Mou. 2003. "Comparative Effectiveness of Augmented Reality in Object Assembly." In Proceedings of the Conference on Human Factors in Computing Systems, SIGCHI Society, 73–80. ACM, New York. doi:10.1145/642611.642626

Verleger, R., P. Jaśkowski, and E. Wascher. 2005. "Evidence for an Integrative Role of P3b in Linking Reaction to Perception." *Journal of Psychophysiology* 19 (3): 165–181. doi:10.1027/0269-8803.19.3.165.

Viola, F. C., J. Thorne, B. Edmonds, T. Schneider, T. Eichele, and S. Debener. 2009. "Semi-Automatic Identification of Independent Components Representing EEG Artifact." *Clinical Neurophysiology* 120 (5): 868–877. doi:10.1016/j.clinph.2009.01.015.

Wallace, J. C., and S. J. Vodanovich. 2003. "Can Accidents and Industrial Mishaps Be Predicted? Further Investigation into the Relationship between Cognitive Failure and Reports of Accidents." *Journal of Business and Psychology* 17 (4): 503–514. doi:10.1023/A:1023452218225.

Wascher, E., H. Heppner, and S. Hoffmann. 2014. "Towards the Measurement of Event-Related EEG Activity in Real-Life Working Environments." *International Journal of Psychophysiology* 91 (1): 3–9. doi:10.1016/j.ijpsycho.2013.10.006.

Yamada, F. 1998. "Frontal Midline Theta Rhythm and Eyeblinking Activity during a VDT Task and a Video Game: Useful Tools for Psychophysiology in Ergonomics." *Ergonomics* 41 (5): 678–688. doi:10.1080/001401398186847.

Young, M. S., and N. A. Stanton. 2007. "Back to the Future: Brake Reaction times for Manual and Automated Vehicles." *Ergonomics* 50 (1): 46–58. doi:10.1080/00140130600980789.

Musculoskeletal disorders as a fatigue failure process: evidence, implications and research needs

Sean Gallagher and Mark C. Schall Jr.

ABSTRACT

Mounting evidence suggests that musculoskeletal disorders (MSDs) may be the result of a fatigue failure process in musculoskeletal tissues. Evaluations of MSD risk in epidemiological studies and current MSD risk assessment tools, however, have not yet incorporated important principles of fatigue failure analysis in their appraisals of MSD risk. This article examines the evidence suggesting that fatigue failure may play an important role in the aetiology of MSDs, assesses important implications with respect to MSD risk assessment and discusses research needs that may be required to advance the scientific community's ability to more effectively prevent the development of MSDs.

Practitioner Summary: Evidence suggests that musculoskeletal disorders (MSDs) may result from a fatigue failure process. This article proposes a unifying framework that aims to explain why exposure to physical risk factors contributes to the development of work-related MSDs. Implications of that framework are discussed.

1. Introduction

Musculoskeletal disorders (MSDs) are widespread throughout the world, and are associated with enormous financial and societal costs. MSDs are the second-greatest cause of disability globally, having increased 45% since 1999 (Horton 2010). They account for roughly one-third of all workplace injuries and illnesses in the United States annually (BLS 2015). In 2004, the estimated direct cost of treatment for MSDs in the United States was estimated at $510 billion, equivalent to 4.6% of the gross domestic product (GDP) (AAOS 2008). Indirect costs were estimated to add $339 billion more, for a total cost for MSDs of $849 billion, or 7.7% of the GDP (AAOS 2008).

Our understanding of the aetiology of MSDs has advanced considerably over the past few decades. Importantly, epidemiological studies have identified several physical risk factors for work-related MSDs common to both upper extremity disorders and low back pain (LBP) (Punnett et al. 2005). These include exposure to tasks requiring: (1) high-force exertions, (2) highly repetitive tasks, (3) adoption of non-neutral postures and (4) exposure to whole-body or hand-arm vibration (NIOSH 1997; NRC-IOM 2001; Punnett et al. 2005). However, lacking in previous analyses of these risk factors has been the development of an underlying theoretical framework that could explain how and why these risk factors are associated with the development of MSDs.

Recently, however, a systematic review of the literature identified a consistent statistical interaction between the risk factors of force and repetition with respect to risk of a wide variety of MSDs including low back disorders (LBDs), carpal tunnel syndrome (CTS), lateral epicondylitis, shoulder pain and many others (Gallagher and Heberger 2013). The authors noted that the pattern of interaction observed in the reviewed studies was indicative of the presence of a fatigue failure process in musculoskeletal tissues and may provide a unifying framework to explain the effects of all of the physical MSD risk factors noted above. If MSDs are indeed the result of a fatigue failure process, numerous important implications arise with respect to how risk of MSDs should be assessed and how prevention efforts should be designed. This paper describes the evidence that suggests MSDs may be the result of a fatigue failure process, explores important implications in terms of exposure and risk assessment and discusses recommendations for future research.

2. Background

It has long been recognised that materials experience failure through either: (1) application of a one-cycle high-magnitude stress (at the so-called 'ultimate stress' [US] of the material), or (2) repeated application of loads at some percentage of the material's US (Peterson 1950). The American Society for Testing and Materials (ASTM) has defined the latter failure mode, known as fatigue failure, as:

> ... the process of progressive localized permanent structural change occurring in a material subjected to conditions that produce fluctuating stresses and strains at some point (or points) and that may culminate in cracks or complete fracture after a sufficient number of fluctuations. (ASTM 2000)

The rate of damage propagation in a material is a function of several loading characteristics and the number of cycles experienced at various loads. The relationship between applied stress and the number of cycles to failure is exponential in nature and is typically described in an S–N diagram, which describes the manner in which the number of cycles to failure (N) varies with respect to a constant cyclic stress (S). An example S–N diagram is provided in Figure 1. As can be seen in this figure, higher levels of loading will result in failure in fewer cycles and lower levels of loading will last an exponentially larger number of cycles. In fact, millions of cycles may be necessary to create failure in low load situations, and for many materials there exists a fatigue (or endurance) limit (usually around 30% of the material's US) where failure will not occur no matter the number of cycles experienced in fully reversed loading conditions (Ashby, Shercliff, and Cebon 2010).

The traditional domain of fatigue failure analysis has been in evaluation of components and/or engineered structures such as bridges, aircraft and automobile parts, and nuclear pressure vessels. However, biological tissues are also materials, and would be expected to incur damage in accordance with the same principles, though with some important differences due to the fact that these materials reside in a complex physiological environment.

While it has been generally recognised by the field that MSDs result from a progression of cumulative damage (as the term 'cumulative trauma disorders' implies), application of fatigue failure principles has not been apparent in the design of current MSD risk assessment tools or recent

epidemiological studies. Nor have the foremost reviews of the MSD literature evaluated risk in accordance with the tenets of fatigue failure (e.g. NIOSH 1997; NRC-IOM 2001). However, as described below, evidence from a variety of sources strongly suggests that fatigue failure is occurring in musculoskeletal tissues, and may be an important etiological mechanism in the development of MSDs.

3. Evidence of a fatigue failure process in musculoskeletal tissues

Several lines of evidence support the notion that MSDs might be the results of a fatigue failure process in musculoskeletal tissues. These include *in vitro* testing of musculoskeletal tissues, animal studies of tissue loading and the epidemiological studies mentioned above.

In vitro studies have been performed on tendons (Schechtman and Bader 1997; Wang, Ker, and Alexander 1995), ligaments (Lipps, Ashton-Miller, and Wojtys 2014; Lipps, Wojtys, and Ashton-Miller 2013; Thornton, Schwab, and Oxland 2007), cartilage (Bellucci and Seedhom 2001) and spinal motion segments, both in compression (Brinckmann, Biggemann, and Hilweg 1988), shear (Cyron and Hutton 1978), and combined compression and shear loading (Gallagher et al. 2005, 2007). Irrespective of the material studied, all studies have demonstrated an exponential relationship between the stress applied and the number of cycles to material failure. In this respect, musculoskeletal tissues are shown to be no different from other (non-biological) materials.

Additional support for the fatigue failure hypothesis of MSD causation can be found in data from a rat model where Sprague-Dawley rats were exposed to one of the following conditions: low-force, low-repetition, low-force, high-repetition, high-force, low-repetition or high-force, high-repetition exertions (Barbe et al. 2013). Tissue pathology results for tendon damage, cartilage damage and bone volume and cytokine responses to applied loading, all demonstrated statistically significant force–repetition interactions of the pattern consistent with an underlying fatigue failure process. Furthermore, Andarawis-Puri and Flatow (2011) showed in an *in vivo* mouse model that fatigue-loaded tendons demonstrated a structural damage progression that started with fibre kinking for low level fatigue loading, to development of a widened interfibre space in tendons with moderate fatigue loading, to a severe matrix disruption with high-level fatigue loading. These results provide evidence that fatigue failure is not just a response observed in *in vitro* studies – musculoskeletal tissues also experience fatigue failure *in vivo*.

Finally, as mentioned previously, epidemiological studies that have examined a force–repetition interaction have shown a pattern of risk consistent with a fatigue failure

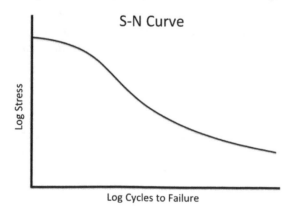

Figure 1. Example of an S–N diagram, relating the level of stress (S) to the number of cycles to failure (N).

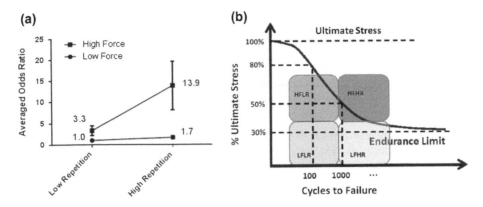

Figure 2. (a) Averaged Odds Ratios for seven studies examining quadrants of risk for force and repetition (Gallagher and Heberger 2013), and (b) fatigue failure curve.
Note: The pattern of risk observed in (a) would be anticipated if MSD development was the results of a fatigue failure process (b).

process (Figure 2). MSDs demonstrating this pattern include CTS, tendinitis, epicondylitis, hand pain and LBDs (Gallagher and Heberger 2013). Recent findings of a large prospective epidemiological study indicating that 'forceful repetition' was the loading variable most associated with development of CTS comports with fatigue failure theory and provides additional support for our model (Harris-Adamson et al. 2015). The additional information provided by these epidemiological studies is that human tissues also appear to be experiencing the same fatigue failure process during the development of MSDs. Given the available evidence, it would seem prudent to consider the implications of fatigue failure theory in terms of MSD risk assessment and prevention efforts, which are discussed below.

4. Implications of MSDs as a fatigue failure process

Despite mounting evidence that fatigue failure may be important in MSD aetiology, there are currently no examples to the authors' knowledge that any ergonomics risk assessment tools have used fatigue failure principles as a basis for ascertaining MSD risk. If MSDs are indeed the result of a fatigue failure process, several fundamental implications related to assessment of risk and methods of prevention for these disorders must be considered. The following sections discuss both general implications and important analysis principles that should be considered when assessing MSD risk from a fatigue failure perspective.

4.1. A unifying framework for MSD risk factors

MSD risk factors have traditionally been assumed to function in a statistically independent manner *vis a vis* MSD risk. For example, several comprehensive reviews of the literature (e.g. NIOSH 1997; NRC-IOM 2001) did not evaluate the potential for an interaction of force and

repetition. However, both *in vitro* and *in vivo* evidence strongly suggests that a consistent statistical interaction exists between force and repetition with respect to MSD risk. This suggests that these two risk factors cannot be treated independently, but instead have an important dependency wherein *the impact of repetition is highly dependent on the forces imposed on the tissues*. If fatigue failure were indeed etiologically significant in MSDs (and a force–repetition interaction exists), examining the main effects of force and repetition would provide unreliable estimates of risk, as main effects are uninterpretable in the presence of an interaction (Meyer 1991). The implication is that force and repetition must be considered in tandem, not in isolation, when assessing MSD risk.

A fatigue process in musculoskeletal tissues clearly would alter the manner in which we would consider the risk factors of force and repetition, but what about the other risk factors for MSDs? For example, consider the adoption of non-neutral postures, another important physical risk factor for MSDs. It should be recognised that adoption of awkward or non-neutral postures often leads to imposition of increased stress on musculoskeletal tissues (in some form or fashion). According to the force × repetition interaction paradigm discussed previously, any increased force demand that may result from the use of non-neutral postures would also be expected to lead to a more rapid escalation of MSD risk. For example, compared to an upright posture, bending the trunk forward into full flexion can triple the stress experienced by the lumbar spine (Nachemson 1976). Studies have demonstrated a dose–response relationship as non-neutral trunk postures becomes more extreme (Punnett et al. 1991), following an expected increase in spinal forces as more extreme neutral postures are adopted. Similarly, working with a deviated wrist posture may increase frictional forces on tissues of the tendons that may also lead to tissue frictional fatigue and, ultimately, CTS (Armstrong and Chaffin 1979).

Adoption of non-neutral postures may have an important role with respect to increasing the mean stress on musculoskeletal tissues, which has an important impact on the fatigue life of tissues, as discussed below. Clearly, awkward postures can have other impacts such as impeding blood flow (Chaffin, Andersson, and Martin 1999) or increasing ligament laxity (Solomonow et al. 2003), which may affect tissue health and/or injury potential. However, a major reason posture emerges as a risk factor for MSDs may simply be due to the increased tissue loads that result from adoption of awkward or non-neutral postures, and the effects of these increased loads in the fatigue failure paradigm.

Finally, it should be recognised that vibration exposure is a combination of force and repetition. When engineers evaluate the effects of vibration on the life of a particular component, they often utilise the techniques of fatigue failure analysis (Sarkani and Lutes 2004). These techniques include cycle counting using the rainflow algorithm (Matsuishi and Endo 1968) and summation of damage incurred by the vibration using the Palmgren–Miner (Miner 1945; Palmgren 1924) technique, both traditional tools of fatigue failure analysis. Thus, it may also be worth evaluating effects of vibration exposure on musculoskeletal tissues from a fatigue failure context as well.

4.2. Validated methods for assessing risk of cumulative damage

The ability to quantify cumulative damage related to repetitive loading of musculoskeletal tissues has long been a holy grail of the ergonomics community. Fortunately, fatigue failure theory has validated methods to predict damage accumulation associated with the variable loading regimens typically experienced by musculoskeletal tissues in occupational (and non-occupational) settings.

The variable amplitude loading typically experienced by musculoskeletal tissues is often termed 'spectrum' loading in the fatigue literature, and the term 'cumulative damage' refers to fatigue effects of non-uniform repeated loading events (Stephens et al. 2001). The most commonly used method of assessing or predicting damage resulting from spectrum loading is the linear cumulative damage rule for fatigue life from spectrum loading was proposed by Palmgren (1924) and Miner (1945):

$$c = \sum_{i}^{k} \frac{n_1}{N_1} + \frac{n_2}{N_2} + \cdots + \frac{n_k}{N_k} \qquad (1)$$

where c is a constant (often set at 1, but which can vary), $n_i \ldots$ equal the number of exposure cycles experienced at force levels at which $N_i \ldots$ cycles would result in fatigue failure. When the right-hand sum is equal to one, the material would be expected to fail. However, for different materials the value of c has been shown to vary above and below this number.

A rudimentary example of the Palmgren–Miner technique is provided below. In this illustration, the need to examine both force and repetition in combination should become apparent. Suppose that a worker performs a task that stresses a tendon at 15 cycles at 60% of ultimate tensile stress (UTS), 100 cycles at 50% UTS and 700 cycles at 40% UTS. Suppose also that the cycles to failure for 60, 50 and 40% UTS are 1000, 10,000, and 100,000 cycles, respectively. The Palmgren–Miner technique would calculate the cumulative damage (Dt) by summing the quotients of the number of cycles experienced at each stress level divided by their respective cycles to failure. The result of the calculation (seen in Figure 3) is 0.032, suggesting that this load would create damage of approximately 3% of the material's fatigue life.

Notice that if one were to focus on the risk factor of repetition in isolation, as has often been done in the past, it would appear that the 700 repetitions at 40% UTS would be the condition of primary concern. However, when examined in relation to the number of cycles to failure at each % UTS, it can be seen from the Palmgren–Miner equation that the 15 cycles at 60% UTS actually result in more than twice the cumulative damage compared to 700 cycles at 40% UTS. This example should illustrate clearly that the impact of repetition changes dramatically as a function of the stress imposed on tissues, and that examining repetition in isolation may lead to improper conclusions regarding the nature of MSD risk experienced by a worker.

The Palmgren–Miner rule often provides a useful approximation of the accumulation of fatigue damage in a material; however, it must be understood that it is an approximation. Numerous factors can influence the development of fatigue failure and these factors can influence the fatigue failure process in a manner not captured by the simple linear relationship expressed above. For example, localised stress concentrations in a material may lead to microstructural failure that causes a portion of the material to become unable to support a load, leading to

% Ult. Strength	Cycles to Failure	Cycles experienced
60%	1000	15
50%	10000	100
40%	100000	700

$$D_{(t)} = \frac{15}{1000} + \frac{100}{10000} + \frac{700}{100000}$$

$$= 0.015 + 0.01 + 0.007 = 0.032$$

Figure 3. Example of cumulative damage calculation using the Palmgren–Miner rule.

a decrease in fatigue life. On the other hand, there may be situations where changes in molecular orientation may actually slow down the fatigue failure process (Stephens et al. 2001). Such factors are not taken into account by the Palmgren–Miner rule, but can have large influences on fatigue life (Roylance 2001). Despite these limitations, the Palmgren–Miner rule remains a useful method with which to estimate cumulative damage in variable loading conditions and may be helpful in ascertaining MSD risk.

Other methods of evaluating cumulative loading metrics for LBP have included measures such as the area under the loading curve (Norman et al. 1998), which makes intuitive sense. Such techniques, however, do not address many important issues that are important in the development of cumulative damage from a fatigue failure perspective (Stephens et al. 2001). The following section discusses some of the issues related to cumulative damage estimation based on loading situations experienced by musculoskeletal tissues.

4.3. The critical roles of stress range and mean stress on cumulative damage

Loading cycles on musculoskeletal tissues can vary in many ways, one of which is the amount of time spent in the loaded phase versus the unloaded phase of a cycle (otherwise known as the duty cycle). To examine the effect of duty cycle, a few basic techniques of fatigue failure

analysis must be covered. To begin, consider a standard method of performing loading in fatigue failure studies, specifically the condition known as completely reversed loading using a sinusoidal loading pattern. Completely reversed loading represents a loading condition where an object is subjected to alternating tensile and compressive stresses and where the mean stress is 0. Figure 4(a) illustrates such a loading pattern. As can be seen in the figure, several characteristics of the loading pattern can be defined. σ_a represents the average of the maximum minus the minimum load of the cycle, σ_m represents the mean loading associated with the cycle, which in the case of fully reversed loading is equal to 0.

The standard S–N curve for a material is developed assuming a fully reversed loading cycle (i.e. where $\sigma_m = 0$). However, if σ_m is not equal to 0, certain specific loading conditions are known as either repeated stress or fluctuating stress (Figures 4(b) and (c)). Repeated stress is defined as a loading pattern where the minimum stress is zero and cycles to some positive (tensile) or negative (compressive) value. This loading pattern is representative of the type of loading experienced by tendons or ligaments. Fluctuating stress is when the minimum stress is non-zero and cycles to a stress of larger absolute magnitude. An example of this would be the loading pattern experienced by a worker repeatedly lifting bags off of a conveyor. The worker begins by standing upright (nothing in the hands) with a load of approximately 500 N on the spine then repeatedly lifts

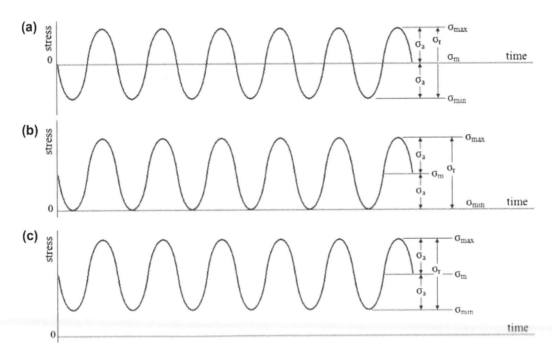

Figure 4. Examples of a (a) fully reversed sinusoidal loading, (b) repeated stress, and (c) fluctuating stress, where σ_a = stress amplitude, σ_r = stress range, σ_{max} = maximum stress, σ_{min} = minimum stress and σ_m = mean stress.
Note: Deflections below zero represent compressive loading, while deflections above zero represent tensile loading (Figure from http://www.engineeringarchives.com/les_machdes_cyclicloading.html).

bags off of the conveyor increasing the compressive load on the spine (say, to 3000 N), which returns to 500 N when the load is released. In this case, the spine is always experiencing compression which cycles from some non-zero number to a larger magnitude compressive stress (i.e. the stress on the spine is never reduced to zero). Both repeated and fluctuating stresses will result in a non-zero mean stress on the tissues, which may shift the fatigue failure curve down, meaning that fewer loading cycles would be needed reach failure compared to a fully reversed loading condition (Figure 5).

Fatigue failure theory has methods for calculating safety factors and expected cycles to failure for materials subjected to repeated or fluctuating stress. These techniques are the Goodman line (Goodman 1899) and the Gerber (1874) criterion, with the Goodman criterion serving as the more conservative of the two. Actual experimental data on fatigue life tend to reside between these two lines. For both criteria, every point on the line corresponds to failure in 10^6 cycles. Thus, combinations of stress amplitude and mean stress that reside beneath the lines would be said to have 'infinite life' and points above the curves would have finite life.

In materials engineering applications, these design techniques often will be used to design materials or parts for 10^6 cycles to failure (considered to be 'infinite life'). In other cases, however, engineers may design for finite life (<10^6 cycles to failure). In such cases, it must be estimated how many cycles would be expected until failure when infinite life conditions are exceeded, given a certain stress amplitude (σ_a) and mean stress (σ_m). Under the completely reversed loading situation, the following equation would be used:

$$N = \left(\frac{\sigma_a}{a}\right)^{\frac{1}{b}} \qquad (2)$$

where N represents cycles to failure, σ_a is the stress amplitude and a and b are:

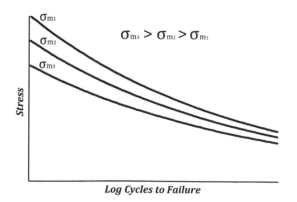

Figure 5. The influence of mean stress on S–N curves. As mean stress increases, cycles to failure will decrease at a give level of stress.

$$a = \frac{(f \cdot S_{ut})^2}{S_e} \qquad (3)$$

$$b = -\frac{1}{3}log\left(\frac{f \cdot S_{ut}}{S_e}\right) \qquad (4)$$

where f is the fatigue strength fraction (approximation of the fatigue strength at 10^3 cycles), S_{ut} is the ultimate tensile strength and S_e is the Stress at the Endurance Limit.

However, in the fluctuating stress situation, we cannot use σ_a in Equation (2) above as this only pertains to the completely reversed loading. In the case of fluctuating stress, we must instead calculate σ_{rev} that represents an equivalent value for completely reversed stress under repeated or fluctuating stress, and will replace σ_a in Equation (2). Thus, for fluctuating stress conditions (using the Goodman design criterion) our equation for N cycles to failure becomes:

$$N = \left(\frac{\sigma_{rev}}{a}\right)^{\frac{1}{b}} \qquad (5)$$

where σ_{rev} is:

$$\sigma_{rev} = \frac{\sigma_a}{1 - \frac{\sigma_m}{S_{ut}}} \qquad (6)$$

and for the Gerber relation σ_{rev} is:

$$\sigma_{rev} = \frac{\sigma_a}{1 - \left(\frac{\sigma_m}{S_{ut}}\right)^2} \qquad (7)$$

The influence of mean stress in this relationship has relevance to the issue of duty cycle in occupational tasks. Figure 6 below shows an example of the influence of mean stress that will be used to calculate expected cycles to failure for each instance. Figure 6(a) illustrates loading with a 40% duty cycle, while Figure 6(b) represents a duty cycle of 70%. The mean stress (σ_m) for the former condition would be 13.93 MPa, while the latter has a mean stress of 25.87 MPa.

There are several methods of predicting cycles to failure given the loading amplitude and mean stress. Two of the most popular are the Goodman equation and the Gerber equation. Of the two, the Goodman criterion tends to result in relatively conservative estimates of fatigue life and the Gerber curve results in a more generous fatigue life prediction. Test data tend to fall between the two predictions (Stephens et al. 2001).

Calculation of the expected cycles to failure according to the Goodman equation suggests a fatigue life of 5700 cycles to failure for the 40% duty cycle condition and 1423

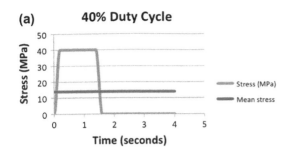 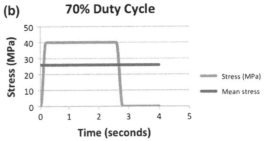

Figure 6. Comparison of the mean stress in (a) a 40% duty cycle (13.93 MPa) versus (b) a 70% duty cycle (25.9 MPa).

cycles to failure for the 70% duty cycle condition. Using the Gerber criterion, the 40 and 70% duty cycle conditions are predicted to last 19,300 and 12,150 cycles, respectively.

4.4. Various types of rest and musculoskeletal health

When considering the impact of rest on MSDs, a distinction may be made between short-term rest (short cyclic loading 'gaps' in the midst of a sequence of repetitive tissue loading) and long-duration rest (tissue unloading) experienced over more extended periods during which significant tissue loading is sparse. With respect to short-term loading gaps, it is likely that the effect on tissue health may be *primarily the influence that the rest period has on the* ***mean stress*** *experienced by tissues during the loading process*. As demonstrated previously, short duty cycles result in a greater fatigue life compared to longer duty cycles due to the lower mean tissue stress. However, remodelling and repair of tissue damage is a time-consuming process taking weeks, months or years, and short-term loading gaps (say, a few seconds in length) are unlikely to provide the opportunity for meaningful tissue healing (Sharma and Maffulli 2005). Tissue unloading lasting extended periods would seem more conducive to repair and remodelling of tissues. Thus, increased rest both in loaded and non-loaded states may be protective against MSDs, but in different manners. Short-term loading (stress) gaps in the midst of frequent tissue loading would decrease the mean stress experienced during loading and increase fatigue life, while long-term stress relief would facilitate tissue remodelling and repair. It should be noted that short-term breaks may be beneficial in terms of other sorts of musculoskeletal function (for example, reducing development of muscle fatigue), but the actual healing and regeneration of tissues during short periods of rest would be minimal at best.

4.5. Counting cycles in variable amplitude loading

Musculoskeletal loading patterns in occupational settings tend to exhibit substantial variability (Mathiassen 2006; Mathiassen, Möller, and Forsman 2003). One commonly used technique to evaluate highly variable loading patterns in the context of a fatigue failure process is known as 'rainflow analysis' (Matsuishi and Endo 1968). The 'rainflow' term refers to an analogy of rain dripping off of the roof of a pagoda. This analysis technique takes a complex set of varying stress exposures and breaks it down into a series of stress reversals. Once these stress reversals are obtained, one can apply the Goodman and/or Gerber relationships and the Palmgren–Miner technique to estimate the amount of cumulative damage in the material of interest. An example of the rainflow analysis technique is provided below. It is important to note that this technique assumes all loads are independent from one another and that there are no sequence effects.

The example in Figure 7(a) presents a stress-loading curve as might be experienced by a tendon during work. Figure 7(b) provides a rainflow analysis of the curve presented in Figure 7(a). The black line represents the variable stress loading experienced by the material, and the blue lines represent the half cycles. The horizontal lengths of the blue lines are dictated by whether you encounter either: (1) a valley lower than or equal to the one at which you started, or (2) a peak greater than or equal to the one at which you started. As can be seen with some of the lines in Figure 7(b), stress reversals may be defined not by the first peak or valley encountered, but subsequent peaks or valleys as the 'rain' drips off of different levels of the pagoda 'roof' (if one were to imagine the figure rotated 90 degrees clockwise). In such circumstances, notice that there are short lines that account for the part of the loading not involved with the line dripping off multiple pagoda roof levels. In this manner, every part of the loading signal is accounted for in the rainflow analysis (Downing and Socie 1982). Table 1 presents the breakdown of stress reversals resulting from this example and calculates predicted cycles to failure (N) and damage per cycle (1/N) using the Goodman criterion. For these analyses, the fatigue strength fraction (f) (i.e. S_f/S_{ut} at 10^3 cycles) was assumed to be 0.42 and the S_e (S_f at 10^6 cycles) was estimated at 10% S_{ut} based on data from Schechtman and Bader (1997) and Thornton, Schwab, and Oxland (2007).

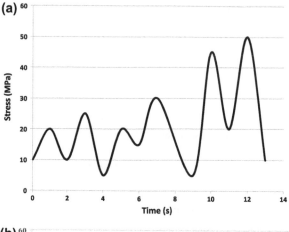

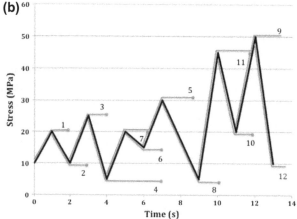

Figure 7. (a) Example of loading curve as might be experienced by a tendon. (b) Rainflow analysis defining half cycles for the loading sequence in (a).

in variable amplitude loading than measures such as maximum stress.

This finding has significant consequences in terms of assessing risk of MSDs in occupational settings. Obviously, it would seem important to evaluate the service load histories of musculoskeletal tissues when exposed to occupational (or non-occupational) stressors, as are commonly performed for bridges, aircraft wings and the like. Materials engineers have the benefit of being able to mount strain gages to critical components that need to be analysed for failure risk or design purposes. For the biomechanist, things are not so simple, especially when *in vivo* loads need to be ascertained. Instead, load histories will usually have to be estimated using biomechanical analysis and modelling techniques. Using biomechanical modelling techniques, estimates of the important variables of fatigue failure analysis can be reasonably estimated for musculoskeletal tissues, for example, the stress range and the mean stress. A technique sometimes used to quantify a service load history (in this instance, perhaps one workday of exposure) is to summarise the results of a rainflow analysis as in Table 2. As can be seen in this table, (half) cycles are broken down into their stress range/mean stress components and tabulated. If one can also estimate the US, fatigue strength fraction and endurance limit of the material, all of the necessary variables are available to estimate the accrual of cumulative damage.

In an ideal world, we would be able to assess the totality of loading on an individual. However, in the real world, it may be necessary to estimate a worker's cumulative loading on the basis of careful work sampling. This would not only include the main tasks performed, but inclusion of infrequent tasks as well. Methods such as Predetermined Time & Motion might be used to derive a statistically representative daily cumulative load. If job rotation were performed, each task would have to be analysed and cumulative loads from each task summed to determine the cumulative daily load.

Inspection of Table 1 provides several insights in terms of the damage associated with different portions of the loading sequence shown in Figure 7. One important insight from this analysis is the importance of the stress range in the development of material damage. As can be seen from this table, the two half cycles resulting in the vast majority of damage (86%) from this loading sequence are the ones possessing the greatest stress range. In fact, stress range can have a bigger impact on fatigue damage

Table 1. Results of rainflow analysis of loading cycles from Figure 7.

Reversal #	Starting Stress (MPa)	Ending Stress (MPa)	Stress Range (MPa)	Mean Stress (MPa)	Predicted N (Goodman)	Damage per cycle (1/N)	% of total damage
1	10	20	10	15	12860623	0.000000053	0.03
2	20	10	10	15	12860623	0.000000053	0.03
3	10	25	15	17.5	1582133	0.000000398	0.19
4	25	5	20	15	457361	0.000001478	0.71
5	5	30	25	17.5	135325	0.000004657	2.25
6	20	15	5	17.5	313226248	0.000000002	0.00
7	15	20	5	17.5	313226248	0.000000002	0.00
8	30	5	25	17.5	135325	0.000004657	2.25
9	5	50	45	27.5	4291	0.000110456	53.44
10	45	20	25	32.5	51509	0.000007965	3.85
11	20	45	25	32.5	51509	0.000007965	3.85
12	50	10	40	30	6388	0.000069026	33.39
Total						0.000206712	100.0

Table 2. Hypothetical example of the number of cycles at various stress range/mean stress combinations from a loading spectrum (subjected to rainflow analysis) as might be experienced in an occupational workday.

Stress range (MPa)	Mean stress (MPa)														
	0	2.5	5	7.5	10	12.5	15	17.5	20	22.5	25	27.5	30	32.5	35
5		539	370	201	152	94	80			42					
10		163	312	49	215	42					25				
15			148	95	427	233	142	37		13		17			
20					12	127	64	99	109	40	31				
25							12	72	49	63	25		19	12	
30								17		15	7	13			5
35												9	13		
40												6			
45													3		
50															

4.6. Individual characteristics and risk

An important precept of fatigue failure theory is that the effect of a given load *is indexed to the US of the exposed tissue.* It is important to recognise that each individual's tissue US will be unique, and variability between individuals may be immense. Given the same load and rate of repetition, different tissue strengths can lead to vastly different rates of damage. For example, a spinal load of 3 kN will cause more damage per cycle to a spine whose US is 6 kN than one of 12 kN. Of course, each individual's unique tissue strength profile will be heavily influenced by factors such as age, gender and anthropometry.

A common tactic of industry for physically demanding jobs is that of worker placement (e.g. getting younger, stronger individuals to work more physically demanding jobs). Designing such that 75% of females and 90% of males have the capability to perform a job should be protective of workers; however, it bears consideration that it may be overprotective for individuals with higher tissue strengths. These individuals may be able to operate quite safely at a higher absolute level of loading that the population-based design criteria would suggest. In this respect, it may be said that design according to population guidelines as above may lead to an unnecessarily strict constraint and a loss of work capacity for industry.

Design according to individual tolerances is not without peril, however, and would need to be very carefully managed. Even individuals with high tissue strength can be overloaded, and develop cumulative damage. However, if MSDs are the result of a fatigue failure process, individualised 'safe loads' may be possible and may make sense as a method of controlling MSDs without being unnecessarily restrictive. Of course, that is not to say that the approach above is to be preferred – ergonomic design of the workplace to eliminate the lift or otherwise reduce the spinal load is clearly a superior approach. However, in circumstances where it is difficult or cost prohibitive to implement an ergonomics fix, having workers with stronger tissues perform the more physically demanding jobs does make some sense from a fatigue failure perspective.

Moreover, the large differences in tissue strength strongly suggest that recommendations for controlling the incidence of MSDs should take age and gender issues into account. As individuals age, the US of musculoskeletal tissues will experience significant weakening and acceptable loads for repetitive activities should decrease correspondingly. It must also be recognised that large differences in tissue strengths are present between the genders, and acceptable loads for repetitive activities are going to differ substantially between males and females. A third individual characteristic that might have influence in terms of acceptable loads is anthropometry. Larger individuals will generally exhibit larger tissue size and strength, which would also significantly influence acceptable loads for repetitive tasks. Again, muscle strength may be a good surrogate measure for tissue strength, and may be useful to monitor from a design standpoint, i.e. designing jobs at some percentage of a maximum voluntary contraction (% MVC) may reduce fatigue failure development.

4.7. Maintenance of tissue homeostasis

In biological systems, both tissue damage and tissue repair processes are continually in progress. The key to prevention of MSDs is to try to ensure that the amount of damage accrued in tissues does not exceed the capacity of the repair mechanisms to heal. The unfortunate truth is that tissue damage can develop relatively rapidly, while the repair process is time-consuming process that can take weeks, months or years. Even when the repair process is complete, many types of tissues (such as tendons, ligaments and cartilage) never regain their original strength and/or tissue quality. Clearly, maintaining a modest degree of damage is critical to maintenance of tissue homeostasis.

Nash (1966) put forth a generalised model of fatigue failure for self-healing biological materials, which can be expressed as follows:

$$D(t) = D_D(t) + D_A(t) + D_S(t) - H(t) \qquad (8)$$

where $D(t)$ represents total tissue damage over a time frame, $D_D(t)$ is damage due to disease, $D_A(t)$ is damage due to aging, $D_s(t)$ is damage due to imposed stresses on tissues and $H(t)$ represents healing over the time frame of interest. If one is considering a time frame where disease is not present and the effect of aging is not significant, then this equation can be simplified:

$$D(t) = D_S(t) - H(t) \qquad (9)$$

where $D(t)$ represents total tissue damage over a time frame, $D_s(t)$ is damage due to imposed stresses on tissues and $H(t)$ represents healing over the time frame of interest. The term $D_s(t)$ can be estimated using the Palmgren–Miner rule, as described above. To do this one would have to know or estimate the ultimate strength of the tissue in question and be able to quantify the stress experienced by the tissue. One would then simply have to know the healing rate of the stressed tissue over time to quantify the total amount of cumulative damage accrual in the tissue.

Healing of tissues requires periods of unloading (rest) and some controlled loading so that damage accumulation can cease and repair mechanisms can have the time necessary to heal the tissue. But how much rest is needed? This is a question that does not yet have a clear answer. However, the model described above would suggest that the amount of rest necessary for healing would have to be related to the amount of damage incurred in some (as yet undetermined) manner. It might be possible, however, to examine the relationship of cumulative tissue loads (developed using fatigue failure techniques) with respect to the amount of rest available to determine whether certain ratios of cumulative loading to rest result in lower MSD rates, while others lead to increased risk.

Poorly vascularised tissues would be expected to be more susceptible to fatigue loading compared to highly vascularised tissues, due to the reduced availability of the biochemical and nutritional elements required for healing. It is worth noting that most of the tissues associated with MSDs are generally poorly perfused and slower to heal, likely permitting increased damage accumulation as the tissue labours to heal.

4.8. Research needs

If a fatigue failure process is involved with the development of MSDs, there may be great opportunity to improve our understanding of the aetiology of these disorders, to better assess risk and to develop more effective interventions. Doing so will require a significant investment in research to acquire the knowledge and techniques necessary to drive this effort forward. The research implications

are substantial, and a few of these are considered below. In general, there are three main areas where research is needed: (1) improved characterisation of the US and fatigue life of musculoskeletal tissues; (2) accurate determination of stresses experienced at the tissue level; and (3) improved characterisation of the dynamic properties of the musculoskeletal system, specifically tissue healing, remodelling and atrophy.

In an ideal world, the risk of cumulative tissue damage for an individual could be accurately predicted, given knowledge of the strength of the tissues being stressed and the magnitude, distribution and frequency of loading on these tissues. First, we would have knowledge of the US (*in vivo*) of the tissue being stressed, and would have precise knowledge of the distribution of tissue strain due to repetitive loading experienced by the tissue. Further, we would have accurate knowledge of the healing rate and capabilities of the involved tissue for the individual in question, and would understand the proper balance in terms of activity and rest for healing to be optimised. We would understand how adoption of different postures would change the stress distribution in the tissues and areas of stress concentration. We would understand the effects of ageing and relevant disease processes on these tissues. The impact of prior tissue damage would be understood. The understanding of how psychological stresses impact tissue loading and healing would also be fully comprehended.

It may be possible that technology eventually will provide methods to better understand some of these issues, though for some aspects, it may be in a rather distant future. However, we are currently missing some rather basic information necessary to evaluate the fatigue failure process *in vivo*. The following sections describe some of the areas where research is needed to better understand fatigue failure processes in the human body, which will hopefully lead not only to better design of jobs, but better overall musculoskeletal health through life.

4.8.1. Improved characterisation of musculoskeletal tissue properties

Data on fatigue failure of musculoskeletal tissues remain relatively sparse, and a much greater exploration of the responses of musculoskeletal tissues is warranted. As an example, there is scant evidence evaluating long-cycle fatigue (>10,000 loading cycles) in spinal motion segments. An improved understanding of long-cycle responses would be important in developing the *S–N* curves that define these tissue responses. Effects of load rate, variable loading amplitudes, damage nucleation and propagation, and fatigue of aging tissues are not sufficient. Viscoelastic responses are still not well understood, particularly the effects of variable load rates during a loading process. Furthermore, developing an improved understanding of load sharing in tissues would be helpful, and how and where stress concentrations develop and how these are affected by changes in posture.

Statistical variability is an important issue with respect to fatigue testing, and using an appropriate sample size is a critical aspect of developing *S–N* curves. The study by Schechtman and Bader (1997) is instructive as a model of how the relationship between loading and cycles to failure is necessary for constructing an *S–N* curve. As these authors demonstrate, it may necessary to measure the fatigue life of 10–20 specimens at each of 10 or so load levels to define the *S–N* curve properly. Obviously, development of a full *S–N* curve is tedious; however, understanding the fatigue responses of musculoskeletal tissues involved with MSDs is critical to our understanding and to MSD prevention efforts. For many tissues involved with specific MSDs, we have little to no data at all. In ligaments, for example, we have only found only two studies of fatigue failure – one in the human anterior cruciate ligament (Lipps, Wojtys, and Ashton-Miller 2013) and one in the rabbit medial collateral ligament (Thornton, Schwab, and Oxland 2007). Further research in this area is necessary.

Furthermore, it should be recognised that damage to viscoelastic tissues can result from both fatigue failure and creep loading. It has been demonstrated that the former will lead to more rapid tissue damage than the latter (Thornton, Schwab, and Oxland 2007); however, it should be recognised that damage is likely the result of the combination of creep and fatigue loading. The nature of the relationship between these two loading modalities in terms of damage accrual is unknown at the current time, and deserves attention in future research studies.

4.8.2. Estimating tissue strength in vivo

Understanding tissue strength is a critical piece of information in assessing MSD risk. Fortunately, there are many technologies that may be very helpful to estimate tissue strengths *in vivo*. An example is dual-energy X-ray absorptiometry, which can provide information on bone densitometry that may be extremely helpful in understanding the risk of vertebral end-plate fractures in the spine. Bone mineral content has been shown to be highly correlated with the ultimate strength (Biggemann, Hilweg, and Brinckmann 1988) and fatigue life (Gallagher et al. 2007) of spinal motion segments *in vitro*, a relationship that is surely present *in vivo* as well. This may be helpful when comparing demands of the job to worker capacity and establishing risk. Other useful imaging techniques include MRI and ultrasound that can help establish the dimensions and estimated strength of tissues important in the determination of stress.

Other methods may also be useful in aiding our understanding of tissue strength. For example, isometric muscle strength appears to be one of the best metrics of recovery from injury (Prasartwuth, Taylor, and Gandevia 2005). Stronger muscle strength implies stronger tissues throughout the system – muscles, tendons and bone. Changes in muscle strength throughout life undoubtedly reflect changes in the integrity of musculoskeletal tissues as a whole. As a result, muscle strength may be a reasonable method with which to estimate tissue strength of individuals, and it may be possible to prevent fatigue failure in tissues.

4.8.3. Understanding dynamic properties of the musculoskeletal system

The unique physiological setting of fatigue failure for musculoskeletal tissues *in vivo* complicates an already complex process. Musculoskeletal tissues are highly dynamic, and can respond relatively quickly to changing demands. The ability of biological tissues to self-heal is a unique aspect of the fatigue failure process in humans that needs to be better understood with respect to its role in MSD risk. In particular, the relationship of rest necessary to heal various amounts of cumulative damage must be better studied. There is a baseline rate of tissue healing (not well quantified presently) that occurs every day and which is apparently able to keep pace with minor damage that occurs when loading is not excessively forceful and repetitive. When loading on tissues leads to a substantial increase in cumulative damage, however, the healing process accelerates. This acceleration, however, varies dramatically between tissues that are highly perfused and those with poor levels of perfusion. Highly perfused tissues such as muscle and bone can heal relatively quickly when a minor amount of cumulative damage is experienced. However, low metabolism tissues such as tendon, ligament and cartilage do not heal as rapidly. Mechanical loading that results in tendon overuse injury can initiate a repair process but, after failed initial repair, non-resolving chronic attempted repair appears to lead to a 'smouldering' fibrogenesis (Thornton and Hart 2011).

4.8.4. Beneficial versus detrimental tissue loading

Musculoskeletal tissues must experience stress to maintain healthy function. This is abundantly clear from the experience of astronauts living in microgravity. Studies have demonstrated that astronauts experience up to a 20% loss of muscle mass on spaceflights lasting five to 11 days, and may lose up to 2% of their bone density per month if bone and muscle are not purposefully stressed (NASA 2001, 2015). Although no other working population is subjected to the dramatic environmental conditions of space, atrophy of tissues will also occur in normal gravity if individuals are not active (though at a less rapid pace).

It is also clear that a limited number of high-stress exertions, combined with sufficient rest, can lead to significant gains in strength of tissues, including muscle, tendons and ligaments. For example, Arampatzis, Karamanidis, and Albracht (2007) showed increased Achilles tendon stiffness during a 14-week exercise protocol involving strain of 4.5%, but not for exercise involving strain of 3%, though exercise frequency and volume were equal. Kongsgaard et al. (2007) also demonstrated increased tendon stiffness (in the patellar tendon) with high-resistance load, but not for a light resistance regimen. These data suggest that a certain loading magnitude threshold must be exceeded to elicit an anabolic tissue response. However, the difference between conditions of high-magnitude stress that lead to anabolic responses and those leading to catabolic tissue changes appears not to be tremendously different (Heinemeier and Kjaer 2011). Understanding these thresholds is an important research issue and poorly understood presently. It would be very useful to increase our understanding of conditions leading to optimal musculoskeletal tissue health compared to those that lead to progressive accumulation of damage.

Furthermore, we need to better understand the negative and positive impacts of loading during the tissue repair process. When tissue becomes damaged, the damaged region becomes an area of stress concentration in the tissue and, therefore, vulnerable to additional fatigue damage even in the presence of lower magnitude loads (Gallagher and Heberger 2013). However, it has been shown that controlled loading during the healing process (for example, in tendons) can help to improve the synthesis and alignment of collagen fibres, leading to an improved repair outcome (Kellett 1986). Thus, tissue loading can be both a hindrance and a benefit in the tissue repair process. The timing and magnitude of loading that should be experienced in the repair process to (1) decrease damage and/or (2) facilitate healing are poorly known presently but important to the preservation of optimum musculoskeletal tissue health.

4.8.5. Risk assessment in epidemiological studies

Past epidemiological studies have not examined MSD risk in a manner congruent with the precepts of fatigue failure. If MSDs are indeed the result of a fatigue failure process, important modifications would be needed with respect to how physical MSD risk factors are assessed. For example, it will be important to evaluate repetition as a function of force level experienced by participants, due to the highly variable impact of repetition at different force levels. Current tools used to assess force and repetition do not appear to appropriately weight the impact of repetition as force levels vary. Nor have important aspects such as the mean stress experienced by tissues been evaluated.

Furthermore, as mentioned above, fatigue failure theory would indicate that a key element of risk assessment is evaluation of the load experienced by an individual relative to their tissue strength, which may be highly influenced by age, gender and anthropometry.

The development of new technologies holds great promise in helping to better quantify the physical exposures of work and may prove extremely beneficial in terms of evaluating MSD risk from a fatigue failure perspective. Wireless wearable motion sensors, for example, are innovative, unobtrusive devices that may be combined with force-sensing technologies (e.g. pressure sensing insoles or gloves, force plates) to obtain biomechanical loading estimates for joints of interest (Faber et al. 2016; Kim and Nussbaum 2014). Wireless electromyography systems, pressure mapping systems and other miniature force and torque measurement products may also be useful in this regard. Furthermore, new video-based technologies and advanced digital human modelling capabilities have shown considerable promise in assessing the data necessary for fatigue failure-based analysis (Chaffin 2005; Chen et al. 2015).

Despite recent advancements, additional research is needed to further improve and evaluate these emerging technologies, particularly for use in dynamic work environments. While several wireless wearable motion sensor systems have been observed to exhibit good accuracy and stability in both field and laboratory settings (e.g. Bauer et al. 2015; Kim and Nussbaum 2013; Schall et al. 2015), the accuracy of these devices has been known to degrade when work activities involve complex, dynamic motions (Brodie, Walmsley, and Page 2008; Godwin, Agnew, and Stevenson 2009) or when measurements are taken in the presence of magnetically distorted fields (Schiefer et al. 2014). Sensor fusion algorithms such as Kalman filters represent one approach to improving the accuracy of motion sensor devices under variable conditions (Bergamini et al. 2014; Ligorio and Sabatini 2015). However, systematic evaluations of the effects of dynamic motion and magnetic distortion (and their interaction) on the accuracy of these technologies are still necessary to further improve estimates of workplace exposure (e.g. Pasciuto et al. 2015). Moreover, evaluation of these technologies for use on complex body segments (such as the wrist) are needed to more effectively study prevalent musculoskeletal conditions that may be a result of a fatigue failure process such as CTS.

"Efficient estimation of the physical demands of work remains somewhat limited by the need for multiple sensors" (Schall, Fethke, and Chen 2016, 107). The creation of additional technologies capable of simultaneously capturing several components of work would allow workers to move more naturally, thereby improving estimates of workplace exposure while further increasing the cost-efficiency of direct measurement (Trask et al. 2014). Moreover, the development of standardised, non-proprietary metrics and procedures for using these new technologies is needed (e.g. Faber et al. 2013; Palermo et al. 2014). Conversion algorithms intended to relate or synthesise workload estimates from various studies may also be useful for efficiently evaluating MSD risk and relating it to fatigue failure theory.

5. Summary

All materials (including biomaterials) have been demonstrated to incur damage via the process of fatigue failure. Recent evidence strongly suggests that a fatigue failure process may be etiologically significant in the development of MSDs. However, as of this writing, the implications of an underlying (and potentially causal) fatigue failure process in MSD development have generally not been considered in prior epidemiological studies, MSD risk assessment tools or MSD prevention strategies.

If this evidence is correct, there are many important implications that need to be considered. These include understanding important interactions between MSD risk factors, the ability to develop improved cumulative loading estimates on tissues, the importance of individual characteristics and MSD risk and perhaps improved understanding of the relationship between tissue damage and healing. The authors hope that the concept that MSDs may be caused (at least in part) by a process of fatigue failure may provide fertile ground for research in the quest to reduce the pain and disability associated with these burdensome health conditions.

References

AAOS [American Academy of Orthopaedic Surgeons]. 2008. The Burden of Musculoskeletal Diseases in the United States: Prevalence, Societal, and Economic Cost. Executive Summary. Rosemont, IL: AAOS.

Andarawis-Puri, N., and E. L. Flatow. 2011. "Tendon Fatigue in Response to Mechanical Loading." Journal of Musculoskeletal and Neuronal Interactions 11: 106–114.

Arampatzis, A., K. Karamanidis, and K. Albracht. 2007. "Adaptational Responses of the Human Achilles Tendon by Modulation of the Applied Cyclic Strain Magnitude." Journal of Experimental Biology 210 (Part 15): 2743–2753.

Armstrong, T. J., and D. B. Chaffin. 1979. "Some Biomechanical Aspects of the Carpal Tunnel." Journal of Biomechanics 12: 567–570.

Ashby, M. F., H. Shercliff, and D. Cebon. 2010. Materials: Engineering, Science, Processing, and Design. 2nd ed. Oxford: Elsevier Butterworth Heinemann.

ASTM [American Society for Testing and Materials]. 2000. Standard Terminology Relating to Fatigue and Fracture Testing. ASTM Designation E1823, Vol. 03.01. West Conshoshoken, PA: ASTM; 1034.

Barbe, M., S. Gallagher, V. Massicotte, M. Tytell, S. Popoff, and A. Barr-Gillespie. 2013. "The Interaction of Force and Repetition on Musculoskeletal and Neural Tissue Responses and Sensorimotor Behavior in a Rat Model of Work-Related Musculoskeletal Disorders." BMC Musculoskeletal Disorders 14: 1–26.

Bauer, C. M., F. M. Rast, M. J. Ernst, J. Kool, S. Oetiker, S. M. Rissanen, J. H. Suni, and M. Kankaanpää. 2015. "Concurrent Validity and Reliability of a Novel Wireless Inertial Measurement System to Assess Trunk Movement." Journal of Electromyography and Kinesiology 25 (5): 782–790.

Bellucci, G., and B. B. Seedhom. 2001. "Mechanical Behaviour of Articular Cartilage under Tensile Cyclic Load." Rheumatology 40 (12): 1337–1345.

Bergamini, E., G. Ligorio, A. Summa, G. Vannozzi, A. Cappozzo, and A. M. Sabatini. 2014. "Estimating Orientation Using Magnetic and Inertial Sensors and Different Sensor Fusion Approaches: Accuracy Assessment in Manual and Locomotion Tasks." Sensors 14 (10): 18625–18649.

Biggemann, M., D. Hilweg, and P. Brinckmann. 1988. "Prediction of the Compressive Strength of Vertebral Bodies of the Lumbar Spine by Quantitative Computed Tomography." Skeletal Radiology 17: 264–269.

Brinckmann, P., M. Biggemann, and D. Hilweg. 1988. "Fatigue Fracture of Human Lumbar Vertebrae." Clinical Biomechanics 3 (Suppl. 1): S1–S23.

Brodie, M. A., A. Walmsley, and W. Page. 2008. "Dynamic Accuracy of Inertial Measurement Units during Simple Pendulum Motion." Computer Methods in Biomechanics and Biomedical Engineering 11 (3): 235–242.

BLS [Bureau of Labor Statistics]. 2015. "Nonfatal Occupational Injuries and Illnesses Requiring Days Away from Work, 2014." November 19, 2015. Accessed January 12, 2016.

Chaffin, D. B. 2005. "Improving Digital Human Modelling for Proactive Ergonomics in Design." Ergonomics 48 (5): 478–491.

Chaffin, D. B., G. B. J. Andersson, and B. J. Martin. 1999. Occupational Biomechanics. 3rd ed. New York: John Wiley and Sons.

Chen, C. H., D. P. Azari, Y. H. Hu, M. J. Lindstrom, D. Thelen, T. Y. Yen, and R. G. Radwin. 2015. "The Accuracy of Conventional 2D Video for Quantifying Upper Limb Kinematics in Repetitive Motion Occupational Tasks." Ergonomics 58 (12): 2057–2066.

Cyron, B. M., and W. C. Hutton. 1978. "The Fatigue Strength of the Lumbar Neural Arch in Spondylolysis." Journal of Bone and Joint Surgery (British) 60-B, 234–238.

Downing, S. D., and D. F. Socie. 1982. "Simple Rainflow Counting Algorithms." International Journal of Fatigue 4 (1): 31–40.

Faber, G. S., C. C. Chang, I. Kingma, J. T. Dennerlein, and J. H. van Dieën. 2016. "Estimating 3D L5/S1 moments and ground reaction forces during trunk bending using a full-body ambulatory inertial motion capture system." Journal of biomechanics 49: 904–912.

Faber, G. S., C. C. Chang, P. Rizun, and J. T. Dennerlein. 2013. "A Novel Method for Assessing the 3-D Orientation Accuracy of Inertial/Magnetic Sensors." Journal of Biomechanics 46 (15): 2745–2751.

Gallagher, S., and J. R. Heberger. 2013. "Examining the Interaction of Force and Repetition on Musculoskeletal Disorder Risk: A Systematic Literature Review." Human Factors: The Journal of the Human Factors and Ergonomics Society 55: 108–124.

Gallagher, S., W. S. Marras, A. S. Litsky, and D. Burr. 2005. "Torso Flexion Loads and the Fatigue Failure of Human Lumbosacral Motion Segments." Spine 30: 2265–2273.

Gallagher, S., W. S. Marras, A. S. Litsky, D. Burr, J. Landoll, and V. Matkovic. 2007. "A Comparison of Fatigue Failure Responses of Old versus Middle-Aged Lumbar Motion Segments in Simulated Flexed Lifting." *Spine* 32: 1832–1839.

Gerber, H. 1874. "Bestimmung derzulässigen Spannungen in Eisenkonstructionen." *Zeitschrift des Bayerischen Architeckten und IngenieurVereins* 6: 101–110.

Godwin, A., M. Agnew, and J. Stevenson. 2009. "Accuracy of Inertial Motion Sensors in Static, Quasistatic, and Complex Dynamic Motion." *Journal of Biomechanical Engineering* 131 (11): 114501.

Goodman, J. 1899. *Mechanics Applied to Engineering*. London: Longmans Green.

Harris-Adamson, C., E. A. Eisen, J. Kapellusch, A. Garg, K. T. Hegmann, M. S. Thiese, A. M. Dale, B. Evanoff, S. Burt, S. Bao, B. Silverstein, L. Merlino, F. Gerr, and D. Rempel. 2015. "Biomechanical Risk Factors for Carpal Tunnel Syndrome: A Pooled Study of 2474 Workers." *Occupational and Environmental Medicine* 72 (1): 33–41.

Heinemeier, K. M., and M. Kjaer. 2011. "In vivo Investigation of Tendon Responses to Mechanical Loading." *Journal of Musculoskeletal and Neuronal Interaction* 11 (2): 115–123.

Horton, R. 2010. "Understanding Disease, Injury, and Risk." *Lancet* 2012 (380): 2053–2054.

Kellett, J. 1986. "Acute Soft Tissue Injuries – A Review of the Literature." *Medicine and Science in Sports and Exercise* 18: 489–500.

Kim, S., and M. A. Nussbaum. 2013. "Performance Evaluation of a Wearable Inertial Motion Capture System for Capturing Physical Exposures during Manual Material Handling Tasks." *Ergonomics* 56 (2): 314–326.

Kim, S., and M. A. Nussbaum. 2014. "Evaluation of Two Approaches for Aligning Data Obtained from a Motion Capture System and an in-Shoe Pressure Measurement System." *Sensors* 14 (9): 16994–17007.

Kongsgaard, M., S. Reitelseder, T. G. Pedersen, L. Holm, P. Aagaard, M. Kjaer, and S. P. Magnusson. 2007. "Region Specific Patellar Tendon Hypertrophy in Humans following Resistance Training." *Acta Physiologica* 191 (2): 111–121.

Ligorio, G., and A.M. Sabatini. 2015. "A Linear Kalman Filtering-Based Approach for 3D Orientation Estimation from Magnetic/Inertial Sensors." Paper presented at the Multisensor Fusion and Integration for Intelligent Systems (MFI), 2015 IEEE International Conference, San Diego, CA, 77–82.

Lipps, D. B., J. A. Ashton-Miller, and E. M. Wojtys. 2014. "Morphological Characteristics Help Explain the Gender Difference in Peak Anterior Ligament Strain during a Simulated Pivot Landing." *American Journal of Sports Medicine* 40 (1): 32–40.

Lipps, D. B., E. M. Wojtys, and J. A. Ashton-Miller. 2013. "Anterior Cruciate Ligament Fatigue Failures in Knees Subjected to Repeated Simulated Pivot Landings." *American Journal of Sports Medicine* 41: 1058–1066.

Mathiassen, S. E. 2006. "Diversity and Variation in Biomechanical Exposure: What is it, and Why Would we like to Know?" *Applied Ergonomics* 37 (4): 419–427.

Mathiassen, S. E., T. Möller, and M. Forsman. 2003. "Variability in Mechanical Exposure within and between Individuals Performing a Highly Constrained Industrial Work Task." *Ergonomics* 46 (8): 800–824.

Matsuishi, M., and T. Endo. 1968. "Fatigue of Metals Subjected to Varying Stress." Paper presented to the Japanese Society of Mechanical Engineers, Fukuoka, Japan, March.

Meyer, D. L. 1991. "Misinterpretation of interaction effects: A reply to Rosnow and Rosenthal." *Psychological Bulletin* 110: 571–573.

Miner, M. A. 1945. "Cumulative Damage in Fatigue." *Journal of Applied Mechanics* 67: A159–A164.

Nachemson, A. 1976. "The Lumbar Spine: An Orthopaedic Challenge." *Spine* 1: 59–71.

Nash, C. D. 1966. *Fatigue of Self-Healing Structure: A Generalized Theory of Fatigue Failure (ASME Publication 66-W a/BHF-3)*. New York: American Society of Mechanical Engineers.

NASA [National Aeronautics and Space Administration]. 2001. "Space Bones." NASA Science News. October 1, 2001. Accessed December 30, 2015.

NASA [National Aeronautics and Space Administration]. 2015. "Muscle Atrophy." NASA Information. Accessed December 30, 2015.

NIOSH [National Institute for Occupational Safety and Health]. 1997. *Musculoskeletal Disorders and Workplace Factors: A Critical Review of Epidemiologic Evidence for Work-Related Musculoskeletal Disorders of the Neck, Upper Extremity and Low Back*. Washington, DC: US Govt. Printing Office.

NRC-IOM [National Research Council & Institute of Medicine]. 2001. *Musculoskeletal Disorders and the Workplace: Low Back and Upper Extremities*. Washington, DC: National Academy Press.

Norman, R., R. Wells, P. Neumann, J. Frank, H. Shannon, M. Kerr, and the Ontario Universities Back Pain Study (OUBPS) Group. 1998. "A Comparison of Peak Vs Cumulative Physical Work Exposure Risk Factors for the Reporting of Low Back Pain in the Automotive Industry." *Clinical Biomechanics* 13: 561–573.

Palermo, E., S. Rossi, F. Marini, F. Patanè, and P. Cappa. 2014. "Experimental Evaluation of Accuracy and Repeatability of a Novel Body-to-Sensor Calibration Procedure for Inertial Sensor-Based Gait Analysis." *Measurement* 52: 145–155.

Palmgren, A. 1924. "Die Lebensdauer von Kugellagern." *Zeitschrift Des Vereins Deutscher Ingenieure* 68: 339–341.

Pasciuto, I., G. Ligorio, E. Bergamini, G. Vannozzi, A. M. Sabatini, and A. Cappozzo. 2015. "How Angular Velocity Features and Different Gyroscope Noise Types Interact and Determine Orientation Estimation Accuracy." *Sensors* 15 (9): 23983–24001.

Peterson, R. E. 1950. "Discussion of a Century Ago concerning the Nature of Fatigue, and Review of Some of the Recent Researches concerning the Mechanism of Fatigue." *ASTM Bulletin* 164: 50–56.

Prasartwuth, O., J. L. Taylor, and S. C. Gandevia. 2005. "Maximal Force, Voluntary Activation and Muscle Soreness after Eccentric Damage to Human Elbow Flexor Muscles." *Journal of Physiology* 567: 337–348.

Punnett, L., L. J. Fine, W. M. Keyserling, G. D. Herrin, and D. B. Chaffin. 1991. "Back Disorders and Nonneutral Trunk Postures of Automobile Assembly Workers." *Scandinavian Journal of Work, Environment & Health* 17: 337–346.

Punnett, L., A. Pruss-Ustun, D. I. Nelson, M. A. Fingerhut, J. L. Leigh, S. Tak, and S. Phillips. 2005. "Estimating the Global Burden of Low Back Pain Attributable to Combined Occupational Exposures." *American Journal of Industrial Medicine* 48: 459–469.

Roylance, D. 2001. *Fatigue*. Accessed December 20, 2015. http://ocw.mit.edu/courses/materials-science-and-engineering/3-11-mechanics-of-materials-fall-1999/modules/fatigue.pdf

Sarkani, L. D., and S. Lutes. 2004. *Random Vibrations Analysis of Structural and Mechanical Systems* ([Online-Ausg.] ed.). Amsterdam: Elsevier.

Schechtman, H., and D. L. Bader. 1997. "In vitro Fatigue of Human Tendons." *Journal of Biomechanics* 30: 829–835.

Schall, M. C., N. B. Fethke, and H. Chen. 2016. "Evaluation of Four Sensor Locations for Physical Activity Assessment." *Applied Ergonomics* 53: 103–109.

Schall Jr, M. C., N. B. Fethke, H. Chen, S. Oyama, and D. I. Douphrate. 2015. "Accuracy and Repeatability of an Inertial Measurement Unit System for Field-Based Occupational Studies." *Ergonomics* 10 (8): 10–23.

Schiefer, C., R. P. Ellegast, I. Hermanns, T. Kraus, E. Ochsmann, C. Larue, and A. Plamondon. 2014. "Optimization of Inertial Sensor-Based Motion Capturing for Magnetically Distorted Field Applications." *Journal of Biomechanical Engineering* 136 (12): 121008.

Sharma, P., and N. Maffulli. 2005. "Tendon Injury and Tendinopathy: Healing and Repair." *Journal of Bone and Joint Surgery (American)* 87 (1): 187–202.

Solomonow, M., R. V. Baratta, A. Banks, C. Freudenberger, and B. H. Zhou. 2003. "Flexion–Relaxation Response to Static Lumbar Flexion in Males and Females." *Clinical Biomechanics* 18: 273–279.

Stephens, R. I., A. Fatemi, R. R. Stephens, and H. O. Fuchs. 2001. *Metal Fatigue in Engineering*. New York: John Wiley and Sons.

Thornton, G. M., and D. A. Hart. 2011. "The Interface of Mechanical Loading and Biological Variables as They Pertain to the Development of Tendinosis." *Journal Musculoskeletal and Neuronal Interaction* 11 (2): 94–105.

Thornton, G. M., T. D. Schwab, and T. R. Oxland. 2007. "Cyclic Loading Causes Faster Rupture and Strain Rate than Static Loading in Medial Collateral Ligament at High Stress." *Clinical Biomechanics* 22: 932–940.

Trask, C., S. E. Mathiassen, J. Wahlstrom, and M. Forsman. 2014. "Cost-Efficient Assessment of Biomechanical Exposure in Occupational Groups, Exemplified by Posture Observation and Inclinometry." *Scandinavian Journal of Work Environment & Health* 40 (3): 252–265.

Wang, X. T., R. Ker, and M. Alexander. 1995. "Fatigue Rupture of Wallaby Tail Tendons." *The Journal of Experimental Biology* 198: 847–852.

The field becomes the laboratory? The impact of the contextual digital footprint on the discipline of E/HF

Sarah Sharples ⓘD and Robert J. Houghton ⓘD

ABSTRACT

The increasing prevalence of affordable digital sensors, ubiquitous networking and computation puts us at what is only the start of a new era in terms of the volume, coverage and granularity of data that we can access about individuals and workplaces. This paper examines the consequences of harnessing this data deluge for the practice of E/HF. Focusing on what we term the 'contextual digital footprint', the trail of data we produce through interactions with many different digital systems over the course of even a single day, we describe three example scenarios (drawn from health care, distributed work and transportation) and examine how access to data directly drawn in considerable volume from the field will potentially change our application of design and evaluation methods. We conclude with a discussion of issues relevant to ethical and professional practice within this new environment including the increased challenges of respecting anonymity, working with n = all data-sets and the central role of ergonomists in promulgating positive uses of data while retaining a systems-based humanistic approach to work design.

Practitioner Summary: The paper envisions the impact of new and emerging sources of data about people and workplaces upon future practice in E/HF. We identify practical consequences for ergonomics practice, highlight new areas of professional competence likely to be required and flag both the risks and benefits of adopting a more data-driven approach.

Introduction

As an applied science, Ergonomics and Human Factors (E/HF) has traditionally used data obtained from experimental and real-world settings to inform our understanding of the way in which people work, and thus the way in which we should design work systems, technologies and environments. While there has often been vigorous debate regarding what form these data should take and how and where it is collected and analysed (see Wilson and Sharples 2015a), a general consensus exists that good ergonomics will tend to take a strong focus on investigating the domain of interest and collecting relevant information about activities within that domain, whether in the field or perhaps through relevant laboratory work. Commensurate with this, considerable effort within the discipline has been devoted to the development of an extensive range of data collection and analysis methods (e.g. Salvendy 2012; Stanton et al. 2013; Wilson and Sharples 2015b).

It is well-recognised, and indeed, considered a matter of pride, that as new technologies have appeared and society itself has changed through aspects such as increased automation, the appearance of the service industry sector,

globalisation or the emergence of the environmental sustainability agenda, E/HF has responded by extending the scope and nature of its domain interests. However, it is perhaps less often noted that these same changes have also considerably altered the practice of the discipline itself (although see Moray 2008). It is clear that, for example, the availability of desktop computers has radically changed the ease with which E/HF laboratory experiments can be undertaken; similarly, advances in visualisation and communications technology, as well as development of advanced data analysis tools, has put complex simulations and statistical analyses within the reach of nearly all practitioners.

Today, the most pervasive changes in technology and perhaps society centre on the emergence of the practical implications of widespread networked computation (National Research Council 2014). The advent of mobile and ubiquitous technologies and novel, embedded sensing technologies, alongside distributed data storage, has contributed to the development of the concept of the 'contextual digital footprint'. The contextual digital footprint can be described as the data which we produce throughout our everyday lives, and represents a 'cradle to grave' collection of explicitly and implicitly produced data about ourselves, our families, our interactions, thoughts, behaviours and work. This footprint is a construct that describes a wide range of current and future forms of data collection that may or may not be discoverable by any given individual; as such it defies further formal definition. However, early work into the properties of data-sets containing information that can be tied to given individuals has demonstrated it typically has certain characteristics, including for example sparsity and higher dimensionality (that is, individuals can be uniquely identified across related data-sets on the basis of specific features, so-called 'jigsaw identification', see Brynjolfsson, Hu, and Smith 2003; Narayanan and Shmatikov 2008). This paper considers the implications of the contextual digital footprint for E/HF science and practice. Up until now, the focus of research into the contextual digital footprint has been conducted primarily within the fields of computer science and human computer interaction. However, the world of business and industry is increasingly becoming aware of and investing in the concept, reflected in initiatives such as customer loyalty schemes or data-driven approaches to personnel management. We consider which types of data are of relevance for E/HF design and analysis; the characteristics of that data and the way in which it is produced, collected and stored of which we need to be aware; and the ethics of using the contextual digital footprint, considering utopian and dystopian views of the practice, and thus how we as E/HF practitioners should embark on responsible use of this data to design effective and safe work systems. Whether this constitutes a new paradigm within E/HF is open to debate. One might argue that E/HF has always been a data-driven discipline and that consequently simply having more data represents merely a change of *degree* in practice rather than a change of *kind*. In the present paper, we make the case that when data about workers and work environments come to exist in sufficient volume, and are increasingly ubiquitous, there is the potential for significant changes not only in the work we study, but also in how we study it and the depth of understanding that can be achieved. At the same time, however, we will also argue that whether we use the digital footprint to 'do better things' or merely 'do things better', this will be done most effectively not by rejecting the values that already guide our practice, but rather by reacquainting ourselves with what our purposes and values actually are.

Contextual digital data in our work and lives

Data exist and are produced throughout our lives. In our personal lives, we might refer to major life events on social media, so that data are stored about when and where we were born, where we went to school and who the members are of our social networks. We might share our location, distributing information about our travelling preferences, choice of leisure activities or consumer selections. We might also likely interact with commercial systems such as shops or online banks. The 'Future Identities' Foresight report by Government Office For Science 2013 refers to the 'wealth of personal data which can be mined'.

In our working lives, data about our movements may be collected or sensed to influence the temperature of the buildings we work in, allow us entry to secure areas of work, or provide us with access to IT such as printers. In addition, formal records of our working lives will record our continuous personal development, safety training, pay level and sickness record. Other formal service providers such as our doctors' surgeries, utility companies and transport organisations will also hold information about our health, habits and behaviours.

We term these data 'contextual digital data'. These data are both rich and imperfect. They represent a tremendous set of business opportunities, and are already used in some forms to support applications such as personalised marketing campaigns. There are also examples of the creeping unification of these data sources – for instance, there have been debates about whether social media information should be referred to when considering an individual for a new employment role, and court cases have referred to social media sources when determining a person's eligibility for disability benefits (The Guardian 2012).

As E/HF specialists, we are concerned with how systems that we use in our work and lives are designed in order to ensure comfort, satisfaction, usability and effectiveness. Beyond the wider ethical, policy and privacy debate, we should consider the implications of the existence, use and value of these data for E/HF methods and practice.

In order to support the consideration of these implications, we present three example scenarios in which the contextual digital footprint is of relevance to E/HF practice and interventions. The aim of these descriptive scenarios is: (a) to highlight some different circumstances in which we might encounter contextual digital data, and consider the different technologies that both currently and in the future will enable this data to be collected, stored and interpreted; (b) to provide a basis from which the positive and negative aspects of using contextual digital data to support E/HF analysis can be identified.

The three scenarios are:

- Situated work: An example of where contextual digital data can be used in a confined workplace. The selected example workplace is a hospital, and the collection of data about clinical work using a range of sensing and systems technologies is presented.

- Distributed work: An example where contextual digital data is used in a work setting where employees are geographically dispersed. The selected example is of crowdsourced human computation (HC), where individuals contribute to an overall task by interacting with separate tasks and systems.
- User experience: An example where contextual digital data is used to capture, manage and enhance the experience of users. The selected context for this is travel, where the user will interact with a number of different official and unofficial systems to help them manage and enjoy their journey.

Example 1: Situated work – health care

Many jobs in many workplaces have routinely been monitored or logged. From air traffic control strips recording agreed changes to flight paths, to voice communications in rail control being recorded, it has been accepted as being reasonable (from the employee perspective) and valuable (from the employer and legislative perspectives) to record many types of decisions and actions at work. The introduction of technologies such as efficient data storage, mobile smartphones, location tracking technologies and movement sensors means that the extent to which it is now possible to monitor and record a wide range of aspects of work is vast.

A particular context in which this is considered is the out-of-hours care hospital setting. Around 75% of time in hospitals is classified as 'out of hours' care, where a small number of clinicians, many of them often quite junior, are responsible for patients on a range of medical wards. Often these wards are geographically distributed over a wide area. Until recently, if a patient needed to be attended by a clinician, the clinician would be alerted via a mobile pager device. On receipt of this short message, the clinician would speak on the telephone to a nurse coordinator, who would relay the message about the patient, including details of their condition and location. This voice and pager system has in many hospitals now been replaced by a task allocation system that uses smartphones to send and display details of tasks to doctors.

These smartphones link with a system (Blakey et al. 2012) that records the number of tasks performed in a shift. As a doctor is required to 'accept' a task, there is also normally an indication of the current task that is being undertaken. However, a single task, such as replacing a catheter, could actually involve a number of distinct actions that are completed in different parts of a ward. Developments in location-based technologies, both in

advanced and discrete sensors that can be worn on the person, mean that the task allocation software can now be combined with location tracking to increase the amount of detail collected about tasks that are completed, building on the knowledge about *what* tasks are done, to also consider *where* they are done and *when*.

The above technology is being implemented in a basic form for research purposes now (Brown et al. 2014). It is reasonable to assume that the capability of these technologies will increase (e.g. wifi coverage will become more reliable, location tracking more accurate) and new technologies will have the potential to be introduced into this context.

Therefore in the future, in addition to detailed monitoring and tracking of movements around a hospital, we may also be able to monitor, in real time, physiological indices of clinicians, which could provide indications of when a doctor is becoming highly stressed or fatigued for example. We may also be able to record conversations or communications, or provide the ability to allow remote support for diagnosis. This has the potential to introduce efficiencies to patient care, for example, by introducing demand-driven staffing (Brown et al. 2014). The data obtained from such technologies can also be used to support staff training, both through development of best practice guidelines for task allocation, as well as providing a reflective tool for clinicians to review their performance on shifts and understand for themselves which types of work strategy are most effective.

Example 2: Distributed work – Human Computation

Human Computation (HC) can be described as using the internet population to perform tasks and provide data to address difficult problems that either cannot be solved by computer algorithms alone (Ma et al. 2009). From an E/HF perspective, this paradigm is recognisable as an implication of Fitt's List (Fitts 1951) that asserts humans and machines have different relative strengths (see also de Winter and Dodou 2014, for a contemporary discussion of the applicability of Fitt's insights). The larger part of HC research concerns commercial HC offerings, principle amongst them the Amazon Mechanical Turk that allows workers ('turkers') to participate in online tasks for micropayment (Vakharia and Lease 2013). These tasks are typically referred to as Amazon Mechanical Turk Tasks (AMTs) and tasks range from image labelling, participation in surveys and undertaking university research experiments through to language translation, carrying out searches and even content generation (such as 'write a short paragraph explaining why hotels are important to business travellers'). Key to the design of HC is that there is a digital platform that both distributes and manages the work at hand, issuing new work and assessing the quality and aggregating completed work units (typically by cross-checking multiple redundant completions of the same work unit).

Several issues have been raised around use of the Mechanical Turk with regard to its commercial aspects, such as the fair pricing of AMTs for workers, questions about ethical practices and whether aggregate earnings for turkers are reasonable. Some have referred to the Mechanical Turk as a 'digital sweatshop' whereas others have preferred the view that AMT provides remunerated diversions people can undertake in their spare time that are not supposed to replace the typical job (see Kittur et al. 2013 for a discussion). From an E/HF perspective we might also be concerned that implementations of HC represent something of a slippage back from lessons already learned about the design of harmonious and productive sociotechnical work because the driver for work design in this sector is not so much what humans can do, as much as what machines so far cannot. This pattern of work allocation is referred to as left-over automation (see Bainbridge 1983). This pattern of work allocation should concern us as it is inimical to the design of satisfying, meaningful work (see, e.g. Hackman and Oldham 1975, 1980; Vicente 1999). While HC might be seen as a relatively niche form of work at present, it seems reasonable to wonder how much further this paradigm might extend in line with developments in computational intelligence. Might it, for example, be possible to break down the work of a legal professional into a set of small bite-sized chunks that are then crunched by a legal rulebase to render an opinion (indeed, the ergonomics expertise in task decomposition techniques might be crucial to this venture). Kittur et al. (2013) have taken the view that HC may transcend its current limits and indeed actually constitute the future of work itself and ask: 'what will it take for us, the stakeholders in crowd work – including requesters, workers, researchers – to feel proud of our own children when they enter such a work force?' (12).

Example 3: User experience – transport

Transport businesses and infrastructure providers are increasingly becoming aware that there is value in capturing travellers' end-to-end journeys. We know, for example, that the 'last mile' is often the barrier to modality change (Rehrl, Bruntsch, and Mentz 2007). Increasingly, people are planning journeys using technologies, and monitoring status of transport infrastructure in real time, making dynamic journey choices (e.g. to walk or take the bus/tube).

The data available to support these activities can be drawn from both official and unofficial sources, and a key technical challenge here is the integration of data from a range of sources in a range of forms. Examples of

technologies currently in place are real-time train system status, on-board booking of onward journey steps and use of formally delivered messages in case of disruption. In addition, unofficial sources, such as twitter feeds from passengers travelling on a disrupted service, can be of value both to transport users and providers, but, of course, the quality of data obtained from these informal sources has little or no official verification and can be variable.

Already there has been a move to make more real-time transport information open, and this has led to a series of entrepreneurial apps that have supported user experience. This therefore yields a 'personal digital travel footprint', where an individual's data about their travel preferences, behaviours and movements can be collected over a long period of time. This not only offers the potential to provide personalised information form and content, but also supports business models such as geographically targeted advertising or varying ticket prices.

This more business- and experience-led example highlights some distinct issues. There are clear opportunities to use such personal footprint data to make travel more efficient and satisfying – two underlying goals of E/HF design. But there are clear questions regarding ownership of data and transparency of decision-making – for example, it may be possible for a transport provider to note that too many people are planning to take a single, congested route, and therefore encourage some individuals to take a longer journey. Should individuals know the rationale behind these decisions? Should they be aware that their suggested route will take them longer to travel?

This example scenario also highlights the need for E/HF to not only consider the work and organisational context of changes to technologies and systems, but also to understand the business implications. Increasingly, we are moving away from dedicated and constrained work settings as the home of E/HF analysis, to more distributed, less controlled contexts. This highlights the need for a new paradigm to our research and practice.

As Table 1 demonstrates, each of these scenarios have the potential for E/HF interventions supported by the contextual digital footprint. These scenarios also being to highlight issues that our E/HF methods need to overcome to ensure that interventions are ethical and effectively influence workplace systems/design.

We now use an alternative framework to consider the challenges and opportunities for the contextual digital footprint in E/HF practice. The following section considers the impact of contextual digital data for five different types of E/HF method that routinely form part of our E/HF toolkit.

Table 1. Dystopian and Utopian elements of the example scenarios.

Case	Description	Potential data-driven E/HF interventions	Utopian vision	Dystopian vision
Situated work	Measure workload and SA at high level of granularity, intervene when WL high	Support for staff training, both through development of best practice guidelines for task allocation, as well as providing a reflective tool for clinicians to review their performance on shifts and understand for themselves which types of work strategy are most effective	Early warning prevention and prognostics Influencing locum spend, decision support, real-time information support	Use for 'blame culture' – unhelpful competition
Distributed work	Combine efforts of distributed workforce to complete large task requiring human computation	Ability to intervene in task/work design through the direct manipulation task-flow, user interfaces etc. and to run *experiments* as embedded *trials*, seamlessly contiguous with or alongside the work itself. Performance can be measured at the level of individual clicks, keyboard presses and time-on-task	More efficient task completion	Reduction of job enrichment and variety
			Removal of need for individuals to complete repetitive work for long periods of time by sharing out tasks Workers can better control their work environment, hours and participation	Radical decomposition of work leads to piece-work rates or similar systems of reward that militate against financial security Increase in prevalence of WRULDs due to repetitive nature of tasks Limited oversight of workstations in individual homes Potential removal of social benefits of work
User experience	Capture details of user behaviours and use to provide a personalised service	Contextual understanding of individual user experience permitting fine-tuning of system design to reduce anxieties, implement affective/emotional design interventions, and the use of personalisation to design for populations beyond the 90%th or 95th percentile that are responsive to the individuals own life changes (e.g. change in work and hence commute)	Optimised data integration leads to seamless use of services such as health care and transport, leading to improved personal experience and efficiencies in service delivery	Failure in data security leads to financial and data loss
			Transparency in routing information or suggestions provided by transport operators, influencing decisions made by users	Failure to accurately profile individual leads to delivery of actively unsuitable services Data used to discriminate against certain user groups outside profitable target demographic or user profiles Annoyance from interruptions from targeted advertising and messages, distracting from primary task

Contextual digital data and E/HF methods

In this section, we explain a series of characteristics of contextual digital data, and specifically consider the implications of these characteristics for E/HF methods.

We use an existing classification of E/HF methods (Wilson and Sharples 2015a) to ground this discussion. Wilson and Sharples outline five types of methods, beyond 'general methods' (a grouping that includes generic techniques such as observation, interviews and experiments) that embody the different approaches that an E/HF researcher or practitioner might wish to employ. These are:

(1) Collection of information about people, equipment and environments
(2) Methods to support Analysis and design
(3) Evaluation of user and system performance
(4) Evaluation of demands on and effects experienced by people
(5) Management and implementation of ergonomics.

In the following sections, we outline the different ways in which contextual digital data can potentially be used by E/HF practitioners, and begin to consider the specific

Table 2. Characteristics of contextual digital data for different E/HF method types.

E/HF method	Relevant characteristics of contextual digital data	Impact of characteristics on E/HF approach	
		Advantages	Disadvantages
Collection of information about people, equipment and environments	Need to consider appropriate and required level of data granularity/ sensitivity	May be able to measure larger samples with greater accuracy than previously	May 'over-measure', leading to unnecessary expense in deployment and time waste in analysis
	Data collection changes from being explicit and targeted activity to embedded and continuous	May reduce 'Hawthorne effect' of behaviour change due to awareness of being observed	May present ethical issues in ensuring participants are aware data is being collected, and able to give informed consent
	Data is stored digitally	May be able to apply algorithmic techniques to speed up and increase power of analysis, as well as combine distinct data-sets	Essential to ensure that data is stored securely and ethical requirements for data collection, storage and use are met
Methods to support analysis and design	Potential to collect data to inform design of specific tasks over longer period of days or weeks	More likely to capture variation in task performance, and impact of unusual events	Cost of data analysis, and no clear guidance on how much data would be 'enough'
	Multiple, discrete, data-sets about a single task are more feasible to collect	Potential for new insights from combined data-sets	Potential for misleading co-ocurrence of data (correlation ≠ causation) and difficulty in assessing reliability of varied data types
	Potential to increase sample sizes	More likely to capture individual differences in task completion strategies and design preferences	Required samples may still be quite large to achieve required power, and larger samples will present time and cost implications
Evaluation of user and system performance	More detailed and varied types of information about task completion can be collected	Potential to increase detail and quantity of task data than previously feasible	Task data (e.g. counts) may not represent performance without contextual information also being captured
Evaluation of demands on and effects experienced by people	Lower intrusion measures of physiological and psychological response	Richer and less intrusive data collected from real world setting	Important to understand meaning of physiological data with respect to E/ HF concepts (e.g. workload)
	Reduced reliance on subjective data reporting in real world context	Opportunity to capture changes in experience at higher resolution than possible with subjective data	Measured physiological changes may not be meaningful or of concern to the individual
Management and implementation of ergonomics	Ability to monitor the long-term effectiveness of E/HF interventions	Potential to collect evidence to support cost benefit analysis of E/HF	Need to understand the role of the E/ HF intervention (as opposed to other workplace changes/behaviours) when interpreting data
	Opportunity for workforce team members to review data in an open and transparent way, and reflect on their own performance and actions	Potential improvement in motivation and commitment to job role from workers	Need to ensure that appropriate and relevant data are captured, understood and used effectively by managers

implications in this paradigm shift. Table 2 summarises the issues for contextual digital data for these five different types of E/HF methods, and feeds into recommendations for the use of contextual digital data to enhance E/HF that we present in the conclusion to this paper.

Collection of information about people, equipment and environments

E/HF has a history of using and developing many specific instruments and approaches to measure the characteristics of people, equipment and environments. These measures can be physical (e.g. anthropometry), physiological (e.g. heart rate), environmental (e.g. lightmeter) or perceptual/ cognitive (e.g. visual acuity tests). Over the past 50 years, the instrumentation to support these measures have been increasing in accuracy and decreasing in their intrusive nature. It is clear that these trends in instrumentation are continuing. This presents two interesting challenges that represent a paradigm shift, as opposed to an incremental change. Firstly, the sensitivity of instrumentation can be

argued to be increasing well beyond the levels required for E/HF design intervention or evaluation. For example, we are already able to measure movement in centre of gravity (used to detect experience of sickness such as might be experienced after a period of using Virtual Reality (see Cobb and Nichols 1999)) to at least a resolution of 0.1 mm. But, realistically, to be distinguishing between an individual who is experiencing symptoms of motion sickness to an extent that it affects their well-being or ability to work, it may well be the case that only measurements to an accuracy of 1 mm are required. Similarly, within the situated work example given above, we may be able to deploy technology that could measure the position of a clinician within a hospital to ±1 cm accuracy, but in fact, only ±1 m accuracy may be needed to inform the design of ward layouts.

Secondly, in the past the collection of data about people, equipment and environments was an explicit and targeted activity. Indeed, phenomena such as the Hawthorne effect demonstrated that the explicit nature of data collection had the consequence of potentially changing participant behaviours once people were aware that they were being measured or recorded. Whilst this might have had an undesirable consequence in terms of the validity of data collected, this has a (perhaps unintended) positive side effect in that participants were clearly aware that data were being collected, and therefore an E/HF practitioner could be confident that the principle of 'informed consent' was being upheld.[1] Now, technologies such as embedded sensors in buildings, or personal devices such as smartphones, mean that participants may not be aware of the presence of sensors, due to their integration into the building infrastructure or technologies that they are routinely using for other purposes. This has the positive consequence of ensuring that collected information is more naturalistic, but presents ethical challenges. We have an underlying ethical principle of ensuring that all participants in research and data collection are able to give 'informed consent' – in other words, that they are able to understand the purpose of the data collection, and consent to their participation in data collection. The ability to capture data about people, equipment and environments using our contextual digital footprint therefore demands more formal and explicit confirmation of 'informed consent' to data collection.

Methods to support analysis and design

Methods to support analysis and design include approaches such as task analysis, modelling and expert evaluation. They traditionally depend on the collection of data about a work task or interaction, and either real-time or off-line analysis conducted by an expert, sometimes using tools such as digital human modelling, to evaluate the workplace, its requirements or design implications. Contextual digital data offers the potential for a richer data-set to be used as the basis for this analysis. For example, rather than observing the tasks completed by an individual supermarket checkout operator, all the interactions with the different systems being used could be collected at several workstations over a period of several weeks. In addition, the increased variety in types of data that can be collected offer the opportunity for combining data-sets and making analytical inferences that are only possibly when two sets of data are combined (for example, the relationship between number of interactions required on a till during a shift could be combined with data on absenteeism for a sequence of shifts). In other contexts, this data analytics approach has been used by organisations such as Google to predict the onset of flu outbreaks (with varying success – see Lazer et al. 2014). In E/HF previous theoretical work has demonstrated that many of the phenomena in which we are interested, such as work-related upper limb disorders (Armstrong et al. 1993), work-related stress (Cox and Griffiths 2005) or influences on human performance (Edwards 2014) are multifactorial. Currently, methods that allow us to examine such phenomena include: expert-led qualitative methods, such as structured critical decision-making interviews; knowledge elicitation methods such as card sorts; or laboratory scenarios with multivariable manipulations. However, in the case of the laboratory examples, power analysis often reveals that to obtain data with a reasonable likelihood of detecting any effects that exist, large participant sample sizes are required. This not only presents time and cost implications, but in many cases it is appropriate for those laboratory study participants to in fact be expert operators themselves (e.g. air traffic controllers); thus they are drawn from a limited participant pool. Contextual digital data presents the potential to gather data where we are interested in analysing multifactorial phenomena from real world data. However, as with all analysis of this type, it is not always straightforward to obtain the appropriate metrics that directly map on to the influences of interest; in the case of Google flu analytics, there were data such as search terms that could be used as indicators of the experiences of users.

As E/HF specialists, we need to consider whether there are equivalent types of contextual digital data that occur during workplace performance that can be used as indicators of multifactorial phenomena. If we are successful in identifying sources of factorial data such as this, we can begin to move away from the constraints presented by multifactorial ANOVA design towards more dynamic epidemiological modelling techniques to understand the development of workplace experiences and effects such as comfort, stress and workload that are at the heart of E/HF.

Evaluation of user and system performance

The challenge of obtaining valid and reliable measures of user and system performance has yielded many methods in E/HF, such as cognitive work analysis (Vicente 1999) or human reliability analysis (Kirwan 1994). There are still however many situations in which there is not a clear and unambiguous measure of work 'performance', and many laboratory tasks are subject to criticism either that they are too artificial and do not reflect the complexity of real world jobs, or that they are subject to classic experimental artefacts such as the traditional speed/accuracy trade-off. Contextual digital data therefore provides an opportunity to deliver new measures of work performance. For example, as in the situated work example provided earlier, research we are conducting in the hospital context is allowing us to collect data relating to workplace tasks (through the smartphone job allocation system) and track clinician movement around the hospital. This will yield a much larger data-set than would be practicable through traditional methods such as observation, and more reliable data than methods such as diary methods. It is not, however, a perfect measure. For example, observations of the technology in-use have shown that the users appropriate the technology to help them manage their work tasks. When a clinician is allocated a job, they are then required to 'accept' the task that has been allocated to them, when they begin that task. This interaction tends to be a fairly reliable indicator of the task being started. However, in order to remove a task from the list, clinicians need to 'complete' a task. Our observations have shown that the task is often in fact left on the smartphone list long after it has actually been completed. This is not because clinicians are deliberately making the system think that a task is taking longer than it actually is, but instead is due to the fact that a clinician will often keep a task on the list as a reminder, perhaps to check later on in their shift on the result of some medical tests that they have ordered. Therefore, it is vital that we accompany interpretation of contextual digital data with a clear understanding of how the technology that yields that data is appropriated by its users.

Evaluation of demands on and effects experienced by people

As noted earlier, one particular technological development that is of value to the practice of ergonomics is that advances in sensor technologies make them much less intrusive than in the past. Therefore, it is realistic to imagine, for example, an entire factory workforce wearing devices such as a 'fitbit' that records physical activity through the day. In addition, technologies such as eye tracking now have the potential to be embedded into standard head gear, and the size of physiological monitors such as heart rate monitors has decreased to such an extent that they can now realistically be worn throughout the working day without being noticeable to the user – therefore hopefully reducing the extent to which participants are aware of the device and therefore changing their behaviour due to their awareness of being monitored. However, it remains critical that we are not seduced by these vast sets of quantitative data that perhaps represent an Aladdin's cave of previously unobtainable data. It is still important that we understand the validity and meaning of such data; whilst these technologies may enhance the accuracy and availability of such data, there is still a challenge in understanding the meaning of these measures (Parasuraman 2003). Science and engineering colleagues often refer to the concept of 'ground truth' when establishing the accuracy of measures (e.g. in developing new Global Navigation Satellite System technologies); in E/HF it is very rare that we have a 'ground truth'. We tend to rely on triangulation to overcome this limitation, but we must acknowledge that whilst this

triangulation shows that our methods are agreeing with each other, it does not necessarily help us to interpret the meaning of such data. This is the classic correlation/causation dilemma – for example, if a participant experiences an increase in heart rate as they report high workload, this could mean that the heart rate is as a consequence of workload, or that both heart rate and workload are influenced by the same external phenomenon. Non-intrusive physiological monitoring undoubtedly offers significant potential but it is important that E/HF practitioners understand the validity of what data are being collected, and how they relate to the multifactorial phenomena that have previously been established.

This type of data is not necessarily solely physiological. For example, if we consider either the distributed work or the user experience examples given above, the level of stress experienced by a participant could potentially be inferred by typing speed or number of errors made whilst providing input to either a mechanical turk work system or an app being used whilst travelling. This is potentially extremely powerful and sensitive data, but it is essential that its meaning in terms of E/HF concepts such as stress is understood and managed appropriately. For example, in the case of distributed work, we would want an E/HF intervention to be focused around work demand management and support for the worker, rather than a punitive or monitoring regime that in fact increases the stress experienced by the individual.

Management and implementation of ergonomics

We also use methods such as Human Factors Integration (Cullen 2007) to support the effective implementation of E/HF in workplace contexts. Contextual digital data offer two opportunities in this area – (a) the ability to monitor the long-term effectiveness of E/HF interventions and (b) the opportunity for workforce team members to review data in an open and transparent way, and reflect on their own performance and actions.

The first opportunity, to monitor long-term effectiveness, is critical to supporting the cost/benefit analysis of E/HF interventions. This has for a long time been something that the discipline has grappled with, as so many ultimate indicators, such as staff turnover, or absenteeism due to illness, are low in frequency and long term in their development. Contextual digital data make obtaining shorter term indicators of overall workplace realistic. Examples of such data might be detailed analysis of frequency of task completion, or lengths of rest breaks taken. This data does of course have to be very carefully managed and interpreted, and clear strategies for management of privacy and use developed. For example, if a group of workers are noted to be taking frequent short breaks during a task, does this indicate that they are not complying with their work requirements, or is it an indicator of a potential problem with the way in which the task is designed. It could in fact be the case that the regular, short, unplanned breaks are an example of good practice, leading to shared intelligence about the status of work, or team building activities. For example, if a doctor whose movement is being tracked is apparently spending frequent periods of time in the mess room, is this an indicator of an inefficient rota, or a sign that the doctor is able to take sufficient rest during their shift, and thus less likely to become stressed?

Secondly, it is already seen in manufacturing and transport contexts that rapid feedback of performance data is possible. Rather than this being perceived as a 'big brother' concept that intimidates and disenfranchises workers, there is potential for the workforce to themselves to take ownership of these data and use the data positively to change their own performance and actions. This therefore requires careful implementation of such data within an organisational structure as formative, rather than summative, feedback.

Characteristics of contextual digital data

As mentioned previously, a principle of research ethics and ethical E/HF practice is that for any data we collect, the participant(s) must be aware that data collection is taking place, and how that data is going to be used and stored. In an experimental context, this is usually quite straightforward to ensure, through the use of consent forms, and influenced by the expectations of the participants, who are aware that they are taking part in a formal process of data collection. E/HF practice has traditionally grappled with the acknowledgement that people change their behaviour in the field once they are aware that they are being monitored or observed. As noted earlier, this was first reported as the Hawthorne effect (Landsberger 1958; but see also Levitt and List 2011) and is a phenomenon that has persisted. Contextual digital data extends the nature of data that can be collected about an individual whilst completing their work or using a system, and presents new challenges for how we ensure that the principle of 'informed consent' is maintained (Eden, Jirotka, and Stahl 2013). Therefore, in addition to the specific impact of the use of contextual digital data within E/HF methods, there are some general, contextual issues of which we need to be aware of as follows:

The blurring of the work–life boundary

Traditionally within E/HF we tend to look at people within a particular context, situation or place of work. Contextual digital data present the opportunity to examine and use data from an individual's life more generally, and these data might span multiple workplaces, or combine their work and home life. If we consider an individual who is experiencing symptoms of an upper limb disorder for example, we may be able to collect data not only from the way in which they interact with systems and equipment at work, but also their activities at home – for example, an individual's frequent interaction with a personal smartphone or games technology may be combining with a physical task they do at work to produce the symptoms that are presenting. This results in a much more sophisticated and extended form of our existing concept of 'archival data'.

Anonymous or not?

The collection of large volumes of contextual data poses significant challenges to practitioner in terms of both security (that is, making sure the data remain confidential) and in terms of maintaining anonymity. The specific issue of security is outside the scope of this paper, but we will note that barely a week passes without reports of a significant security breach concerning personal data, be it the result of deliberate hacking (as in the case of leaks of photographs stored on the cloud) or some form of simple error (losing a datastick). Indeed, with human error reported as being implicated at some level in 95% of cyber incidents (IBM Global Technology Services 2014), this is in itself an important new facet of safety science research. As the potential custodians and users of contextual data, E/HF practitioners will have increasingly onerous responsibilities in this area.

Maintaining anonymity will also be challenging. Normal ethical practices in this area typically include removing personal identifying information (such as names or addresses) and perhaps coding respondent's data by job role or simply with an index number. It is increasingly clear that such practices will not be sufficient when dealing with contextual data; indeed, risks to anonymity will exist even where no ostensive identifying data was ever collected by an individual researcher or team. The key to this difficulty lies in the specific characteristics of contextual data both linked to its sheer volume as a result of its automated collection. First, we may fail to fully appreciate what identifying information is hidden in our data-set. It may for example contain metadata[2] we are unaware of. Second, it may be possible to carry out inferences over data we did not directly foresee, possibly because the data are captured at a higher spatial or temporal resolution than we were aware of. For example, analysis of time-load data from energy monitoring devices can be used to identify time-varying appliance load signatures that identify when specific electrical items were being used (National Institute of Standards and Technology 2010, 13). This may in itself be embarrassing or could be used then to produce further inferences about who was at home and their pattern of life. Relatedly, it has been established that the phenomena of *higher dimensionality* and *sparsity* tend to exist in data-sets that contain so-called micro-records of individual behaviour as an inevitable consequence of the long-tail distribution (Brynjolfsson, Hu, and Smith 2003; Narayanan and Shmatikov 2008). The practical consequence of this is that once an individual record is considered from multiple dimensions in terms of all the attributes it may contain, far from blending into the crowd, individual specific records can be easily located on the basis of specific diagnostic features (i.e. there will be one or more dimensions in a given micro-record that will disambiguate it from other similar micro-records). Third, even if (a) our data-set is clean of metadata and (b) sparsity and higher dimensionality do not lead practically to 'jigsaw' identification, considerable risks are posed by the existence of other data-sets, particularly where many different types of data-set are already publicly available. A definitive example of this occurred relatively early in the modern history of contextual data where a Massachusetts hospital discharge database was deanonymised by correlating it with a public voter database via postal codes (Sweeney 1997). Other famous examples include correlating a publicly released data-set of films watched by supposedly anonymous individuals on Netflix with film reviews posted to the Internet Movie Database: "Using the Internet Movie Database as a source of background knowledge, we successfully identified the Netflix records of known users, uncovering their apparent political preferences and other potentially sensitive personal information" (Narayanan and Shmatikov 2008). This type of anonymity breach is not limited just to online databases, for example locational privacy can also be broken with reference to social networks as a correlate (Srivatsa and Hicks 2012) or with reference to other kinds of sensor data that modern smartphones collect like accelerometer, magnetometer and gyro data (Lane et al. 2012).

One might however note that the risk of deanonymisation also implies something positive about the qualities of contextual data in that it shows that these data tend to be highly specific and thus in principle, more data means potentially more information. More generally, this possibility can be seen simply as a corollary of the power of data mining to produce insights, albeit in this case ethically dubious ones. The mining of hospital data together with voter records could just as easily generate epidemiological data. The challenge lies in using data ethically and in an informed manner. Useful tools in this area may include ways of categorising data in a risk-based manner based not upon absolute security, but the amount of effort that would be required to deanonymise a data-set and the

nature of the likely intent (identifying a specific individual known to be in the data-set, attempting to identify a specific individual who might be in the data-set, attempting to identify as many people as possible in a data-set) (El Emam 2010).

Another aspect of practice will be to educate users and workers effectively as to the nature of the risks involved and to offer them appropriate levels of control over their own data where possible (ENSIA 2011). In the workplace this would however require an appropriate managerial and cultural viewpoint on whether workers are indeed allowed this kind of privacy or ownership in the first place.

Beyond the sample

Traditionally, we are able to look at constrained contexts, samples which are governed by physical rules. For example, when completing a Cognitive Work Analysis (Vicente 1999), we establish an abstraction hierarchy that outlines the context to which our analysis will apply. Contextual digital data, in principle, allows us to look at complete populations. This presents a tremendous opportunity in terms of coverage of a range of user types, but we (a) must really be sure we have captured the whole population, and (b) need to acknowledge that the examination of a complete population represents a change from what is sometimes our normal good practice – sometimes we deliberately do not try to look at the whole population but consider specific user groups and their needs, on the basis that if their needs are met, others are automatically met (e.g. door height for tall people, button size for big fingers, visibility of contrast for those with visual impairments etc.), (c) this makes the concept of statistical significance tricky. We already see this in the context of correlations when we have large samples, where we need to be cautious in the inferences that we make from statistical tools such as correlation, and remember what the numbers produced from applying a statistical test actually mean. For example, a correlation of 0.3 for a sample of $N = 50$ will be considered 'significant' at a level of $p < 0.05$ (i.e. the correlation would only have occurred by chance 5% of the time). However, there is also a meaning to the correlation coefficient of 0.3 – by converting it to the coefficient of determination, we know that 9% of the variance in one variable is explained by knowing the other variable. Whilst 9% may be 'significant', the meaningfulness of having explained only 9% of the variance in a variable needs to be acknowledged (and methods used to help capture the nature of the other 91%!).

The challenge for E/HF is therefore to a) be able to interpret the correlations in massive data-sets in ways which are meaningful and b) to pose hypotheses or offer explanations which can exploit these data-sets. (For further discussion of the notion that $n = N$ in big data contexts, see Drury 2015).

The uses of E/HF in a contextual data environment

In addition to practical considerations as to how we might use contextual data in E/HF, there is also the significant issue of how E/HF would in turn itself be used in organisations and what part we will play within these ventures. At the present time, the use of such 'business intelligence' has arguably happened ahead of substantial efforts by ergonomists to understand it. Of course, using data within management is hardly new and has not been without its proven benefits and equally, its discontents, particularly when linked to 'targets culture'. Timecards, for example, have long been a form of employee tracking. Indeed, the overbearing 'big brother' manager (Chaplin's vision predates Orwell's) who tracked his employees even up to the point of tracking and intruding on their bathroom breaks was famously parodied in the Charlie Chaplin film 'Modern Times' (1936). One might feel that a trajectory from registers to punched time cards through to swipe cards and then employee location tracking is merely a quantitative change in the fidelity with which employees can be tracked. However, a key development is that this tracking data is just one of a range of measures that can now be easily applied, and most importantly, the development of computational intelligence to track employees (e.g. Parenti 2001). Recent media attention has been focused on the use of location and activity tracking data as part of employee monitoring at mail-order warehouse and fulfilment centres (BBC 2013) with several workers expressing unhappiness at their perceived lack of control within their workplace: 'Workers are treated more as robots than humans' (Streitfeld 2013).

In wider focus, one of the biggest challenges ergonomists will have to face regards the potential for improvement specifically in terms of production. We have, in a sense, been here before. One of the first responses to having accurate information about employee behaviour (in the form of artefacts like time and motion methods and Frank Gilbreth's filming of the workplace) was so-called scientific management (Taylor 1911). In an echo of the present situation, F.W. Taylor himself was surprised to find that the Ford motor company had implemented methods of scientific management ahead of the involvement of experts, including himself (Sorensen 1956). While the sociotechnical turn corrected for this tendency (e.g. Trist 1981), there is a risk that with the lure of data-driven improvements in efficiency, lessons learned at great cost are once again forgotten leading to a 'neo-Taylorist' future. At the same time of course there is fantastic potential for the contextual footprint to serve what we might regard as sound sociotechnically informed ends such as permitting job enlargement or even allowing employees newfound

flexibility in how they transact their labours by providing them with rich sources of data. Indeed, at an extreme, exploitation of the digital contextual footprint could permit the removal of management functions in favour of self-synchronising teams within the workplace as has been envisioned as a consequence of ubiquitous data sharing in military domains (e.g. Alberts and Hayes 2006). The implication here is that ergonomists may occupy the role of ombudsman with regard to effect of new technologies on the workplace (Meister 1999; see also Hancock 2009 for an extended discussion). It may be increasingly necessary that we stick to our guns with regard to what we understand as the appropriate ways to design work, and ensure that we understand how contextual digital data can be used to support this dialogue.

Discussion

The emergence of contextual digital data, like most technological developments, presents both opportunities and hazards to the discipline of E/HF at several different levels of analysis. Traditionally, E/HF has been something of a data hungry discipline where practitioners may often have found data collection expensive and time-consuming even assuming easy access to relevant sites and Subject Matter Experts, and sometimes have to accept that their resources may not stretch as far as they might want. The potential for a deluge of rich and seemingly unlimited data about individuals and work systems has clear appeal, signalling the potential to become more confident about the effects of work design on a wider population, and reducing the time and financial costs of data collection. At the same time, E/HF as a newly 'data rich' venture presents numerous fresh challenges in terms of the interpretation of these data, the practical and ethical handling of large data-sets and ultimately, determining how it fits in with the concerns of our discipline and how it should actually be used and what could and should change in the world as a result.

Abstracted empiricism and 'the ergonomic imagination'

Although the contextual digital footprint is a new phenomenon with several distinctive features, this is not the first time a discipline has had to consider its reaction to the availability of a flood of new data and in it is instructive to examine the lessons that were learned. In 1959, the American sociologist C. Wright Mills expressed concerns that have a familial similarity to our own in this paper. Mills had noted that the then-emerging technology of electric computers meant that survey research on public opinion could be rapidly coded onto Hollerith punch cards 'which are used to make statistical runs by means of which relations are sought. Undoubtedly this fact, and the consequent ease with which the procedure is learned by any fairly intelligent person, accounts for much of its appeal' (Mills 1959, 50). The concern expressed by Mills was that this technology would lead to the distortion of sociology as an academic discipline towards 'abstracted empiricism' where data and method were not appropriately contextualised or integrated with theory with an endgame emerging where sociology degenerates into the analysis of opinion polling rather than retaining focus on understanding social structures and phenomena. A further corollary of Mills' concerns was that the easy availability of data leads to a potential confusion between what is important and what is easy to measure. Mills' 'abstract empiricism' of the Hollerith card has much in common with contemporary critiques of big data that emphasise its likewise theory-free interrogation of data privileging correlational statistics over hypothesis-testing inferential methods and several of the concerns expressed in the present paper. Mills' response to this was to invoke the notion of the 'sociological imagination', essentially a call for a sociology that took a three-dimensional, holistic view of society combining macro- and micro-perspectives such that individual experience could be understood in terms of larger, interlinked phenomena, a view not dissimilar – at least by analogy – to the systems ergonomics perspective in E/HF (e.g. Wilson 2014) most typically expressed through ideas such as the onion model (see Wilson and Sharples 2015a) or ergonomics as 'reflective practice' (see Sharples and Buckle 2015). In view of this, we have no apparent need at the present time for a putative New Ergonomics but it is perhaps ironic that in the consideration of a new paradigm within E/HF, our attention is drawn back to the key pre-existing foundations of our discipline. Ultimately the safe, positive and effective accommodation of the contextual footprint within our subject will require a recommitment to our core values and concerns.

Using the contextual data footprint to enhance E/HF

Contextual digital data already exists, and is here to stay. As E/HF specialists, it is our responsibility to understand how these data can be used ethically and responsibly to improve the way in which we design systems, technologies and work.

We require at least the following:

- Training in methods to handle large data-sets, and retaining a fundamental understanding of statistical inference, so that colleagues are aware of the way in which statistical tests behave with large data-sets.
- Training in and provision of appropriate techniques to ensure data security, coupled with methods to ensure that ethical requirements are met.
- Developing methods to store and dispose of digital data
- Making sure procedures are in place to ensure 'informed consent' is feasible whenever data are used as part of ergonomics analysis
- Developing approaches that allow us to maintain participant anonymity, being particularly aware of the hazard of jigsaw identification.

But, in addition to these recommendations regarding the ethical and responsible use of contextual digital data, we should not be blinkered, and should embrace the opportunities presented by these data. Contextual digital data may well provide us with the opportunity to have new insights and advance our theories about causation and response to stimuli. The contextual data footprint, if used responsibly and ethically, has the potential to transform the nature of E/HF analysis and track the impact of design changed informed by E/HF analysis over days, months and years. We can move beyond concerns about the transferability of data from the laboratory to the field and, consider the possibility of the field becoming the laboratory.

Notes

1. Whilst in laboratory studies or formal activities such as interviews or focus groups, a standard consent form will be used to confirm informed consent, in many less formal workplace observations, participants do not always give written consent to data being collected, but the E/HF practitioner will clearly verbally explain the reason for their presence and the types of data (e.g. written notes) that will be collected.
2. Additional, explanatory data that is attached to the primary data value.

Acknowledgements

The authors would like to thank all of their collaborators in these contributory projects, as well as the anonymous reviewers who gave constructive feedback that improved the flow and argument presented in this paper.

Disclosure statement

No potential conflict of interest was reported by the authors.

Funding

This work was supported by the Health Foundation, Transport Systems Catapult and EPSRC Horizon Digital Research Institute [grant numbers EP/M02315X/1, EP/G065802/1].

ORCID

Sarah Sharples iD http://orcid.org/0000-0003-0288-915X
Robert J. Houghton iD http://orcid.org/0000-0001-7334-8567

References

Alberts, D. S., and R. E. Hayes. 2006. *Understanding Command and Control.* Washington, DC: DODCCRP.

Armstrong, T. J., P. Buckle, L. J. Fine, M. Hagberg, B. Jonsson, A. Kilbom, I. A. A. Kuorinka, B. A. Silverstein, G. Sjogaard, and E. R. A. Viikari-Juntura. 1993. "A Conceptual Model for Work-related Neck and Upper-limb Musculoskeletal Disorders." *Scandinavian Journal of Work, Environment and Health* 19: 73–84.

Bainbridge, L. 1983. "Ironies of Automation." *Automatica* 19 (6): 775–779.

BBC. 2013. "Panorama: Amazon, the Truth behind the Clicks." Broadcast, November 29.

Blakey, J. D., D. Guy, C. Simpson, A. Fearn, S. Cannaby, P. Wilson, and D. Shaw. 2012. "Multimodal Observational Assessment of Quality and Productivity Benefits from the Implementation of Wireless Technology for out of Hours Working." *BMJ Open* 2: 2:e000701.

Brown, M., J. Pinchin, J. Blum, S. Sharples, D. Shaw, G. Housley, S. Howard, S. Jackson, M. Flintham, K. Benning, and J. Blakey. 2014. "Exploring the Relationship between Location and Behaviour in Out of Hours Hospital Care." *HCI International 2014-Posters, Part II CCIS 435,* 395–400.

Brynjolfsson, E., Y. Hu, and M. Smith. 2003. "Consumer Surplus in the Digital Economy." *Management Science* 49 (11): 1580–1596.

Cobb, S. V. G., and S. C. Nichols. 1999. "Static Posture Tests for the Assessment of Postural Instability after Virtual Environment Use." *Brain Research Bulletin* 47 (3): 459–464.

Cox, T., and A. Griffiths. 2005. "The Nature and Measurement of Work-related Stress: Theory and Practice." In *Evaluation of Human Work,* edited by J. R. Wilson and E. N. Corlett, 3rd ed, 553–572. Boca Raton, FL: CRC Press.

Cullen, L. 2007. "Human Factors Integration–Bridging the Gap Between System Designers and End-users: A Case Study." *Safety Science* 45 (5): 621–629.

Drury, C. 2015. "Human Factors/Ergonomics Implications of Big Data Analytics: Chartered Institute of Ergonomics and Human Factors Annual Lecture." *Ergonomics* 58 (5): 1–15.

Eden, G., M. Jirotka, and B. Stahl. 2013. "Responsible Research and Innovation: Critical Reflection into the Potential Social Consequences of ICT." In *2013 IEEE Seventh International Conference on Research Challenges in Information Science (RCIS),* pp. 1–12. Paris: IEEE.

Edwards, T. E. 2014. "Human Performance in Air Traffic Control." PhD thesis, University of Nottingham.

El Emam, K. 2010. "Risk-Based De-identification of Health Data." *IEEE Security & Privacy* 8 (3): 64–67.

ENSIA (European Network and Information Security Agency). 2011. *To Log or Not to Log? Risks and Benefits of Emerging Life-logging Applications.* Heraklion: European Union.

Fitts, P. M. 1951. *Human Engineering for an Effective Air-navigation and Traffic-control System.* Oxford: National Research Council.

Foresight Future Identities. 2013. *Final Project Report.* The Government Office for Science, London.

Hackman, J. R., and G. R. Oldham. 1975. "Development of the Job Diagnostic Survey." *Journal of Applied Psychology* 60 (2): 159–170.

Hackman, J. R., and G. R. Oldham. 1980. *Work Redesign.* Reading, MA: Addison-Wesley.

Hancock, P. A. 2009. *Mind, Machine and Morality: Towards a Philosophy of Human-technology Symbiosis.* Farnham: Ashgate.

IBM Global Technology Services. 2014. *IBM Security Services 2014 Cyber Security Intelligence Index: Analysis of Cyber Attack and Incident Data from IBM's Worldwide Security Operations.* Somers, NY: IBM Corporation.

Kirwan, B. 1994. *A Guide to Practical Human Reliability Assessment.* London: Taylor & Francis.

Kittur, A., J. V. Nickerson, M. S. Bernstein, E. M. Gerber, A. Shaw, J. Zimmerman, M. Lease, and J. J. Horton. 2013. "The Future of Crowd Work." *16th ACM Conference on Computer Supported Collaborative Work (CSCW '13),* San Antonio, TX, 1301–1318.

Landsberger, H. A. 1958. *Hawthorne Revisited.* Ithaca, NY: Cornell University.

Lane, N. D., J. Xie, T. Moscibroda, and F. Zhao. 2012. "On the Feasibility of User De-anonymisation from Shared Mobile Sensor Data." *Phonesense '12: Proceedings of the Third International Workshop on Sensing Applications on Mobile Phones.* Toronto, Canada, 1–5

Lazer, D., R. Kennedy, G. King, and A. Vespignani. 2014. "The Parable of Google Flu: Traps in Big Data Analysis." *Science* 343 (March): 1203–1205.

Levitt, S. D., and J. A. List. 2011. "Was There Really a Hawthorne Effect at the Hawthorne Plant? An Analysis of the Original Illumination Experiments." *American Economic Journal: Applied Economics* 3 (1): 224–238.

Ma, H., R. Chandrasekar, C. Quirk, and A. Gupta. 2009. "Improving search engines using human computation games." In *Proceedings of the 18th ACM conference on Information and knowledge management (CIKM '09),* 275–284. New York, NY, USA: ACM.

Meister, D. 1999. *The History of Human Factors and Ergonomics.* Mahwah, NJ: Lawrence Erlbaum Associates.

Mills, C. W. 1959. *The Sociological Imagination.* Oxford: Oxford University Press.

Moray, N. 2008. "The Good, the Bad, and the Future: On the Archaeology of Ergonomics." *Human Factors: The Journal of the Human Factors and Ergonomics Society* 50 (3): 411–417.

Narayanan, V., and H.-J. Shmatikov. 2008. "Robust De-anonymisation of Large Sparse Datasets." *Proceedings of the 2008 IEEE Symposium on Security and Privacy,* Oakland, California, 111–125.

National Institute of Standards and Technology. 2010. *Guidelines for Smart Grid Cyber Security, Volume 2: Privacy and the Smart Grid.* NISTIR 7628. http://www.nist.gov/smartgrid/upload/nistir-7628_total.pdf

National Research Council. 2014. *Complex Operational Decision Making in Networked Systems of Humans and Machines: A Multidisciplinary Approach.* Washington, DC: National Academies Press.

Parasuraman, R. 2003. "Neuroergonomics: Research and Practice." *Theoretical Issues in Ergonomics Science* 4 (1–2): 5–20.

Parenti, C. 2001. "Big Brother's Corporate Cousin: High-tech Workplace Surveillance is the Hallmark of a New Digitial Taylorism." *The Nation* 275 (5): 26–30.

Rehrl, K., S. Bruntsch, and H.-J. Mentz. 2007. "Assisting Multimodal Travelers: Design and Prototypical Implementation of a Personal Travel Companion." *IEEE Transactions on Intelligent Transportation Systems* 8: 1, 31–42.

Salvendy, G. 2012. *Handbook of Human Factors and Ergonomics.* Hoboken: Wiley.

Sharples, S., and P. Buckle. 2015. "Ergonomics and Human Factors as Reflective Practice." In *Evaluation of Human Work*, edited by J. R. Wilson and S. Sharples, 4th ed, 975–982. Boca Raton, FL: CRC Press.

Sorensen, C. E. 1956. *My Forty Years with Ford*. New York: Norton.

Srivatsa, M., and M. Hicks. 2012. "Deanonymising Mobility Traces: Using Social Network as a Side-channel." *Proceedings of the 19th ACM Conference on Computer and Communications Security*, Raleigh, NC, USA 628–637.

Stanton, N. A., P. Salmon, G. H. Walker, C. Baber and D. A. Jenkins. 2013. *Human Factors Methods: A Practical Guide for Engineering and Design*. 2nd ed. Farnham: Ashgate.

Streitfeld, D. 2013. "Amazon Workers in Germany Strike Again." *New York Times*. Accessed January 2, 2015. http://bits.blogs.nytimes.com/2013/12/16/amazon-strikers-take-their-fight-to-seattle/

Sweeney, L. 1997. "Weaving Technology and Policy Together to Maintain Confidentiality." *Journal of Law, Medicine and Ethics* 25: 2–3.

Taylor, F. W. 1911. *The Principles of Scientific Management*. Harper & Brothers.

The Guardian. 2012. "Are Sun Readers Ready to 'Beat the Cheat' Armed with the Facts?" Accessed March 1, 2014. http://www.theguardian.com/commentisfree/2012/mar/07/sun-beat-the-cheat-benefits

Trist, E. 1981. *The Evolution of Socio-technical Systems: A Conceptual Framework and an Action Research Program*. Ontario: Ontario Quality of Working Life Centre Occasional Papers.

Vakharia, D., and M. Lease. 2013. Beyond AMT: An Analysis of Crowd Work Platforms. *ArXiv*:1310.1672. Accessed November 2014. http://arxiv.org/abs/1310.1672

Vicente, K. 1999. *Cognitive Work Analysis: Towards Safe, Productive and Healthy Computer-based Work*. Mahwah, New Jersey, London: Lawrence Erlbaum Associates.

Wilson, J. R. 2014. "Fundamentals of Systems Ergonomics/Human Factors." *Applied Ergonomics* 45 (1): 5–13.

Wilson, J. R., and S. Sharples. 2015a. "Methods in the Understanding of Human Factors." In *Evaluation of Human Work*, edited by J. R. Wilson and S. Sharples, 1–36, 4th ed. Boca Raton, FL: CRC Press.

Wilson, J. R., and S. Sharples. 2015b. *Evaluation of Human Work*. 4th ed. Boca Raton, FL: CRC Press.

de Winter, J. C. F., and D. Dodou. 2014. "Why the Fitts List Has Persisted Throughout the History of Function Allocation." *Cognition, Technology & Work* 16 (1): 1–11.

Imposing limits on autonomous systems

P. A. Hancock

ABSTRACT

Our present era is witnessing the genesis of a sea-change in the way that advanced technologies operate. Amongst this burgeoning wave of untrammelled automation there is now beginning to arise a cadre of ever-more independent, autonomous systems. The degree of interaction between these latter systems with any form of human controller is becoming progressively more diminished and remote; and this perhaps necessarily so. Here, I advocate for human-centred and human favouring constraints to be designed, programmed, promulgated and imposed upon these nascent forms of independent entity. I am not sanguine about the collective response of modern society to this call. Nevertheless, the warning must be voiced and the issue debated, especially among those who most look to mediate between people and technology.

Practitioner Summary: Practitioners are witnessing the penetration of progressively more independent technical orthotics into virtually all systems' operations. This work enjoins them to advocate for sentient, rational and mindful human-centred approaches towards such innovations. Practitioners need to place user-centred concerns above either the technical or the financial imperatives which motivate this line of progress.

1. Introduction

Visions of the future of technology abound. They range from utopian ideals of machines as perfect servants to the dystopian nightmares of machine usurpation of all control. In their essence, each of these polarised perspectives expresses contrasting visions of the future direction of our own human narrative concerning our personal and collective freedom. Like all prognostications about the future (Bartlett 1962; Hancock 2008), neither of these absolutes is liable to represent our coming reality. Instead, we will witness an evolving palimpsest in which the residual remnants of past incarnations of automated technologies will be overlaid by the often chimerical advances, touted of the very latest offerings. The principle that there is nothing new under the sun (*nihil sub sole novum*) will still mostly predominate. However, like the regally apocryphal cloud 'no bigger than a man's hand', there looms upon our horizon a potentially game-changing saltation in the evolution of technological capacity. The more lurid anticipation of this transformation has pre-occupied film-makers just as much as it has futurists, scientists and philosophers (cf., De Chardin 1955/1959; Kurzweil 2005). But, to what degree are these differing presentiments likely to come to pass? In

what follows, I consider this question in the hope of shedding some light on what might more realistically represent the future of our human–machine hybrid species and our ergonomic discipline which researches it.

2. Automation vs. autonomy?

The foundation for each of my observations depends upon an understanding of both automation and autonomy. Therefore, I here define automation as: *automated systems are those designed to accomplish a specific set of largely deterministic steps, most often in a repeating pattern, in order to achieve one of an envisaged and limited set of pre-defined outcomes.* In apparent contrast, and I emphasise the word apparent here, I define autonomous systems as: *autonomous systems are generative and learn, evolve and permanently change their functional capacities as a result of the input of operational and contextual information. Their actions necessarily become more indeterminate across time.* While we can cite many examples of automation, there are far fewer of these latter autonomous systems in current operation. What is very clear however, from the vector of our present progress is that the nascent growth of autonomy is necessarily predicated upon increasing levels of automation. In consequence, the two terms are certainly not necessarily in any form of contrast or mutual contradiction, one of the other. Rather, they look to represent differing serial stages of computational evolution. Like the evolutionary rates of all species, some facets of automation will rapidly change into autonomy. In other contexts, simple automation will continue to suffice. In these latter cases, little impetus will drive such specific systems to any higher level of complexity. But we should make no mistake; the transition from automation to autonomy is rapidly approaching. The landscape of our own natural world provides a facile analogue of this overall mimetic, inter-species existence (Hancock 2014a). The only fundamental difference between the evolutionary conceptions of Lamarck and Darwin lie in the passage of time. The apparently haphazard exploration of possible developments in the realm of life via the engine of spontaneous mutation (Darwin) applies to humans. The more intentional and notionally, rational strategy of iterative change (Lamarck) applies to machine evolution. However, each process actually serves to explore the exact same fundamental realms of possibility. The fact that technological evolution most closely approximates Lamarck's vision does not mean species evolutionary principles are not also applicable in this latter technological case. Such technical systems simply change more quickly and, supposedly, in more purposive and logical sequences.

In their first phase of incarnation, autonomous systems will be fundamentally characterised by an increase in the indeterminacy of their action. As these progressively more independent systems 'explore' the myriad operational phase spaces available to them, the degree of certainty implicit in our reductionistic perspective (and characteristics of what we now conceive of as a major feature of automation) will progressively dissolve. We see pale anticipations of this forthcoming state in circumstances characterized for example, as 'mode error awareness' (Sarter and Woods 1995). Here, individuals in control of a system confuse its current state (i.e. mode of operation) to their potential detriment. The analogy here with growing autonomy is not exact, but the behavioural confusion will be very similar. Should a machine possess a simple bifurcated mode (e.g. on vs. off), then mode confusion may be minimal (but see Hancock 2011). However, even common systems, such as personal vehicles, can now be supported by millions of lines of operational code (Zax 2012). Increasing operational degrees of freedom necessarily lead to mode proliferation. In combination with operational speed, it is already the case that many 'insensate' automated systems cycle through states at such a rate that no human controller (individually or team) can ever keep pace. While the system at hand may be operating with perfect reliability

and perfect predictability it may still appear opaque to the attendant operator. This latter, effective indeterminacy will become ever more worrisome. We have already seen some foreshadowing of such effects. For example, one advertising agency recently apologised for inappropriate sexual content which 'popped-up' on a popular children's smartphone app. The company lamented that they had been unable to identify which network was responsible for the incursion. A spokesman noted that 'we can't guarantee control of third party ads on our site'. The editorial went on to note that:

> The aim of programmatic [i.e., automation-mediated] advertising was to remove human ad buyers on the grounds that they were expensive and didn't always get the best deal. Now that it has emerged that some major sites and apps have lost control over their pages, it looks as if humans might have had their uses after all. (Editorial 2015) [parentheses mine]

While such events may prove embarrassing and even generate a degree of trepidation and trauma in exposed children, it is the self-same concern for unpredictability that has to be expressed in the control strategies for extremely high-risk, high-consequence systems.

For critical systems, such as power plants, aircraft and the like, it may well become effectively impossible, either prospectively or even in retrospect to conduct standard, traditional verification and validation procedures. Indeed, we will have to begin to ask questions as to what verification and validation actually mean for these increasingly autonomous systems. Equally, speed and complexity threaten to defeat standard test and evaluation processes, at least conducted by any human agency. In the same way that our current vehicles are robots which are built by robots, we may well have to construct either an effective 'praetorian guard' or more appealingly a set of 'guardian angels' who will act to indemnify us from the indeterminate actions of such evolving autonomous systems (and note the use of the personal pronoun here). We see some such frissoms of concern in respect to human–robot interaction where the issues of trust and machine (robot) transparency are becoming of ever greater importance (Hancock et al. 2011; Sanders et al. 2015). But do such autonomous systems truly yet exist? And how would we know when and if they come 'on-line?'

3. Revisiting Turing

Perhaps one of the most important questions for all of humanity then concerns the potentiality for sentience in our iatrogenic, (i.e. self-created), technological systems (Hancock 2013). This question is an ancient one and, over the years, has occupied the minds of some of our very best. One standard approach has been to adopt the criteria advocated by Alan Turing (1950). His test is formally equivalent to the solution suggested to the philosophical conundrum of induction. Induction had become, in Broad's prescient words, 'the glory of science and the scandal of philosophy' (Broad 1926). It was a riddle to which Karl Popper provided an apparently acceptable resolution. It was a resolution that Fisher encoded and operationalised for the behavioural sciences in the form of null hypothesis testing (Fisher 1935). We are only now emerging from the shadow of this enshrined scientific shibboleth. Turing himself derived his own test from a Cambridge Parlour Game in which an interlocutor (tester) sought to distinguish the sex of two 'players' who could not be seen, but to whom a series of queries could be posed. If after the requisite time, or limited number of questions, the tester could not distinguish which was the male and which the female, then they lost. Successful distinction however, connoted victory. Stated in its simplest terms, the test requires us to distinguish a 'significant' difference between the streams of data emerging from each player and then link those streams to the purported characteristics (e.g. male vs. female; then sentient vs. insensate judgments) of the specific originating entity. In this manner, Turing asserted that if we were not able to distinguish human from machine (i.e. there were no significant differences derived from the intrinsically applied t-test), then the consciousness and intentionality that we attributed to humans must also be granted to that machine. To date, no machine has completely passed the Turing test, although a number have veered close to success (and see e.g. Parsons 1990; Moor 2003). But, we must consider here what that success might look like.

Over the years, there have been many critics of Turing's criterion (see e.g. Searle 1990). What often remains obscure is the precise, transactional nature of the interrogation. Presumably, ever more intelligent inquisitors are progressively less liable to 'fall' for the machine's answers and thus render an incorrect verdict. In signal detection terms, the system must exhibit at least as good as, if not superior, d' (and also Beta) capacities than the interlocutor (Hancock, Volante, and Szalma forthcoming). The machine must thus prove to be at least as intelligent if not more so than the questioner. That the attribution of intelligence is contingent upon a priori intelligence of the judge is not simply a self-referential loop, it also implies a non-stationary criterion and not one single and ubiquitous threshold for the affirmation of machine intelligence. This same critique can be applied to the recognition of the threshold for autonomous systems. The fact that the Turing test is hopelessly anthropocentric is well known and there are strong reasons to believe that our present vector of technical progress in computational abilities is not primarily geared to the reproduction of human capacities (Hancock 2007). By this I do not mean that there are not many manifest projects designed to explore and achieve the recreation of human capacities – assuredly there are. What I mean is that our currently predominant computer-based technical architectures are not set up to exactly recreate human processing characteristics. Perhaps they never should be?

4. Caveat Machina

Before we proceed to consider the limits and constraints that we may wish to impose on nascent autonomous systems, it is helpful to understand a comparable mandate with respect to current automated systems. After all, they provide a very concrete existence proof of the potential problems of unbound technical proliferation. Here, we can distil a preview of forthcoming concerns by evaluating some important current trends. Technology and automation are often hailed as the harbingers of progress and 'good', however, such approbation is not universal (Hancock 2014b). As epitomised in the nomothetic calculation of Benthamite[1] Utilitarianism, the fundamental question is: what connotes 'good?' Further, to what degree can we provide a quantitative assessment of that qualitative value; which, after all, is arguably one fundamental *raison d'etre* of ergonomics anyway (see Hancock, Weaver, and Parasuraman 2002). More problematically, how do we distil the number of individuals who each partake in this chimerical good? Let us look at specifics. We might agree that robots now produce cars in a more efficient manner than human beings and thereby render greater value and reliability to customers accordingly. Further, we can argue that this provides higher financial return for investment stake-holders; but what of car-workers? Benthamite calculus renders a static assessment of the trade-off at that one moment in time, but those human beings, now made redundant, live on.

Now let us suppose that the next increment in automation deprives another set of workers of employment, vocation and income. The marginal increase in efficiency is calculated to render a greater collective good, but these newly disenfranchised individuals join their ex-car producing peers as part of an undifferentiated and perhaps unemployed lump. Their collective dolour in losing their preferred occupation nowhere appears in the equation calculating overall 'good'. So what happens when the overall majority are now displaced by automation? No recidivist Luddite policy is liable to be then approved by those in charge to rectify the situation. In short, human happiness resides in memory and prospect as well as immediate sensation; it is a property of our past and our future as much as it is our present. Death by a thousand automated cuts may well prove just as painful, if not more so, than a single extinguishing thrust. It is arguable that technology, and its automated incarnation, is a major driver of profit and therefore influential in the on-going disproportional distribution of wealth. In this essence, automation fuels growth in the Gini Coefficient which illustrates the relative degree of this distribution of wealth (e.g. see Gini 1921; Ansell and Samuels 2014). Automation assists in this process of wealth re-distribution by generating greater wealth for the wealthy and inducing increasing poverty across the rest of the social spectrum. These are not simply polemical political points but arguably necessary sequelae of increasing automation. What of ergonomics in this problematic circumstance? Do our social and professional mores uphold the mantra of ever more efficiency and progress when the value derived therefrom is so manifestly disproportionally distributed? Is it this line of progress that we currently support when we advocate for ever greater automation? Where is individual human *good* considered, for example, when we now embrace such an over-arching 'systems-based' approach to our science (Carayon et al. 2015)?

In a recent and intriguing article, Frey and Osborne (2013) have attempted to plot the susceptibility of a panoply of occupations to automation; what they term computerisation (Frey and Osborne 2013, 2, note 1). Regardless of the arguments about their calculational techniques, the central thrust of their work is obvious. Advances in automation do not threaten human occupations uniformly. Rather, their model serves to identify a number of professions which are in imminent danger from automation's overthrow, candidates such as insurance underwriter, loan officer, and credit analyst are now to the very fore. Of course, it is not that these tasks go away but rather it is the human worker who is forced to migrate to another work sector; regardless of the degree to which they previously enjoyed underwriting analysing or whatever activity it was that is now gone from the human domain (Hancock, Pepe, and Murphy 2005). Perhaps most intriguingly for us, Frey and Osborne's underlying arguments rely on an ergonomic analysis of job content and job description. However, while references to financial considerations and economics feature heavily in their citations, once again formal reference to the science of ergonomics features not at all. This, despite the fact that fundamentally it is an ergonomic study! It provides yet a further testament to both the criticality and invisibility of our discipline.

While established and known evolutionary vectors such as limits on physical resources and the like, will surely direct the nature of future human society, autonomy in technology may well prove to be a saltation; a discontinuity in our gradualistic progress which will catapult us to a new state. Given the foregoing observations upon the power, influence and ubiquity of automation, we must be crucially concerned with this approaching watershed. Thus, we need to formulate design boundaries and constraints to such a leap if we are to exercise a meaningful

teleologic intent, rather than simply suffer the vacuous and happenstance outcome of the ministrations of so-many 'invisible hands'.

5. Morality first?

With respect to envisaged futures, there is a potential danger in over-featuring the dimension of pessimism. Both author and reader stand in danger of metaphorically throwing up their cognitive hands to embrace a dark form of fatalism. I very much wish to counteract such a propensity here (Hancock 1999). Indeed, I would rather observe that now is the moment when we can exert our most profound influence on our evolutionary line of development and hence one major motivation for the present appeal now. However, exhortations without solid recommendations tend to evaporate very quickly. Thus, the first of my recommendations concern the necessity for programmed morality (and see Hancock 2009). What I especially advocate here is that we insist that morality for machines is a design imperative. This is not simply another exhortative expression since the area of research on moral machines is actually quite a broad and burgeoning one (see e.g. Murphy and Woods 2009; Wallach and Allen 2009). What is clear is that 'morality' writ large, is an amorphous construct whose specification has tasked (and some would say eluded) the world's philosophers now for many centuries. But here we might rather consider imposing the requirement for *ethical principles* into machine operations. For example, Asimov's original 'laws of robotics' (Asimov 1942) looked to provide such an overarching set of guidelines which, sadly, have been more honoured in the breach than the observance. The challenge, of course, remains the specific coding and acceptance of each individual 'principle'. However, and in principle, such an endeavour should not be beyond us. What this may also mean is retro-fitting some current operational software with these mandatory ethical components. That morals vary across time and culture is self-evident; that individual, programmable ethical principles may transcend time and geography remains an essential, ergonomic challenge.

It is feasible and perhaps advisable that, like software bus access, these individual and grouped ethical principles become inherent OS characteristics and not simply 'apps' or 'add-on modules' to existing software architecture (e.g. patches). One imperative might well be 'first do no harm' (*primum non nocere*). Of course, we then immediately have to wrestle with the contradiction of the existence of 'killing' machines such as UCAVs. Again, this serves to emphasise that each act of design is an inherently political act, whether the designer is aware of it or not (see Hancock 2005, 2009). But, as scientists and ergonomists we must decide whether we lead or whether we follow; whether we innovate or whether we serve? Such decisions are necessarily always political and raise in us both conscience and concern. Sadly, some scientists finesse the question by protesting that such social issues are beyond the purview of science itself. Ergonomists do not possess that luxury. Yet, with growing autonomy, it may only be a short time in the future when such design decisions are removed from human control completely. And indeed, we cannot neglect to understand that providing operating systems with ethical algorithms is again necessary to introduce at least the appearance of further independence of action. Moreover, the introduction of such algorithms also represents another step along the path to nascent machine intentionality. As has been noted, the road to hell is paved with good intent and we must be very wary that ethicality per se, does not serve as one instrument of progress towards unbounded machine autonomy.

6. Send in the drones

One place where the translation from inner loop, to outer loop, to out-of-the-loop progression is unfortunately, liable to be accelerated, is in the automatisation and eventual autonomy of killing. Our contemporary asymmetric warfare environment brings this progress to the fore. However, this is only one extreme example of the general trend towards unconstrained autonomy. In general, living organisms possess a strong degree of inherent inhibition about destroying members of their own species. Even the titanic struggles of male members of fierce, large, mammalian species most often stop short of such a destructive act. Indeed, killing in this context perhaps traduces the very purpose of the overall process in which rivals, albeit defeated ones, still have an important role to play. The exception is *homo sapiens*. Sadly, and perhaps progressively, via the use of technology, we have become more and more capable of destroying other human beings. The sadness of modern conflict is that often those destroyed are not merely 'innocent' but their collateral destruction is both biologically maladaptive and sociologically counter-productive. Such is the pathological state of our world.

As with all contemporary problems, technology appears to have much to offer. In the case of asymmetric warfare, this is embodied in the highly seductive and antiseptic solution found in the notion of autonomous killing (ICRC 2014). Here, I do not wish to engage in the moral battles about the validity of such killing (but see Benjamin 2013). Neither do I want presently to expand this issue into discussions of the morality of technology in general, since I have addressed such concerns elsewhere (Hancock 2009). I do, briefly want to focus upon the slippery slope that may lead to autonomous destruction (see Stowers et al. forthcoming). Paradoxically, to understand the impetus for autonomous

destruction, we must focus not on those who are destroyed but rather on the putative 'controllers' of current technologies. To the present, relatively little public attention has been directed to the unfortunate individuals whose job it is to perform the patriotic duty of remote killing. By day they live perhaps even an excessively 'normal' life, by night, they stalk the enemies of their state with legitimized assassination at their fingertips. These individuals, unlike their cockpit-bound forebears, witness death close up. They may have 'surveilled' or 'stalked' their target for days, perhaps even weeks or more. They may 'know' them in a way that airborne pilots never could, and they witness their destruction up close and in technicolour. For many individuals, this proves to be a morally ambiguous and overtly stressful circumstance at best. The technical temptation then will be to remove these individuals from the loop completely – hence autonomous killing.

We are not allowed to know the success or failure rates of our present form of drone warfare, nor the ratio of 'targets' destroyed to 'collateral' damage. Pessimists suspect that this ratio might be very low and that the information is kept hidden especially because of this fact. However, it is clear that such actions can, and do, have excoriating effects on normal individuals at both ends of that particular technological chain. Basically, armed drone operators are, on average, not insensitive killers; hence, the slippery slope. For, we have already taken the first step along the purgatorial pathway in removing combatants from direct combat. Each of the following, sequential iterations towards autonomous killing then become progressively easier. Necessarily, the release of lethal force, in its current form, requires a series of steps. The temptation to automate first some, then all of those steps will prove ever harder to resist. After all, isn't efficiency the purpose of progressive technology? We can well imagine this progress. First, automation's help will be invoked to process the prohibitive mass of surveillance data. How can it not? To be actionable, such data must be current and the days or hours of human processing can well defeat this real-time purpose. Next comes flight control. Do we really need the 'stick monkeys,' or are their days also numbered? Now the rate-limiting element becomes the invidious mathematics of destruction. How and when can the 'shot' be taken that maximises success and minimises damage? But this, after all, is only computational re-iteration of Bentham's greater good. It is a signal detection criterion setting exercise and the momentary Beta level assumedly will be much better calculated by machine? Critically, that machine has no conscience to trouble it, no personal morality to traduce, and no post-traumatic stress symptoms to treat. Like all such justifications in general, the populace will be sold on the delusion of democratically chosen destruction. However, we will then have sown the explicit technological seeds of our own destruction. And the resulting harvest will come much more quickly than we may desire. Although the example toward the realisation of autonomous killing is a very stark one, it is only a single exemplary representative of the more general case of the abdication of control to unconstrained autonomous systems. No rational species should be the architect of its own destruction.

7. The 'off' switch

It is obvious that we have created a technological infrastructure that cannot simply be turned 'off'. Despite the siren call of naturalistic Arcadians (cf., Milton 1667; Nietzsche 1878), any aspiration for an untrammelled world of the noble savage remains simply delusional. As Thoreau noted, we have now become tools of our tools and such is the degree of symbiosis between modern humans and technology, we really should regard ourselves as co-dependent species, with modern civilisation as the prime emergent property. That being said, one design constraint for autonomous systems must be that modular elements of the technological infrastructure should always, functionally, be able to be circumvented or at least isolated. Perhaps this facet of control can best be couched in terms of system 'resilience' (see Hoffman and Hancock 2016; Hoffman, Hancock, and Woods forthcoming). Here, we might express our aspiration for 'redundancy' and 'back-up' in the face of elemental failure. However, in reality, this may simply be a form of a human defence mechanism. In various systems we have already witnessed numerous occasions of cascading failure. As systems become more inter-related, the opportunity for, and sheer scale of these cascading failures also necessarily increase. Designing buffering capacities into the overall technical architecture at a more global 'systems' level can therefore facilitate both operational resilience and the retention of human control. Of course, how and what is switched off, and under what circumstances, become context-contingent issues. Manifestly, these 'stop orders' will also traverse across national and institutional boundaries. The spectre of local optimisation for one small constituency leading to global sub-optimisation and even disaster still persists. However, that form of dysfunction is already prevalent on our planet and the case devolves to one of management not of absolute prevention. If we look to create fully autonomous systems to manage these local versus global optimisation concerns, we will have already abrogated our necessary minimal level of human control. What will derive then is, like our own future, indeterminant. However, we shall have to relinquish the illusion of control; and at present our human narrative dictates that we have no choice but to believe in free will.

8. Considering the antithesis

The central thesis of this present commentary is that; (i) automation is ubiquitous and its penetration and influence continue to grow across all of human society; (ii) the evolutionary vector of such automation tends towards increasing autonomy in emerging systems founded upon this burgeoning automation; (iii) while yet to find a current, unequivocal existence proof, such autonomous systems will act (or appear to act) as independent, self-directed entities; (iv) it is advisable that we now work on design principles and imperatives that effectively constrain such systems with respect to their future paths of action. I here take the first two points as essentially incontestable. Point three is much more polemical, but with the included parentheses, I think it is supportable. My antithetical discussion here concerns the final assertion: that of continued human control. As the well-known poem 'Invictus' exhorts us to be the master of our fate and the captain of our soul (Henley 1888), it seems to be self-evident that we humans wish to retain our privileged place at the peak of living things. However, can we truly say that we always know best? If we examine our current record of planetary stewardship it is painfully obvious that we are lacking in both rationality and a necessary standard of care. It may well be possible that globally-interconnected operations are better conducted by quasi- and subsequently fully autonomous systems? What of Ergonomists if this anti-thesis were proved to be either the preferable choice or the de facto default position?

9. Summary and conclusion

In 1975 there was a meeting held at the Asilomar Conference facility in Pacific Grove, California. The purpose was to examine the concerns over research into recombinant DNA and what risks such work could pose to continuing humans existence (see e.g. Kourilsky 1975; Wade 1979). The purpose was to collect the community of experts to solicit their input, ideas, understanding and projections with respect to what was perceived than as an imminent threat. Here, I do not rehearse the subtle and not-so-subtle scientific and political wrangling that derived from this meeting (but see Frederickson 1991). While one can always debate such outcomes, we should at least applaud the fundamental motivation behind the meeting and the holding of it. But the birth and growth of autonomous systems promise a threat and risk at least as great as that of the DNA question if not more so. Indeed, if we believe that memes emerge from genes, and that technology is the physical representation of memes, then autonomy actually represents the same threat but in an emergent form. I believe it is now the time for us to convene an equivalent meeting. The stakes are, I believe, even higher. Whether ergonomists have the prescience, the foresight, or the fortitude to advance and lead this necessary enterprise is the challenge for our own professional future.

Ergonomics and Human Factors are purpose-directed disciplines that seek to mediate between human beings and the machines they create. For most of their existence, tools and technology have served to render human visions into material form in order to facilitate individual and collective well-being. That such technology and much of its associated progress has derived from the conflictive nature of the human character remains problematic. Humans and their propensity toward individual and small group optimisation are often in conflict with the goals and aspirations of some putative 'other', out-group (broadly defined). Such an inherent propensity to conflict has proved to be no great cause for celebration. The wars with our distant kin, other living things, and the very environment that sustains us have seen great expansions of such material visions. Yet today many of these advances stand in direct opposition to and threaten the continued existence of our whole species. Now, the very technological vehicles of that 'progress' are beginning to provide a profound and even existential threat. If we do not react and respond, progress will still continue apace, but to what end? In terms of the chess game of life – it's our move.

Note

1. The assertion 'greatest good for the greatest number' often ascribed to Bentham actually pre-dated him and was, even at that juncture the subject of debate, contention and rightful rejections. (see https://en.wikiquote.org/wiki/Jeremy_Bentham).

Acknowledgments

I am most grateful to the insightful comments of two anonymous reviewers on an earlier version of this work.

Disclosure statement

No potential conflict of interest was reported by the author.

References

Ansell, B. W., and D. Samuels. 2014. *Inequality and Democratization: An Elite-competition Approach*. Cambridge: Cambridge University Press.

Asimov, I. 1942. *I Robot*. New York: Gnome Press.

Bartlett, F. C. 1962. "The Future of Ergonomics." *Ergonomics* 5: 505–511.

Benjamin, M. 2013. *Drone Warfare: Killing by Remote Control*. New York: Verso Books.

Broad, C. D. 1926. *The Philosophy of Francis Bacon*. (p. 67). Cambridge: Cambridge University Press.

Carayon, P., P. A. Hancock, N. Leveson, I. Noy, L. Sznelwar, and G. van Hootegem 2015. "Advancing a Sociotechnical Systems Approach to Workplace Safety – Developing the Conceptual Framework." *Ergonomics* 58 (4): 548–564.

De Chardin, T. 1955/1959. *The Phenomenon of Man*. New York: Harper-Collins.

Editorial. 2015. "Ad Nauseam." *Private Eye* 1408.

Fisher, R. A. 1935. "The logic of inductive inference." *Journal of the Royal Statistical Society* 98 (1): 39–82.

Frederickson, D. S. 1991. *Asilomar and Recombinant DNA: The End of the Beginning*. Accessed January 13, 2016. www.ncbi.nlm.nih.gov/books/NBK234217/

Frey, C. B., and M. A. Osborne. 2013. *The Future of Employment: How Susceptible Are Jobs to Computerisation*. Oxford: Oxford Martin School, University of Oxford.

Gini, C. 1921. "Measurement of Inequality of Incomes." *The Economic Journal* 31 (121): 124–126.

Hancock, P. A. 1999. "Life, Liberty and the Design of Happiness." In *Automation, Technology, and Human Performance: Current Research and Trends*, edited by M. Scerbo and M. Mouloua, 42–47. Mahwah, NJ: Erlbaum.

Hancock, P. A. 2005. "Nature Will Not Be Fooled: Observations on the Interdependence of Science and Politics." Paper presented at the 171st National Meeting of the American Association for the Advancement of Science, Washington, DC, February.

Hancock, P. A. 2007. "On the Nature of Time in Conceptual and Computational Nervous Systems." *KronoScope* 7 (2): 185–196.

Hancock, P. A. 2008. "Fredric Bartlett: Through the Lens of Prediction." *Ergonomics* 51 (1): 30–34.

Hancock, P. A. 2009. *Mind, Machine and Morality*. Chichester: Ashgate.

Hancock, P. A. 2011. "The Key to a Quiet Life … or Death?" *The Ergonomist* 487: 4–5.

Hancock, P. A. 2013. "In Search of Vigilance: The Problem of Iatrogenically Created Psychological Phenomena." *American Psychologist* 68 (2): 97–109.

Hancock, P. A. 2014a. "Autobiomimesis: Toward a Theory of Interfaces." Invited Plenary Keynote Presentation given at the 6th International Conference on Automotive User Interfaces and Interactive Vehicular Applications, Seattle, WA, September.

Hancock, P. A. 2014b. "Automation: How Much is Too Much?" *Ergonomics* 57 (3): 449–454.

Hancock, P. A., D. R. Billings, K. Olsen, J. Y. C. Chen, E. J. de Visser, and R. Parasuraman. 2011. "A Meta-Analysis of Factors Impacting Trust in Human-Robot Interaction." *Human Factors* 53 (5): 517–527.

Hancock, P. A., A. Pepe, and L. L. Murphy. 2005. "Hedonomics: The Power of Positive and Pleasurable Ergonomics." *Ergonomics in Design* 13 (1): 8–14.

Hancock, P. A., W. Volante, and J. L. Szalma Forthcoming. "Defeating the Vigilance Decrement." *IIE Transactions on Occupational Ergonomics and Human Factors*.

Hancock, P. A., J. L. Weaver, and R. Parasuraman. 2002. "Sans Subjectivity, Ergonomics is Engineering." *Ergonomics* 45 (14): 991–994.

Henley, W. E. 1888. *A Book of Verses*. London: Nutt.

Hoffman, R. R., and P. A. Hancock 2016. "Measuring Resilience." Submitted.

Hoffman, R. R., P. A. Hancock, and D. D. Woods. Forthcoming. "Measuring Effective Resilience." *IEEE Intelligent Systems*.

ICRC (International Committee of the Red Cross). 2014. *Report of the ICRC Expert Meeting on 'Autonomous Weapon Systems'*. Geneva, Switzerland.

Kourilsky, P. 1975. "Manipulations génétiques *in vitro*: compterendu de la conference de Pacific Grove [Genetic manipulations in vitro: An account of the Conference at Pacific Grove]." *Biochimie* 57 (2).

Milton. J. 1667. *Paradise Lost*. London: Simmons.

Moor, J. H., ed. 2003. *The Turing Test: The Elusive Standard of Artificial Intelligence*. New York: Springer.

Nietzsche, F. 1878. *Human All Too Human*. Chemnitz: Ernst Schmeitzner.

Kurzweil, R. 2005. *The Singularity is near*. New York: Viking.

Murphy, R. R., and D. D. Woods. 2009. Beyond Asimov: The Three Laws of Responsible Robotics. *IEEE Intelligent Systems*, July/August, 2–6.

Parsons, H. M. 1990. "Turing on the Turing Test." In *Ergonomics of Hybrid Automated Systems*, edited by W. Karwowski and M. Rahimi, 787–793. Amsterdam: Elsevier.

Sanders, T., P. Oppold, K. E. Schaefer, T. Kessler, P. A. Hancock, and J. Y. C. Chen. 2015. "The Influence of System Transparency on User Stress Levels." *Proceedings of the 19th Triennial Congress of the International Ergonomics Association*, Melbourne, Australia.

Sarter, N. B., and D. D. Woods. 1995. "How in the World Did We Ever Get into That Mode? Mode Error and Awareness in Supervisory Control." *Human Factors* 37 (1): 5–19.

Searle, J. R. 1990. "Is the Brain's Mind a Computer Program?" *Scientific American* 262 (1): 26–31.

Stowers, K., A. Leva, G. M. Hancock, and P. A. Hancock. Forthcoming. "Life or Death by Robot." *Ergonomics in Design*.

Turing, A. M. 1950. "Computing Machinery and Intelligence." *Mind* 59: 433–460.

Wade, N. 1979. *The Ultimate Experiment: Man-Made Evolution*. New York: Walker.

Wallach, W., and C. Allen. 2009. *Moral Machines: Teaching Robots Right from Wrong*. Oxford: Oxford University Press.

Zax, D. 2012. "Many Cars Have a Hundred Million Lines of Code." *MIT Technology Review*, December 3. https://www.technologyreview.com/s/508231/many-cars-have-a-hundred-million-lines-of-code/

Nature: a new paradigm for well-being and ergonomics

Miles Richardson, Marta Maspero, David Golightly, David Sheffield, Vicki Staples and Ryan Lumber

ABSTRACT

Nature is presented as a new paradigm for ergonomics. As a discipline concerned with well-being, the importance of natural environments for wellness should be part of ergonomics knowledge and practice. This position is supported by providing a concise summary of the evidence of the value of the natural environment to well-being. Further, an emerging body of research has found relationships between well-being and a connection to nature, a concept that reveals the integrative character of human experience which can inform wider practice and epistemology in ergonomics. Practitioners are encouraged to bring nature into the workplace, so that ergonomics keeps pace with the move to nature-based solutions, but also as a necessity in the current ecological and social context.

Practitioner Summary: Nature-based solutions are coming to the fore to address societal challenges such as well-being. As ergonomics is concerned with well-being, there is a need for a paradigm shift in the discipline. This position is supported by providing a concise summary of the evidence of the value of the natural environment to well-being.

1. Introduction

Ergonomics is concerned with well-being. The second of the two objects of the royal charter of the Chartered Institute of Ergonomics and Human Factors (CIEHF) refers to the promotion of well-being through the use of ergonomics knowledge (CIEHF 2014). This paper argues that the ergonomics knowledge base should include the benefits of nature for human well-being. The health benefits of nature outlined in the present review provide that knowledge and show that nature provides a new paradigm for well-being (European Commission 2015; Stevens 2010), in contrast to the existing biomedical model of health care that essentially views people as separate from the environment and affected by events, with deviation from normal being treated with costly interventions. Upstream nature-based solutions that harness the power of nature to turn challenges into opportunities are coming to the fore to address societal challenges such as well-being (European Commission 2015). As the EU research and policy agenda recognises the human need for nature, disciplines such as ergonomics will also have to transform the solutions they offer, bringing nature into the workplace to address major challenges such as work-related stress and ill-health. At present, workplace health programmes tend not consider nature (Lottrup, Grahn, and Stigsdotter 2013; Trau et al. 2015), despite the health benefits of nature being known for many years (Logan and Selhub 2012; Nisbet, Zelenski, and Murphy 2011). This mirrors, and is perhaps caused by, the wider societal dissociation from nature in an age of rapid global urbanisation (Barnosky et al. 2011; Maller et al. 2009), bringing about increases in mental health issues (Walsh 2011) and lifestyle diseases (Pappachan 2011). This should be noted by ergonomics practitioners, as reduced performance at work and long-term sickness absence are related to mental health issues (Sahlin et al. 2014).

Nature does much more than provide a route to workplace wellness; it provides a new paradigm for ergonomics informing epistemology of the discipline (the first object of the CHIEF royal charter) and wider practice through revealing the integrative character of human experience. The continued loss of biodiversity (EEA 2015) and the links to human well-being (von Hertzen et al. 2015) have brought the concept of connection to nature and reconnecting people with nature to the fore (DEFRA 2011). The concept has been the focus of many high-profile campaigns recently (eg Wild Network 2015), including ones that focus on the workplace (eg 30 × 30 at Work, David Suzuki Foundation 2015). Rather than simple exposure to nature, there is emerging evidence that an affinity or connection to nature is good for well-being to a level similar to established variables such as income and education (Capaldi, Dopko, and Zelenski 2014). The construct of nature connectedness (NC) is seeing one's *self* as part of a wider ecology and has a positive impact on valuable workplace factors such as vitality, creativity and happiness (Capaldi, Dopko, and Zelenski 2014), while also leading to other benefits such as pro-social behaviour and pro-environmental behaviour (eg Frantz and Mayer 2014; Zhang et al. 2014). This concept of a shared place in nature ties into integrative perspectives on ergonomics, and the indivisibility of cognition and environment (Dekker, Hancock, and Wilkin 2013; Flach, Dekker, and Jan Stappers 2008).

In order to support the case that nature provides a new paradigm for ergonomics, the beneficial impact of nature is reviewed to indicate how nature can help deal with workplace well-being, with well-being defined in the review as encompassing variables such as life satisfaction, vitality and mood (Cervinka, Roderer, and Hefler 2012). To support this need and impact on decision-making, the present review provides a concise summary and armoury of the evidence of the benefits of nature to well-being while highlighting the emerging importance of connectedness to nature. In addition to bringing the benefits of well-being and innovation into the workplace, Ergonomics and Human Factors practitioners can, at the same time, contribute to the revival of nature through their efforts (cf. Hanson 2013). As well as bringing nature into the workplace for its benefits to humans, there is an opportunity, indeed a necessity, to understand and promote a connection to nature so that benefits to both humans and the natural world can be realised in order to deliver a sustainable future. The research presented below focuses on everyday exposure to nearby nature (Kaplan and Kaplan 1989); the nature accessible on our journeys to work, in our lunch breaks or even viewable through the window (Richardson, Hallam, and Lumber 2015). It is this everyday nature that will increasingly become where we engage with nature in a progressively urbanised world (Dunn et al. 2006), although it should be noted that there is emerging evidence of links between biodiversity and well-being (eg von Hertzen et al. 2015). Papers were selected based on the themes of nature, health, well-being and restoration in order to provide a selection of peer-reviewed studies indicative of the area from broadly relevant populations. The selected papers were then tabulated to indicate key characteristics such as measures used and design in order to provide an accessible overview. In order to evidence nature as a new paradigm for well-being and ergonomics, the current paper opens by presenting the body of empirical research examining the benefits of exposure to

nature and then the concept of a connection to nature is introduced. Next, the theories regarding the human need for nature are briefly introduced. The wider implications of the concept of NC for a paradigm shift in ergonomics are then considered.

2. Beneficial effects of nature

The following sections introduce the evidence for the role of nature in well-being. Starting with general health it goes on to examine subjective well-being and restoration, providing an evidence base for the ergonomist promoting provision and access to nature in the workplace in order to improve well-being, and ultimately absenteeism and productivity. Going beyond everyday exposure to nature, to activity in nature, there is evidence that 'nature experience', including wilderness experiences, have benefits for health and well-being (eg Hartig, Mang, and Evans 1991). This is not reviewed here but has been comprehensively discussed within a theoretical context by Hartig et al. (2011).

2.1. General health benefits

A cluster of studies investigating the link between nature and health have focused on the health gap between people living in rural and urban locations, providing underlying support to bringing nature to the workplace. Studies in many countries have shown that urban people are more likely to report a poorer health status than their rural counterparts (Verheij et al. 1998; Weich, Twigg, and Lewis 2006). However, recent epidemiological studies have demonstrated that the association between people's perceived health and the availability of green spaces is stronger than the one between health and urbanicity (de Vries et al. 2003; Maas et al. 2006; Verheij, Maas, and Groenewegen 2008). These findings suggest that the urban–rural health gap is not fully accounted for by differences in environmental factors and unhealthy behaviours and is instead mediated by an actual discrepancy in nature availability.

Several other studies add to the growing evidence of the major health benefits provided by the natural environment, which can feed into workplace design and work routine options. For example, a study by Agyemang et al. (2007) established a relationship between the presence and the quality of green spaces in the neighbourhood and lower hypertension rates among residents. Donovan et al. (2013) found evidence to suggest that areas subjected to a loss of trees owing to disease had increased mortality due to lower respiratory-tract and cardiovascular illnesses. Further, in a study by Raanaas, Patil, and Hartig (2012), patients of a residential rehabilitation programme self-reported a better physical and mental health if their bedroom had a view of natural surroundings. Finally, green buildings have higher light levels, greater access to windows, conditions associated with thermal comfort and fewer airborne particulates with occupants reporting a lower frequency of visual and physical discomfort symptoms, better mood and better sleep quality (Newsham et al. 2013). Green spaces have also been argued to facilitate exercise and social contact (Van den Berg, Hartig, and Staats 2007) and a study by Van den Berg et al. (2010) has shown that the amount of nearby green areas moderates the relationship between stress and health, thus suggesting that nature might help preserve health by acting as a buffer against stress. As indicated by the summary presented in Table 1, generic guidelines for work can be derived from such research, namely the availability of, and access to, green space with trees.

2.2. Well-being benefits

A positive association between nature and subjective well-being has also been established, see Table 2. Well-being is a complex construct for which several definitions exist (McMahan and Estes 2011). In this context, it has a wide-ranging meaning, encompassing variables such as mood, life satisfaction, psychological well-being and vitality (Cervinka, Roderer, and Hefler 2012). Availability of green spaces in nearby areas (Gidlöf-Gunnarsson and Öhrström 2007; Groenewegen et al. 2006), natural views from windows (Kaplan 2001; Ulrich 1979) and time spent in nature (Lafortezza et al. 2009; Pretty et al. 2007) have all been demonstrated to increase well-being, including job satisfaction (Leather et al. 1998). Furthermore, exposure to nature, both physically and visually, has been shown to have a positive effect on mental health (Guite, Clark, and Ackrill 2006; Ottosson and Grahn 2008), vitality (Guite, Clark, and Ackrill 2006), mood (Hartig et al. 1996, 2003; Hull 1992; Mayer et al. 2009) and emotional self-regulation (Korpela et al. 2001).

2.3. Restoration

Visual, virtual or actual exposure to nature has been related to improvements in physiological responses (eg heart rate, blood pressure, muscle tension) (Agyemang et al. 2007; Miyazaki et al. 2011; Ottosson and Grahn 2005; Park

et al. 2010; Ulrich et al. 1991), attention capacity (Staats, Kieviet, and Hartig 2003) and affective states (Berto 2005; Hartig et al. 2003; McMahan and Estes 2015), following a stressful event (see Table 3). These effects are referred to as expressions of nature's restorative power (Ulrich 1979); their empirical investigation has started with the pioneering studies by Ulrich (1984) and Kaplan and Kaplan (1989). Ulrich (1984) has shown that patients who had a hospital room with a view of trees recovered more quickly and required fewer painkillers after a gallbladder surgery than patients with a view of a brick wall. Kaplan and Kaplan (1989) have demonstrated nature's ability of restoring mental capacity after prolonged fatigue in a series of studies. Subsequently, numerous studies have replicated these findings, with a body of research utilising various antecedent conditions (mental fatigue, stress, anxiety), assessed variables (physiological, affective or cognitive measures), and type of nature exposure. Positive participant response has been found both after spending time in nature (Hartig, Mang, and Evans 1991; Hartig et al. 2003), and after being exposed to real or virtual natural scenes (Berto 2005; Hartig et al. 1996; Laumann, Gärling, and Stormark 2003; McMahan and Estes 2015; Parsons et al. 1998; Ulrich et al. 1991; Van den Berg, Koole, and van der Wulp 2003); and it should be noted that actual nature gives a stronger response than virtual nature (Kahn et al. 2008). With regard to the work and the workplace, Lee et al. (2015) found that a 40-second view of green roof can restore attention, similarly Chow and Lau (2015) found that people exposed to photos of nature restored their 'inner-strength' after depletion to have greater persistence in logic and reasoning tasks. Lottrup, Grahn, and Stigsdotter (2013) found a significant relationship between decreased stress and workplace attitude, and visual and physical access to workplace greenery. Such findings have informed workplace stress management interventions, for example, Sahlin et al. (2014) reduced long-term sick leave and stress symptoms through a garden and nature-based intervention.

Collectively, results from these studies have confirmed the greater physiological, cognitive and affective restorative power of natural settings. Physiological and affective recovery from viewing a stressful movie (Ulrich et al. 1991), and a video of a drive (Parsons

Table 1. Summary of the nature and general health research considered.

Authors	Location	Sample	Design	Measures	Theme
Weich, Twigg, and Lewis (2006)	UK	7659	Cross-sectional	Self-reported mental health	Urban/rural
de Vries et al. (2003)	Netherlands	17000	Cross-sectional, regression	Symptoms, self-reported health, GHQ	Green space
Maas et al. (2006)	Netherlands	250782	Regression	Self-reported health	Green space
Agyemang et al. (2007)	Netherlands	1286	Regression	Blood pressure	Green space
Donovan et al. (2013)	US	N/A	Regression	Deaths	Trees and health
Raanaas, Patil, and Hartig (2012)	Norway	278	Quasi-exp	Self-reported health, SF-12	Natural Views
Newsham et al. (2013)	US & Canada	2545	Cross-sectional	Self-reported health	Green buildings
Van den Berg et al. (2010)	Netherlands	4529	Regression	Symptoms, self-reported health	Green space

Table 2. Summary of the nature and subjective well-being research considered.

Authors	Location	Sample	Design	Measures	Theme
McMahan and Estes (2011)	Various	2356	Meta-analysis	Positive affect	Natural environment & positive affect
Gidlöf-Gunnarsson and Öhr-ström (2007)	Sweden	500	Cross-sectional	Self-reported well-being	Green space
Groenewegen et al. (2006)	Netherlands	403,000	Cross-sectional	Self-report & illness	Green space & health and wellbeing
Kaplan (2001)	US	188	Cross-sectional	Self-reported well-being	Natural views
Lafortezza et al. (2009)	UK and Italy	800	Cross-sectional	Self-reported well-being	Green space
Pretty et al. 2007	UK	263	Pre–post	Self-reported well-being	Green space Nature activities
Leather et al. (1998)	Europe	100	Cross-sectional	Self-reported well-being	Natural views
Guite, Clark, and Ackrill (2006)	UK	1012	Cross-sectional	Self-reported mental health/well-being	Green space
Ottosson and Grahn (2008)	Sweden	547	Cross-sectional	Self-reported well-being	Green Space
Hartig et al. 1996;	Europe	N/A	Empirical	Self-report	Natural views
Hartig et al. (2003)	Europe	112	Empirical	Blood pressure, self-report positive affect & perfor-mance	Exposure to nature
Hull (1992)	US	108	Pre–post	Self-report mood	Urban park
Korpela et al. (2001)	Europe	199	Qualitative	Qualitative	Green space

et al. 1998) was faster in subjects who were exposed to natural, rather than urban, virtual scenes. Furthermore, exposure to nature stimuli has been shown to restore attention capacity (Berman, Jonides, and Kaplan 2008; Hartig et al. 2003), to foster positive affect (Berto 2005) and to improve mood and concentration (Van den Berg, Koole, and van der Wulp 2003). For instance, a view of nature (Hartig et al. 2003), and the presence of plants in the workplace (Lohr, Pearson-Mims, and Goodwin 1996) have been associated with a more rapid decline in blood pressure after attention demanding tasks, leading to improved worker productivity.

3. Connectedness with nature

Parallel to research investigating the benefits of being exposed to nature above, several recent studies have started to demonstrate the beneficial effects of NC (Mayer and Frantz 2004; Mayer et al. 2009), although the importance of being connected to nature and the involved mechanisms are still unclear (Mayer et al. 2009). Rather than being a connection across some form of artificial human–nature boundary, connectedness to nature is comprised of affective and experiential sense of belonging to the natural world (Mayer and Frantz 2004), and includes the extent to which nature is included within an individual's view of self (Schultz 2002). Individual differences are also important in possessing a connection to nature and involve the affective and experiential factors mentioned previously along with cognitive aspects (Zelenski and Nisbet 2014). Given the evidence presented above on exposure to nature, it would seem likely the level of connectedness to nature is important to health and well-being, and is therefore a potential route to improved health in the workplace.

3.1. The benefits of nature connectedness

While the studies introduced earlier have focused on the association between exposure to nature and well-being, an emerging body of literature, see Table 4, has found a relationship between positive affect and individual differences in connectedness with nature (Cervinka, Roderer, and Hefler 2012; Howell et al. 2011; Nisbet, Zelenski, and Murphy 2011). In particular, a connection to nature has been shown to significantly correlate with life satisfaction (Mayer and Frantz 2004), lower cognitive anxiety (Martyn and Brymer 2014), vitality (Cervinka, Roderer, and Hefler 2012), meaningfulness (Cervinka, Roderer, and Hefler 2012; Howell, Passmore, and Buro 2012; Mayer et al. 2009), happiness (Nisbet, Zelenski, and Murphy 2011) and mindfulness (Howell et al. 2011). These correlations are of a similar magnitude to those found between well-being and other variables, such as marriage and education, whose relationships with well-being are well established (Mayer and Frantz 2004). Further, in a recent meta-analysis, Capaldi, Dopko, and Zelenski (2014) found people with a stronger connection to nature experienced more life satisfaction, positive affect and vitality at levels associated with established predictors such as personal income. There is also emerging evidence of physiological responses to a more embedded experience of nature (eg Park et al. 2010).

A recent campaign that encouraged daily nature contact for one month delivered sustained increases in health and happiness, with improvements in connection to nature mediating that relationship (Richardson et al. 2016). A connection to nature has also been demonstrated to partially mediate the relationship between exposure to nature and well-being; people who are more connected with nature experience greater psychological benefits from contact with nature (Hartig et al. 2011).

Table 3. Summary of the nature and restoration research considered.

Authors	Location	Sample	Design	Measures	Theme
Park et al. (2010)	Japan	288	Pre–post	Heart rate, cortisol, blood pressure	Exposure to nature
Miyazaki et al. (2011)	Japan	420	Pre–post intervention	Physiological	Exposure to nature
Ottosson and Grahn (2005)	Europe	15	Pre–post intervention	Blood pressure & heart rate	Exposure to nature
Ulrich et al. (1991)	Sweden	40	Empirical	Skin conductance	Exposure to sounds of nature
Staats, Kieviet, and Hartig (2003)	Sweden	101	Empirical	Perceived restoration	Virtual nature
McMahan and Estes (2015)	Various	2356	Meta-analysis	Positive affect	Exposure to nature
Berto (2005)	Italy	32, 32, 32	Empirical	Sustained attention test	Virtual nature
Laumann, Gärling, and Stormark (2003)	Norway	28	Empirical	Posner's attention-orienting task	Virtual nature
Parsons et al. (1998)	US	160	Empirical	Blood Pressure, EMG	Virtual nature
Van den Berg, Koole, and van der Wulp (2003)	Netherlands	114	Empirical	Mood & concentration	Virtual nature
Kahn et al. (2008)	US	90	Empirical	Heart Rate	Natural Views
Chow and Lau (2015)	Hong Kong	42, 58, 185	Empirical	Reasoning performance	Exposure to nature
Lottrup, Grahn, and Stigsdotter (2013)	Sweden	439	Cross-sectional	Self reported stress	Green Space
Sahlin et al. (2014)	Sweden	33	Pre-post	Sick leave, symptoms	Nature activities
Lohr, Pearson-Mims, and Goodwin (1996)	US	96	Empirical	Blood pressure, heart rate, emotional state	Workplace plants

More recently, the aspects that mediate the relationship between NCand well-being have been investigated, with spirituality (Kamitsis and Francis 2013) and natural beauty (Zhang, Howell, and Iyer 2014) mediating the relationship between nature connection and psychological well-being. Thus, increasing people's connection to nature is at least as important as increasing the availability and access of green space, particularly in urban locations (Lin et al. 2014). While the emerging benefits of NC are important, there is a need for further understanding of how to facilitate and improve people's connection to nature, and how this might translate to a workplace context. Further, there is a need to understand the pathways by which people connect to nature, for example, through contact, meaning, emotional attachment, compassion and nature's beauty (Richardson et al. 2016). The benefits and routes to nature connection outlined above are summarised in Figure 1.

3.2. Nature–human relationship theory

Several theories have been developed to account for the human need for nature and the beneficial effects of nature. A brief insight of three key theories related to human–nature relationship is useful. Wilson's (1984) biophilia hypothesis is widely acknowledged and provides background and a catalyst for much research into human–nature relations (Hartig et al. 2011) and has informed the biophilic design of buildings (Kellert, Heerwagen, and Mador 2011; Ryan et al. 2014), which has clear links to ergonomics (Thatcher 2013). Biophilia states that humans have an inborn tendency to affiliate with nature (Wilson 1984). This spontaneous affiliation with nature is justified from an evolutionary perspective as humans have lived for most of their existence embedded in natural environments

(Frumkin 2001; Pretty 2002). Our cognitive and emotional apparatus instinctively respond with attraction or aversion to natural stimuli. Human innate affiliation with nature is therefore argued to be an indirect confirmation of its beneficial effects (Wilson 1984). Similarly, Ulrich's (1993) psycho-evolutionary model posits humans' innate affiliation with natural environments. In particular, Ulrich et al. (1991) argues that natural environments induce positive emotions and soothe autonomic arousal. This occurs because humans respond positively to natural environments, in which survival possibilities abound. Hence, natural environments elicit an affective and psycho-physiological restorative effect on humans (Ulrich 1993). Rather than focusing on nature's ability of restoring from stress, Kaplan and Kaplan's (1989) Attention Restoration Theory focuses on nature's role in recovering from mental fatigue, of particular relevance to ergonomics. According to Kaplan (2001), nature's main beneficial effect lies in the effortless attention and pleasurable fascination that natural settings elicit in humans. This provides them with a chance to restore their attention capacity and to recover from the mental fatigue caused by the cognitive tasks of modern society in which prolonged directed attention is required (Kaplan 1995).

From an ergonomics perspective, the cyberneticist Bateson (1972), and more recently Guddemi (2011), propose a systems-orientated approach to the relationship between individuals and nature. In this interpretation, consciousness, which is primarily goal-directed, is only a partial window on our systemic, dynamic relationship with our environment. A closer relationship with nature, which includes greater exposure to and immersion within nature, may facilitate a move away from a purely egocentric and goal-directed interpretation of the world and, allow us to

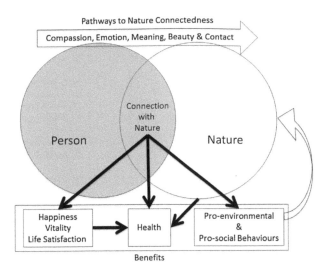

Figure 1. Summary of the benefits of, and pathways to nature connection.

develop a more holistic relationship with our environment in the broadest sense (Bateson 1972). One possibility is such a shift away from goal-driven activity may encourage an emphasis on right-hemisphere over left-hemisphere processing (Guddemi 2011). This shift in processing would also map to a shift between using local-feature processing, located in the left-hemisphere, and more global feature processing, located in the right-hemisphere (Fink et al. 1996), which could explain the restorative effects of nature after a demanding attentional task (Kaplan and Kaplan 1989; Staats, Kieviet, and Hartig 2003).

4. Research into the beneficial effects of nature: critical analysis

The reviewed studies constitute a comprehensive and diverse body of research, demonstrating exposure to nature can reduce hyper-tension, respiratory tract and cardiovascular illnesses; improve vitality and mood; benefit issues of mental well-being such as anxiety; and of particular note for workplace performance, restore attention capacity and mental fatigue. The research provides an evidence base for nature as a new paradigm for ergonomics and support for the ergonomist advocating provision and access to nature in order to improve health, well-being, restoration and ultimately absenteeism and productivity.

The literature presented is characterised by a variety of samples, designs (correlational, quasi-experimental, and empirical) and settings (laboratory and field). Nevertheless, it is not without methodological and conceptual limitations. The predominance of self-reported measures for the health and well-being assessments listed in Tables 1 and 2 is a weakness, as self-reported measures may lack objectivity and introduce reporting biases and artefacts (Braun et al. 2012).

However, physiological and non-self-report measures (including performance measures such as attention which are of interest to ergonomists) do feature strongly in the studies on the restorative benefits in nature (see Table 3), and measures such as heart rate and blood pressure here are precursors and indicators of health and well-being. This research also tends to be empirical, covering both exposure to real and virtual nature, raising confidence in the self-report studies that dominate the health and well-being literature.

Although most of the health-related studies in Table 1 have large sample sizes, a correlational design is used in many of the studies looking to establish the relationship between nature, health and well-being. This approach is repeated by those studies investigating the well-being benefits of the emerging construct of nature connection summarised in Table 4. However, these correlational studies do not allow for causal inferences. Now that the relationship between nature, health and well-being is well established investigating the mechanisms by which nature brings about health becomes paramount as this helps establish a cause-and-effect link (Kuo 2015). Whereas evidence of nature's positive effects abounds, a few studies have attempted to examine the mechanisms underpinning this relationship (eg Kuo 2015; Van den Berg et al. 2010). Although several possible mediators have been identified (eg exercise promotion, social contact facilitation) (Brown and Bell 2007; Kuo 2001; Maas et al. 2008), only nature's restorative power has been extensively researched (Hartig et al. 2003; Ulrich et al. 1991). This lack of investigation into the underpinning mechanisms and moderating factors is frequent in the environmental and eco-psychology literature (Winkel, Saegert, and Evans 2009). Because of the complex interactions among the environment, outcome variables and other psychological and social factors, research designs are often simplified leading to the mechanisms involved to not be extensively investigated. Therefore, a challenge for ergonomists is to disambiguate the direct and indirect factors involved in the benefits brought about by human relationships with nature and consider how these relationships can be deepened within the workplace through both environmental design and behavioural interventions. Ergonomists are well placed to take the holistic perspective required to progress this work, bringing together many disciplines and building on an understanding of the richness of the human–nature relationship revealed by a range of research approaches.

5. Implications for ergonomics

5.1. Applied value

The well-being benefits of nature are often overlooked in reviews and models of workplace well-being (eg Danna

Table 4. Summary of the nature connection research considered.

Authors	Location	Sample	Design	Measures	Theme
Cervinka, Roderer, and Hefler (2012)	Austria	547	Cross-sectional	Self-reported physical & mental well-being, SWLS, WHOQOL-Bref	NC & well-being
Mayer and Frantz (2004)	US	60, 102, 270, 135	Cross-sectional	NC, self-report well-being	NC & well-being
Mayer et al. (2009)	US	76, 92, 64	Empirical	NC, self-report positive affect	NC & well-being
Howell et al. (2011)	Canada	452, 275	Cross-sectional	NC, self-report well-being	NC & well-being
Nisbet, Zelenski, and Murphy (2011)	Canada	184, 145, 170	Cross-sectional	NC, self-report well-being, PANAS, PWB, SWLS	NC & well-being
Martyn and Brymer (2014)	Australia	305	Cross-sectional	NC, self report anxiety	NC & well-being
Howell, Passmore, and Buro (2012)	Canada	311, 227	Cross-sectional	NC, self-report well-being	NC & well-being
Capaldi, Dopko, and Zelenski (2014)	Canada	8523	Meta-analysis	NC, self-report positive affect	NC & well-being
Kamitsis and Francis (2013)	Australia	190	Cross-sectional	NC, self-report well-being – WHOQOL-Bref	NC & well-being
Zhang, Howell, and Iyer (2014)	US	1108, 151	Cross-sectional	NC, self-report life satisfaction	NC & well-being

and Griffin 1999; Wilson et al. 2004) and in guidance on creating healthy workplaces (eg Day, Kelloway, and Hurrell 2014). Similarly, although workplace health promotion is known to be valuable for employee's well-being, the literature is limited beyond traditional approaches such as exercise (eg Kuoppala, Lamminpää, and Husman 2008), despite nature exposure being an easy and inexpensive solution (Trau et al. 2015). Likewise, key texts in ergonomics do not promote the benefits of the natural environment for well-being and restoration of performance (eg Salvendy 2012). Given the importance of the work environment and that stressors at work are associated with ill-health such as common mental disorders (eg Stansfeld and Candy 2006), there is a need to promote the full range of solutions. A nature-orientated approach is taken in urban design for public health (eg Brown and Grant 2005; Tzoulas et al. 2007) and the design of biophillic buildings (eg Ryan et al. 2014), but there is a need for nature-based solutions to become part of ergonomics practice. The strong evidence base for the benefits of nature, and lack of formal guidance, allows the ergonomist freedom to make nature part of the working day in the most straightforward and cost-effective manner, with evaluation of outcomes where possible to build the research base.

For example, good work on the benefits of rest breaks on productivity and well-being (eg Dababneh, Swanson, and Shell 2001) that has informed practice can be enhanced through those breaks including restorative natural environments. There has been a shift from manual to non-manual work, and as the physical and chemical hazards have become more controlled, there has been a greater focus on the psychosocial environment at work, particularly the social environment and the key factors of psychological job demands and decision latitude (Kuoppala, Lamminpää, and Husman 2008). These factors can be used to identify high-strain jobs with greater risk of illnes, anxiety, depression and fatigue (Karasek and Theorell 1992). Just as social support can 'buffer' the impact of these demands (Johnson

and Hall 1988), nature, as evidenced in the literature above, also provides restorative benefits. Recently, Sachita and Ruchi (2015) have found that working in a restorative and green environment is a mediator of the relationship between organisational socialisation and employee happiness. Clearly, the beneficial effects of nature can be included in current models, as a restorative buffer and mediator of workplace well-being. Access to nature at work may be as fundamental as the need for a rest break.

The success of nature-based interventions will be influenced by the environmental context and also the workplace culture, for example, are activities needed to encourage employees to simply spend some time outside each day. This has some interesting implications when considering the constraints that work design may place on the access to nature. One area where this is apparent is shiftwork. There is mounting evidence for the negative consequences of shiftwork (eg Vyas et al. 2012), which can be mediated by quite practical issues, for example, the lack of opportunities for good nutrition, and the need for increased caffeine intake (Amani and Gill 2013). For many shift workers, there will be limited or no access to nature during rest periods simply because it is dark and nearby parks, or the areas surrounding the parks may be unsafe (Bedimo-Rung et al. 2005). If nature is to be implemented within worker well-being, strategies will also need to consider access for shift and night workers.

In addition to supporting a nature as a new paradigm for ergonomics and workplace well-being, the literature on nature's beneficial effects has great applied value for leveraging the value of the natural environment at work and informing practice, with recent examples demonstrating this well. In a correlational study, outdoor, indoor and indirect contact with nature within the workplace was positively related to decreases in stress and related health issues, suggesting contact with nature contributes to a healthy work environment (Largo-Wright et al. 2011). From an intervention perspective, Sahlin et al.

(2014) used nature and gardening activities within a multimodal stress management course. The 12-week course involving gardening and nature walks led to reductions in burnout, long-term sick leave and improved work ability over a 12 month follow-up. Similarly, Tyrväinen et al. (2014) showed that short-term visits to a large urban park during the working day reduced both perceived stress and cortisol levels, while Brown et al. (2014) found a nature-based 'Walks4Work' intervention to be more effective than a built environment walk in improving mental health. Although there is evidence of the benefits of a connection to nature, there is little work on how to improve connection to nature in a sustained manner. From the work presented above, suggestions include noting the good things in nature, such as nature's beauty, though writing or activities such as photography which can give walks from work a purpose. However, the literature is yet to provide a clear set of guidelines for the best nature-based pathways for workplace well-being.

Despite the lack of clear guidance, the breadth of research considered above shows that a great deal of benefit can come in three areas. Simple exposure to nature in the form of green spaces, gardens and trees and even plants in the office. There was evidence that simply having a view of such spaces is beneficial, with windowless workers more likely to want plants and pictures of nature (Bringslimark, Hartig, and Grindal Patil 2011). Once access to nature is established, informal measures can be taken to encourage employees to spend time in nature, both during breaks and as a location for meetings as part of health promotion campaigns. Secondly, and more specifically, given the research on restoration, time in nature can be formalised, particularly for those jobs that place high demands on attention. Thirdly, formal nature-based interventions can be designed to deliver benefits such as reductions in burnout and sick leave.

Finally, as discipline ergonomics should engage with global challenges where it can, such as sustainability, climate change and the state of nature (eg Moray 1993; Thatcher 2013). Strengthening human exposure to, and connection with nature through simple interventions would be not only beneficial for human health and well-being, but for the environment as well, and there is a need for a coalition of disciplines to promote human interaction with nature (Sandifer, Sutton-Grier, and Ward 2015). Literature has shown that, in contrast with negative, alarmist campaigns which can make us feel helpless (PIRC 2013), connectedness with nature (Mayer and Frantz 2004), and exposure to nature (Brown and Kasser 2005; Ewert, Place, and Sibthorp 2005; Hartig, Kaiser, and Bowler 2001) encourage environmentally friendly attitudes and behaviours. It would be anticipated that considerations relevant to other forms of successful health and safety intervention, such as energy and creativity, engagement (Hale et al. 2010) will be as relevant to interventions to bring nature into working practice.

5.2. Further implications

The well-being benefits of nature provide one aspect of a new paradigm for ergonomics in the delivery of well-being. There are, however, wider implications for the discipline related to the reasons nature is beneficial, and our shared place in nature. Connectedness to nature was introduced earlier as a sense of belonging to the natural world (Mayer and Frantz 2004) which includes the extent to which nature is included within an individual's view of self (Schultz 2002). The self is a key construct in Western thinking and the disembodied or independent self is a common notion in modern Western societies (Bragg 1996). This philosophical stance is built upon the dominant Cartesian tradition of modernity where the object is seen as separate from the subject. An alternative is a phenomenological perspective (eg Merleau-Ponty and Lefort 1968) which suggests a shared place in the world.

There has been previous discussion of phenomenology and ergonomics, for example, a phenomenology of human–machine interaction, or coagency, where the machine becomes 'transparent' and part of how the world is experienced (Hollnagel and Woods 2005). The roots of such cognitive integration, where mind and environment operate as a coupled system (Clark and Chalmers 1998; Thompson 2010), can be found in phenomenology and the philosophy of Merleau-Ponty (eg Merleau-Ponty and Lefort 1968); thinking that has developed into embodied cognition (eg Clark 1997; Gallagher 2005; Lakoff and Johnson 1999) and the notion of the extended mind. Concepts such as distributed and extended cognition, discussed previously in the ergonomics literature (eg Hollnagel 2001) are also relevant. These perspectives suggest that the mind extends beyond the body to be embedded in the environment, so, for example, that hand tools become integrated by the mind into body schema and the task becomes deeply integrated into our experience (eg Borghi and Cimatti 2010).

Ergonomics, at its core, is interested in the relationship between the environment and people, although this is often from a positivist perspective (Dekker, Hancock, and Wilkin 2013) involving some 'interface' which suggests a boundary where the task is an external element, something we encounter. However, from nature connection, to cognitive integration, embeddedness and well-being there is value, in a more general integrative perspective as it is difficult to establish where the environment begins and system ends (Dekker, Hancock, and Wilkin 2013). Building upon the concept of connection to nature where self and the external natural world are integrated, the task becomes part of our being. Being is our interaction with the world and the things we do within it, so that fitting the task to the human, goes beyond interaction to our situated state, place and cognition. The philosophical basis underpinning nature connectedness, self and embeddedness provides a different, and fully holistic, perspective for ergonomics – if people are embedded within the natural environment, they are also embedded within the work environment. This viewpoint is represented in Figure 2, where the straightforward 'concentric rings' model of ergonomics which depicts interactions of factors relevant to applied ergonomics is adapted (Grey, Norris, and Wilson 1987; Wilson and Corlett 2005). The adaptation attempts to capture the holistic need to consider nature within the workplace through the encompassing leaf, but the model also references the phenomenological perspective of nature connectedness, task embeddedness and cogntive integration through the larger central figure which shares experiences with all factors directly, rather than across a series of boundaries, therefore providing a straightforward focal point to inform education and scholarship in the epistemology of ergonomics (Dekker, Hancock, and Wilkin 2013); the first of the two objects of chartership being to advance education and knowledge in ergonomics (CIEHF 2014). As noted earlier, a move away from a positivist, goal-directed interpretation of the world to a holistic perspective, may itself be facilitated by a closer relationship with nature (Bateson 1972).

More generally, this embeddedness within the environment provides a new paradigm for well-being. Stevens (2010) presents an ecopsychological view in contrast to existing models of health which essentially view people as separate from the environment and affected by specific events. The 'biomedical' model of medicine is based on a deviation from 'normal' within the individual, with the 'biopsychosocial' model reflecting how biological, psychological and social factors play a significant role in health (Engel 1977). Seeing people as embedded within the environment shifts the emphasis away from the person and their health issues to a consideration of dynamic relationships between people and environment. Given the evidence above, it is time for a wider paradigm shift and an embedded model based on 'biopsychophysis', reflecting how health depends on the unity of biology, psychology and nature.

In summary, the concept of NC is further argument for a paradigm shift in ergonomics, with a move away from purely reductive Cartesian viewpoints (eg Dekker, Hancock, and Wilkin 2013), echoing debate on other core topics for ergonomics such as situation awareness (eg Dekker 2013; Stanton et al. 2014) whereby the role of the

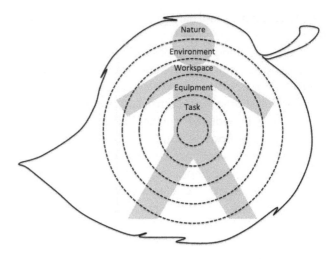

Figure 2. A nature connectedness informed, embedded model of ergonomics.
Notes: Adapted from Grey, Norris and Wilson (1987); Wilson and Corlett (2005).

individual, and their understanding of their environment, is indivisible from the environment as a unit of analysis.

6. Future research

Although the evidence of nature's benefits to health and well-being is extensive, more work is required on the linkages between biodiversity, nature and health. In order to place nature at the centre of human well-being, there is a need for research on mechanisms and quantification of well-being outcomes to drive policy change (Sandifer, Sutton-Grier, and Ward 2015). From an applied perspective, at present, the knowledge about the beneficial effects of nature, and the most beneficial kinds of interaction with nature, are insufficient to be applied in a systematic way in areas related to health promotion (Van den Berg, Hartig, and Staats 2007). Future research should therefore focus on applied studies aimed at exploring ways to translate theoretical notions, such as nature as a restorative environment, into practice and to assess the effectiveness of nature-based interventions in the workplace to inform policy and well-being programmes.

To expand, this research should follow the three themes set out in the applied implications for ergonomics above. Firstly, how does exposure to, and time in, nature impact on employee health, well-being and performance? The research for this broader question is likely to take a cross-sectional and self-report approach, particularly in the first instance before building into intervention-based studies that will also support a causal link. Secondly, informed by wider research into theoretical knowledge of mechanisms, there is a need to tackle applied issues head-on and explore nature interventions as a route to well-being

Figure 3. Moving from theory to workplace guidance.

(cf. Richardson et al. 2016). Such studies would consider the design and evaluation of formal nature-based interventions to deliver benefits such as reductions in burnout, sick leave and improved performance. Such work should do more to empirically examine the value of nature as distinct from other associated factors that might lead to well-being benefits such as exercise or greater exposure to daylight (Mills, Tomkins, and Schlangen 2007). As with the research presented earlier, depending on the outcome being targeted, both subjective and objective measures are possible. Research in the first two themes should also consider associated benefits such as people's connection to nature and pro-environmental behaviours. Thirdly, there is an opportunity for research into how time in nature can be formalised and implemented as a restorative break-based intervention, particularly for those jobs that place high demands on attention. Building on the laboratory and empirical research into restoration and attention, this research should be field based with applied performance measures. Finally, these broad research themes would inform the revision of models of workplace well-being to include nature. This process of moving from theory to concrete guidance is presented in Figure 3.

7. Conclusions

Although there is work to be done to understand the relationship further, the message for the practitioner is straightforward: exposure to nature is beneficial to well-being. There is freedom to bring nature into the work environment in numerous ways, and the opportunity to cement a paradigm shift by evaluating and reporting the impact. As humans are part of nature, there can be no surprise that exposure to nature is beneficial for our well-being. While modern life is preferable to that of our predecessors in many ways, it has created new pressures and recently health experts have started to recognise that a divorce from nature may present high costs, not just in terms of health but also in wider concerns about disrupting the systemic relationship between us and our environment (Bateson 1972; Guddemi 2011). A large body of literature has demonstrated that nature exerts many beneficial effects on humans. People who are exposed to nature, or feel connected with it, seek out natural places

in urban locations and present higher levels of well-being, as nature has been shown to have a restorative effect on both a stressed autonomic nervous system and a depleted attention capacity. Extending and sharing this knowledge has importance in behavioural and work environment interventions in which nature's beneficial effects could be capitalised and a positive attitude towards nature can be encouraged as we work towards a sustainable future. For ergonomics, this presents a new paradigm for its object of promoting human well-being, knowledge that should be incorporated to meet the first of the two objects of chartership, to advance education and knowledge in ergonomics (CIEHF 2014). As nature-based solutions come to the fore, ergonomists should understand the value of nature, and how to accommodate its impact within working environments and working patterns. Moreover, the systemic relationship between us and nature further highlights the relevance of a non-dualistic stance between people and the environment that is applicable to all aspects of ergonomics and socio-technical approaches.

Disclosure statement

No potential conflict of interest was reported by the authors.

References

Agyemang, C., C. van Hooijdonk, W. Wendel-Vos, J. Ujcic-Voortman, E. Lindeman, K. Stronks, and M. Droomers. 2007. "Ethnic Differences in the Effect of Environmental Stressors on Blood Pressure and Hypertension in the Netherlands." *Biomedcentral Public Health* 7 (1): 118. doi:10.1186/1471-2458-7-118.

Amani, R., and T. Gill. 2013. "Shiftworking, Nutrition and Obesity: Implications for Workforce Health – A Systematic Review." *Asia Pacific Journal of Clinical Nutrition* 22 (4): 505.

Barnosky, A. D., N. Matzke, S. Tomiya, G. O. U. Wogan, B. Swartz, T. B. Quental, C. Marshall, J. L. McGuire, E. L. Lindsey, K. C. Maguire, B. Mersey, and E. A. Ferrer. 2011. "Has the Earth's Sixth Mass Extinction Already Arrived?" *Nature* 471: 51–57. doi:10.1038/nature09678.

Bateson, G. 1972. *Steps to an Ecology of Mind*. Chicago, IL: University of Chicago Press.

Bedimo-Rung, A. L., A. J. Mowen, and D. A. Cohen. 2005. "The Significance of Parks to Physical Activity and Public Health: A Conceptual Model." *American Journal of Preventive Medicine* 28 (2): 159–168.

Berman, M. G., J. Jonides, and S. Kaplan. 2008. "The Cognitive Benefits of Interacting with Nature." *Psychological Science* 19 (12): 1207–1212.

Berto, R. 2005. "Exposure to Restorative Environments Helps Restore Attentional Capacity." *Journal of Environmental Psychology* 25 (3): 249–259. doi:10.1016/j.jenvp.2005.07.001.

Borghi, A. M., and F. Cimatti. 2010. "Embodied Cognition and Beyond: Acting and Sensing the Body." *Neuropsychologia* 48: 763–773. doi:10.1016/j.neuropsychologia.2009.10.029.

Bragg, E. A. 1996. "Towards Ecological Self: Deep Ecology Meets Constructionist Self-theory." *Journal of Environmental Psychology* 16: 93–108.

Braun, E., A. Woodley, J. T. E. Richardson, and B. Leidner. 2012. "Self-rated Competences Questionnaires from a Design Perspective." *Educational Research Review* 7 (1): 1–18.

Bringslimark, T., T. Hartig, and G. Grindal Patil. 2011. "Adaptation to Windowlessness: Do Office Workers Compensate for a Lack of Visual Access to the Outdoors?" *Environment and Behavior* 43 (4): 469–487. doi:10.1177/0013916510368351.

Brown, D. K., J. L. Barton, J. N. Pretty, and V. Gladwell. 2014. "Walks4Work: Assessing the Role of the Natural Environment in a Workplace Physical Activity Intervention." *Scandinavian Journal of Work, Environment & Health* 40 (4): 390–399. doi:10.5271/sjweh.3421.

Brown, C., and M. Grant. 2005. "Biodiversity and Human Health: What Role for Nature in Healthy Urban Planning?" *Built Environment* 31 (4): 326–338.

Brown, K. W., and T. Kasser. 2005. "Are Psychological and Ecological Well-being Compatible? The Role of Values, Mindfulness, and Lifestyle." *Social Indicators Research* 74 (2): 349–368. doi:10.1007/s11205-004-8207-8.

Brown, T., and M. Bell. 2007. "Off the Couch and on the Move: Global Public Health and the Medicalisation of Nature." *Social Science and Medicine* 64 (6): 1343–1354. doi:10.1016/j.socscimed.2006.11.020.

Capaldi, C. A., R. L. Dopko, and J. M. Zelenski. 2014. "The Relationship between Nature Connectedness and Happiness: A Meta-analysis." *Frontiers in Psychology* 5: 976.

Cervinka, R., K. Roderer, and E. Hefler. 2012. "Are Nature Lovers Happy? On Various Indicators of Well-being and Connectedness with Nature." *Journal of Health Psychology* 17 (3): 379–388. doi:10.1177/1359105311416873.

Chartered Institute of Ergonomics and Human Factors. 2014. *Royal Charter.* http://iehf.org/ehf/wp-content/uploads/2013/04/CIEHF-Charter-documents.pdf.

Chow, J. T., and S. Lau. 2015. "Nature Gives Us Strength: Exposure to Nature Counteracts Ego-depletion." *The Journal of Social Psychology* 155 (1): 70–85.

Clark, A. 1997. *Being There.* Cambridge, MA: MIT Press.

Clark, A., and D. Chalmers. 1998. "The Extended Mind." *Analysis* 58: 7–19.

Dababneh, A. J., N. Swanson, and R. L. Shell. 2001. "Impact of Added Rest Breaks on the Productivity and Well Being of Workers." *Ergonomics* 44 (2): 164–174.

Danna, K., and R. W. Griffin. 1999. "Health and Well-being in the Workplace: A Review and Synthesis of the Literature." *Journal of Management* 25 (3): 357–384.

David Suzuki Foundation. 2015. *30x30 Nature Challenge.* http://30x30.davidsuzuki.org

Day, A., E. K. Kelloway, and J. J. Hurrell Jr., eds. 2014. *Workplace Well-being: How to Build Psychologically Healthy Workplaces.* Oxford: Wiley.

Dekker, S. W. 2013. "On the Epistemology and Ethics of Communicating a Cartesian Consciousness." *Safety Science* 56: 96–99.

Dekker, S. W., P. A. Hancock, and P. Wilkin. 2013. "Ergonomics and Sustainability: Towards an Embrace of Complexity and Emergence." *Ergonomics* 56 (3): 357–364.

Department for Environment, Food and Rural Affairs. 2011. *The Natural Choice: Securing the Value of Nature (Vol. 8082).* London: The Stationery Office.

de Vries, S., R. Verheij, P. Groenewegen, and P. Spreeuwenberg. 2003. "Natural Environments – Healthy Environments? An Exploratory Analysis of the Relationship Between Greenspace and Health." *Environment and Planning A* 35 (10): 1717–1731.

Donovan, G. H., D. T. Butry, Y. L. Michael, J. P. Prestemon, A. M. Liebhold, D. Gatziolis, and M. Y. Mao. 2013. "The Relationship Between Trees and Human Health: Evidence from the Spread of the Emerald Ash Borer." *American Journal of Preventive Medicine* 44 (2): 139–145. doi:10.1016/j.amepre.2012.09.066.

Dunn, R. R., M. C. Gavin, M. C. Sanchez, and J. N. Solomon. 2006. "The Pigeon Paradox: Dependence of Global Conservation on Urban Nature." *Conservation Biology* 20 (6): 1814–1816.

European Commission. 2015. *Towards an EU Research and Innovation Policy Agenda for Nature-based Solutions & Re-naturing Cities: Final Report of the Horizon 2020 Expert Group on 'Nature-Based Solutions and Re-naturing Cities'.* https://ec.europa.eu/research/environment/pdf/renaturing/nbs.pdf.

European Environment Agency. 2015. *State of Nature in the EU.* http://www.eea.europa.eu/publications/state-of-nature-in-the-eu/at_download/file.

Engel, G. L. 1977. "The Need for a New Medical Model: A Challenge for Biomedicine." *Science* 196 (4286): 129–136.

Ewert, A., G. Place, and J. Sibthorp. 2005. "Early-life Outdoor Experiences and an Individual's Environmental Attitudes." *Leisure Sciences* 27 (3): 225–239.

Fink, G. R., P. W. Halligan, J. C. Marshall, C. D. Frith, R. S. J. Frackowiak, and R. J. Dolan. 1996. "Where in the Brain does Visual Attention Select the Forest and the Trees?" *Nature* 382 (6592): 626–628.

Flach, J. M., S. Dekker, and P. Jan Stappers. 2008. "Playing Twenty Questions with Nature (The Surprise Version): Reflections on the Dynamics of Experience." *Theoretical Issues in Ergonomics Science* 9 (2): 125–154.

Frantz, C. M., and F. S. Mayer. 2014. "The Importance of Connection to Nature in Assessing Environmental Education Programs." *Studies in Educational Evaluation* 41: 85–89.

Frumkin, H. 2001. "Beyond Toxicity: Human Health and the Natural Environment." *American Journal of Preventive Medicine* 20 (3): 234–240. doi:10.1016/S0749-3797(00)00317-2.

Gallagher, S. 2005. *How the Body Shapes the Mind.* Oxford: Oxford University Press.

Gidlöf-Gunnarsson, A., and E. Öhrström. 2007. "Noise and Well-being in Urban Residential Environments: The Potential Role of Perceived Availability to Nearby Green Areas." *Landscape and Urban Planning* 83 (2–3): 115–126. doi:10.1016/j.landurbplan.2007.03.003.

Grey, S. M., B. J. Norris, and J. R. Wilson. 1987. *Ergonomics in the Electronic Retail Environment.* Slough: ICL.

Groenewegen, P., A. van den Berg, S. de Vries, and R. Verheij. 2006. "Vitamin G: Effects of Green Space on Health, Well-Being, and Social Safety." *BMC Public Health* 6 (1): 1. doi:10.1186/1471-2458-6-149.

Guddemi, P. 2011. "Conscious Purpose in 2010: Bateson's Prescient Warning." *Systems Research and Behavioral Science* 28 (5): 465–475.

Guite, H. F., C. Clark, and G. Ackrill. 2006. "The Impact of the Physical and Urban Environment on Mental Well-Being." *Public Health* 120 (12): 1117–1126. doi:10.1016/j.puhe.2006.10.005.

Hale, A. R., F. W. Guldenmund, P. L. C. H. Van Loenhout, and J. I. H. Oh. 2010. "Evaluating Safety Management and Culture Interventions to Improve Safety: Effective Intervention Strategies." *Safety Science* 48 (8): 1026–1035.

Hanson, M. A. 2013. "Green Ergonomics: Challenges and Opportunities." *Ergonomics* 56 (3): 399–408.

Hartig, T., A. Böök, J. Garvill, T. Olsson, and T. Gärling. 1996.

"Environmental Influences on Psychological Restoration." *Scandinavian Journal of Psychology* 37 (4): 378–393. doi:10.1111/j.1467-9450.1996.tb00670.x.

Hartig, T., G. W. Evans, L. D. Jamner, D. S. Davis, and T. Gärling. 2003. "Tracking Restoration in Natural and Urban Field Settings." *Journal of Environmental Psychology* 23 (2): 109–123. doi:10.1177/0013916508319745.

Hartig, T., F. G. Kaiser, and P. A. Bowler. 2001. "Psychological Restoration in Nature as a Positive Motivation for Ecological Behavior." *Environment and Behavior* 33 (4): 590–607. doi:10.1177/00139160121973142.

Hartig, T., M. Mang, and G. W. Evans. 1991. "Restorative Effects of Natural Environment Experiences." *Environment and Behavior* 23 (1): 3–26. doi:10.1177/0013916591231001.

Hartig, T., A. van den Berg, C. Hagerhall, M. Tomalak, N. Bauer, R. Hansmann, A. Ojala, E. Syngollitou, G. Carrus, A. van Herzele, S. Bell, M. T. Camilleri Podesta, and G. Waaseth. 2011. "Health Benefits of Nature Experience: Psychological, Social and Cultural Processes." In *Forests, Trees, and Human Health*, edited by K. Nilsson, M. Sangster, C. Gallis, T. Hartig, S. de Vries, K. Seeland, and J. Schipperijn, 127–168. Dordrecht: Springer. doi:10.1007/978-90-481-9806-1.

Hollnagel, E. 2001. "Extended Cognition and the Future of Ergonomics." *Theoretical Issues in Ergonomics Science* 2 (3): 309–315.

Hollnagel, E., and D. D. Woods. 2005. *Joint Cognitive Systems: Foundations of Cognitive Systems Engineering*. Boca Raton, FL: CRC Press.

Howell, A., H. A. Passmore, and K. Buro. 2012. "Meaning in Nature: Meaning in Life as a Mediator of the Relationship between Nature Connectedness and Well-being." *Journal of Happiness Studies* 11 (1): 1–16. doi:10.1007/s10902-012-9403-x.

Howell, A. J., R. L. Dopko, H. A. Passmore, and K. Buro. 2011. "NatureConnectedness: Associations with Well-being and Mindfulness." *Personality and Individual Differences* 51 (2): 166–171. doi:10.1016/j.paid.2011.03.037.

Hull, R. B. I. 1992. "Brief Encounters with Urban Forests Produce Moods That Matter." *Journal of Arboriculture* 18 (6): 322–324.

Johnson, J. V., and E. M. Hall. 1988. "Job Strain, Work Place Social Support, and Cardiovascular Disease: A Cross-sectional Study of a Random Sample of the Swedish Working Population." *American Journal of Public Health* 78 (10): 1336–1342.

Kahn, P. H., B. Friedman, B. Gill, J. Hagman, R. L. Severson, N. G. Freier, N. Feldman, S. Carrère, and A. Stolyar. 2008. "A Plasma Display Window? — The Shifting Baseline Problem in a Technologically Mediated Natural World." *Journal of Environmental Psychology* 28 (2): 192–199.

Kamitsis, I., and A. J. P. Francis. 2013. "Spirituality Mediates the Relationship Between Engagement with Nature and Psychological Wellbeing." *Journal of Environmental Psychology* 36: 136–143. doi:10.1016/j.jenvp.2013.07.013.

Kaplan, R. 2001. "The Nature of the View from Home: Psychological Benefits." *Environment and Behavior* 33 (4): 507–542. doi:10.1177/00139160121973115.

Kaplan, R., and S. Kaplan. 1989. *The Experience of Nature. a Psychological Perspective*. Cambridge: Cambridge University Press.

Kaplan, S. 1995. "The Restorative Benefits of Nature: Toward an Integrative Framework." *Journal of Environmental Psychology* 15 (3): 169–182.

Karasek, R., and T. Theorell. 1992. *Healthy Work: Stress, Productivity, and the Reconstruction of Working Life*. New York, NY: Basic Books.

Kellert, S. R., J. Heerwagen, and M. Mador. 2011. *Biophilic Design: The Theory, Science and Practice of Bringing Buildings to Life*. Oxford: Wiley.

Korpela, K. M., T. Hartig, F. G. Kaiser, and U. Fuhrer. 2001. "Restorative Experience and Self-regulation in Favorite Places." *Environment and Behavior* 33 (4): 572–589. doi:10.1177/00139160121973133.

Kuo, F. E. 2001. "Coping with Poverty: Impacts of Environment and Attention in the Inner City." *Environment and Behavior* 33 (1): 5–34. doi:10.1177/00139160121972846.

Kuo, M. 2015. "How Might Contact with Nature Promote Human Health? Promising Mechanisms and a Possible Central Pathway." *Frontiers in Psychology* 6: 1093. doi:10.3389/fpsyg.2015.01093.

Kuoppala, J., A. Lamminpää, and P. Husman. 2008. "Work Health Promotion, Job Well-being, and Sickness Absences – A Systematic Review and Meta-analysis." *Journal of Occupational and Environmental Medicine* 50 (11): 1216–1227.

Lafortezza, R., G. Carrus, G. Sanesi, and C. Davies. 2009. "Benefits and Well-being Perceived by People Visiting Green Spaces in Periods of Heat Stress." *Urban Forestry and Urban Greening* 8 (2): 97–108. doi:10.1016/j.ufug.2009.02.003.

Lakoff, G., and M. Johnson. 1999. *Philosophy in the Flesh: The Embodied Mind and Its Challenge to Western Thought*. New York: Basic Books.

Largo-Wright, E., W. W. Chen, V. Dodd, and R. Weiler. 2011. "Healthy Workplaces: The Effects of Nature Contact at Work on Employee Stress and Health." *Public Health Report* 1 (126): 124–130.

Lee, K. E., K. J. Williams, L. D. Sargent, N. S. Williams, and K. A. Johnson. 2015. "40-second green roof views sustain attention: The role of micro-breaks in attention restoration." *Journal of Environmental Psychology* 42: 82–189.

Leather, P., M. Pyrgas, D. Beale, and C. Lawrence. 1998. "Windows in the Workplace: Sunlight, View, and Occupational Stress." *Environment and Behavior* 30: 739–762.

Laumann, K., T. Gärling, and K. M. Stormark. 2003. "Selective Attention and Heart Rate Responses to Natural and Urban Environments." *Journal of Environmental Psychology* 23 (2): 125–134.

Lin, B. B., R. A. Fuller, R. Bush, K. J. Gaston, and D. F. Shanahan. 2014. "Opportunity or Orientation? Who Uses Urban Parks and Why." *PLoS ONE* 9 (1): e87422. doi:10.1371/journal.pone.0087422.

Logan, A. C., and E. M. Selhub. 2012. "Vis Medicatrix Naturae: Does Nature 'Minister to the Mind'?" *BioPsychoSocial Medicine* 6 (1): 11–23. doi:10.1186/1751-0759-6-11.

Lohr, V. I., C. H. Pearson-Mims, and G. K. Goodwin. 1996. "Interior Plants May Improve Worker Productivity and Reduce Stress in a Windowless Environment." *Journal of Environmental Horticulture* 14 (2): 97–100.

Lottrup, L., P. Grahn, and U. K. Stigsdotter. 2013. "Workplace Greenery and Perceived Level of Stress: Benefits of Access to a Green Outdoor Environment at the Workplace." *Landscape and Urban Planning* 110: 5–11.

Maas, J., R. Verheij, P. Spreeuwenberg, and P. Groenewegen. 2008. "Physical Activity as a Possible Mechanism behind the Relationship between Green Space and Health: A Multilevel Analysis." *BMC Public Health* 8 (1): 206–219.

Maas, J., R. A. Verheij, P. P. Groenewegen, S. de Vries, and P. Spreeuwenberg. 2006. "Green Space, Urbanity, and Health: How Strong is the Relation?" *Journal of Epidemiology and Community Health* 60 (7): 587–592.

Maller, C., M. Townsend, L. St Leger, C. Henderson-Wilson, A. Pryor, L. Prosser, and M. Moore. 2009. "Healthy Parks, Healthy People: The Health Benefits of Contact with Nature in a Park Context." *The George Wright Forum* 26 (2): 51–83.

Martyn, P., and E. Brymer. 2014. "The Relationship Between Nature Relatedness and Anxiety." *Journal of Health Psychology*. Advance online publication. doi:10.1177/1359105314555169.

Mayer, F. S., C. M. Frantz, E. Bruehlman-Senecal, and K. Dolliver. 2009. "Why is Nature Beneficial? The Role of Connectedness to Nature." *Environment and Behavior* 41 (5): 607–643.

Mayer, S. F., and C. M. Frantz. 2004. "The Connectedness to Nature Scale: A Measure of Individuals' Feeling in Community with Nature." *Journal of Environmental Psychology* 24 (4): 503–515. doi:10.1016/j.jenvp.2004.10.001.

McMahan, E. A., and D. Estes. 2011. "Measuring Lay Conceptions of Well-being: The Beliefs about Well-being Scale." *Journal of Happiness Studies* 12 (2): 267–287.

McMahan, E. A., and D. Estes. 2015. "The Effect of Contact with Natural Environments on Positive and Negative Affect: A Meta-analysis." *The Journal of Positive Psychology* 10 (6): 507–519.

Merleau-Ponty, M., and C. Lefort. 1968. *The Visible and the Invisible: Followed by Working Notes.* Evanston, IL: Northwestern University Press.

Mills, P. R., S. C. Tomkins, and L. J. Schlangen. 2007. "The Effect of High Correlated Colour Temperature Office Lighting on Employee Wellbeing and Work Performance." *Journal of Circadian Rhythms* 5 (1): 1.

Miyazaki, Y., J. Lee, B. J. Park, Y. Tsunetsugu, and K. Matsunaga. 2011. "Preventive Medical Effects of Nature Therapy." *Japenese Journal of Hyiene* 66 (4): 651–656.

Moray, N. 1993. "Technosophy and Humane Factors." *Ergonomics in Design: The Quarterly of Human Factors Applications* 1 (4): 33–39.

Newsham, G. R., B. J. Birt, C. Arsenault, A. J. Thompson, J. A. Veitch, S. Mancini, and G. J. Burns. 2013. "Do 'Green' Buildings Have Better Indoor Environments? New Evidence." *Building Research & Information* 41 (4): 415–434.

Nisbet, E., J. Zelenski, and S. Murphy. 2011. "Happiness is in Our Nature: Exploring Nature Relatedness as a Contributor to Subjective Well-being." *Journal of Happiness Studies* 12 (2): 303–322. doi:10.1177/0956797611418527.

Ottosson, J., and P. Grahn. 2005. "A Comparison of Leisure Time Spent in a Garden with Leisure Time Spent Indoors: On Measures of Restoration in Residents in Geriatric Care." *Landscape Research* 30 (1): 23–55.

Ottosson, J., and P. Grahn. 2008. "The Role of Natural Settings in Crisis Rehabilitation: How Does the Level of Crisis Influence the Response to Experiences of Nature with Regard to Measures of Rehabilitation?" *Landscape Research* 33 (1): 51–70. doi:10.1080/01426390701773813.

Pappachan, M. J. 2011. "Increasing Prevalence of Lifestyle Diseases: High Time for Action." *Indian Journal of Medical Research* 134 (2): 143–145.

Park, B. J., Y. Tsunetsugu, T. Kasetani, T. Kagawa, and Y. Miyazaki. 2010. "The Physiological Effects of Shinrin-Yoku (Taking in the Forest Atmosphere or Forest Bathing): Evidence from Field Experiments in 24 Forests across Japan." *Environmental Health and Preventive Medicine* 15 (1): 18–26.

Parsons, R., L. G. Tassinary, R. S. Ulrich, M. R. Hebl, and M. Grossman-Alexander. 1998. "The View from the Road: Implications for Stress Recovery and Immunization." *Journal of Environmental Psychology* 18 (2): 113–140.

PIRC. 2013. *Common Cause for Nature – Full Report.* http://valuesandframes.org/download/reports/Common%20Cause%20for%20Nature%20-%20Full%20Report.pdf.

Pretty, J. 2002. *Agri-culture: Reconnecting People, Land, and Nature.* London: Earthscan.

Pretty, J., J. Peacock, R. Hine, M. Sellens, N. South, and M. Griffin. 2007. "Green Exercise in the UK Countryside: Effects on Health and Psychological Well-being, and Implications for Policy and Planning." *Journal of Environmental Planning and Management* 50 (2): 211–231.

Raanaas, R. K., G. G. Patil, and T. Hartig. 2012. "Health Benefits of a View of Nature through the Window: A Quasi-experimental Study of Patients in a Residential Rehabilitation Center." *Clinical Rehabilitation* 26 (1): 21–32. doi:10.1177/0269215511412800.

Richardson, M., A. Cormack, L. McRobert, and R. Underhill. 2016. "30 Days Wild: Development and Evaluation of a Large-scale Nature Engagement Campaign to Improve Well-being." *PLoS ONE* 11 (2): e0149777. doi:10.1371/journal.pone.0149777.

Richardson, M., J. Hallam, and R. Lumber. 2015. "One thousand good things in nature: Aspects of nearby nature associated with improved connection to nature." *Environmental Values* 24 (5): 603–619.

Ryan, C. O., W. D. Browning, J. O. Clancy, S. L. Andrews, and N. B. Kallianpurkar. 2014. "Biophilic Design Patterns: Emerging Nature-based Parameters for Health and Well-being in the Built Environment." *International Journal of Architectural Research: ArchNet-IJAR* 8 (2): 62–76.

Sachita, S. P., and G. Ruchi. 2015. "Happiness and Organizational Socialization: Exploring the Mediating Role of Restorative Environments." *International Research Journal of Social Sciences* 4 (8): 15–20.

Sahlin, E., G. Ahlborg, J. Vega Matuszczyk, and P. Grahn. 2014. "Nature-Based Stress Management Course for Individuals at Risk of Adverse Health Effects from Work-related Stress – Effects on Stress Related Symptoms, Workability and Sick Leave." *International Journal of Environmental Research and Public Health* 11 (6): 6586–6611. doi:10.3390/ijerph110606586.

Salvendy, G. 2012. *Handbook of Human Factors and Ergonomics.* Oxford: Wiley.

Sandifer, P. A., A. E. Sutton-Grier, and B. P. Ward. 2015. "Exploring Connections among Nature, Biodiversity, Ecosystem Services, and Human Health and Well-being: Opportunities to Enhance Health and Biodiversity Conservation." *Ecosystem Services* 12: 1–15.

Schultz, P. W. 2002. "Inclusion with Nature: Understanding the Psychology of Human–Nature Interactions." In *Psychology of Sustainable Development*, edited by P. Schmuck and P. W. Schultz, 61–78. New York: Kluwer.

Staats, H., A. Kieviet, and T. Hartig. 2003. "Where to Recover from Attentional Fatigue: An Expectancy-value Analysis of Environmental Preference." *Journal of Environmental Psychology* 23 (2): 147–157.

Stansfeld, S., and B. Candy. 2006. "Psychosocial Work Environment and Mental Health – A Meta-analytic Review." *Scandinavian Journal of Work, Environment & Health* 32 (6): 443–462.

Stanton, N. A., P. M. Salmon, and G. H. Walker. 2014. "Let the Reader Decide a Paradigm Shift for Situation Awareness in Sociotechnical Systems." *Journal of Cognitive Engineering and Decision Making* Advance online publication. doi:10.1177/1555343414552297.

Stevens, P. 2010. "Embedment in the Environment: A New Paradigm for Well-being?" *Perspectives in Public Health* 130 (6): 265–269.

Thatcher, A. 2013. "Green Ergonomics: Definition and Scope." *Ergonomics* 56 (3): 389–398.

Thompson, E. 2010. *Mind in Life: Biology, Phenomenology, and the Sciences of Mind*. Cambridge: Harvard University Press.

Trau, D., K. A. Keenan, M. Goforth, and V. Large. 2015. "Nature Contacts Employee Wellness in Healthcare." *HERD: Health Environments Research & Design Journal*. Advance online publication. doi:10.1177/1937586715613585.

Tyrväinen, L., A. Ojala, K. Korpela, T. Lanki, Y. Tsunetsugu, and T. Kagawa. 2014. "The Influence of Urban Green Environments on Stress Relief Measures: A Field Experiment." *Journal of Environmental Psychology* 38: 1–9.

Tzoulas, K., K. Korpela, S. Venn, V. Yli-Pelkonen, A. Kaźmierczak, J. Niemela, and P. James. 2007. "Promoting Ecosystem and Human Health in Urban Areas Using Green Infrastructure: A Literature Review." *Landscape and Urban Planning* 81 (3): 167–178.

Ulrich, R. S. 1979. "Visual Landscapes and Psychological Well-being." *Landscape Research* 4 (1): 17–23.

Ulrich, R. S. 1984. "View Through a Window May Influence Recovery from Surgery." *Science* 224 (4647): 420–421.

Ulrich, R. S. 1993. "Biophilia, Biophobia and Natural Landscapes." In *The Biophilia Hypothesis*, edited by S. R. Kellert and E. O. Wilson, 75–137. Washington, DC: Island Press.

Ulrich, R. S., R. F. Simons, B. D. Losito, E. Fiorito, M. A. Miles, and M. Zelson. 1991. "Stress Recovery during Exposure to Natural and Urban Environments." *Journal of Environmental Psychology* 11 (3): 201–230.

Van den Berg, A. E., T. Hartig, and H. Staats. 2007. "Preference for Nature in Urbanized Societies: Stress, Restoration, and the Pursuit of Sustainability." *Journal of Social Issues in Mental Health Nursing* 63 (1): 79–96.

Van den Berg, A., S. Koole, and N. van der Wulp. 2003. "Environmental Preference and Restoration: (How) Are They Related?" *Journal of Environmental Psychology* 23 (2): 135–146.

Van den Berg, A. E., J. Maas, R. A. Verheij, and P. P. Groenewegen. 2010. "Green Space as a Buffer Between Stressful Life Events and Health." *Social Science and Medicine* 70 (8): 1203–1210.

Verheij, R. A., J. Maas, and P. P. Groenewegen. 2008. "Urban-Rural Health Differences and the Availability of Green Space." *European Urban and Regional Studies* 15 (4): 307–316. doi:10.1177/0969776408095107.

Verheij, R. A., H. D. van de Mheen, D. H. de Bakker, P. P. Groenewegen, and J. P. Mackenbach. 1998. "Urban–Rural Variations in Health in the Netherlands: Does Selective Migration Play a Part?" *Journal of Epidemiology and Community Health* 52 (8): 487–493.

von Hertzen, L., B. Beutler, J. Bienenstock, M. Blaser, P. D. Cani, J. Eriksson, and W. M. de Vos. 2015. "Helsinki Alert of Biodiversity and Health." *Annals of Medicine* 47(3): 218–225.

Vyas, M. V., A. X. Garg, A. V. Iansavichus, J. Costella, A. Donner, L. E. Laugsand, and D. G. Hackam. 2012. "Shift Work and Vascular Events: Systematic Review and Meta-analysis." *BMJ* 345: e4800.

Walsh, R. 2011. "Lifestyle and Mental Health." *American Psychologist* 66 (7): 579–592. doi:10.1037/a0021769.

Weich, S., L. Twigg, and G. Lewis. 2006. "Rural/Non-rural Differences in Rates of Common Mental Disorders in Britain: Prospective Multilevel Cohort Study." *The British Journal of Psychiatry* 188 (1): 51–57. doi:10.1192/bjp.bp.105.008714.

Wild Network. 2015. *Project Wild Thing the Wild Network*. http://http://projectwildthing.com/thewildnetwork

Wilson, E. O. 1984. *Biophilia*. Cambridge: Harvard University Press.

Wilson, M. G., D. M. Dejoy, R. J. Vandenberg, H. A. Richardson, and A. L. McGrath. 2004. "Work Characteristics and Employee Health and Well-being: Test of a Model of Healthy Work Organization." *Journal of Occupational and Organizational Psychology* 77 (4): 565–588.

Wilson, J. R., and N. Corlett, eds. 2005. *Evaluation of Human Work*. Boca Raton, FL: CRC Press.

Winkel, G., S. Saegert, and G. W. Evans. 2009. "An Ecological Perspective on Theory, Methods, and Analysis in Environmental Psychology: Advances and Challenges." *Journal of Environmental Psychology* 29 (3): 318–328.

Zelenski, J. M., and E. K. Nisbet. 2014. "Happiness and Feeling Connected: The Distinct Role of Nature Relatedness." *Environment and Behavior* 46 (1): 3–23.

Zhang, J. W., R. T. Howell, and R. Iyer. 2014. "Engagement with Natural Beauty Moderates the Positive Relation between Connectedness with Nature and Psychological Well-being." *Journal of Environmental Psychology* 38: 55–63.

Zhang, J. W., P. K. Piff, R. Iyer, S. Koleva, and D. Keltner. 2014. "An Occasion for Unselfing: Beautiful Nature Leads to Prosociality." *Journal of Environmental Psychology* 37: 61–72.

Index

Milton Keynes UK
Ingram Content Group UK Ltd.
UKHW050131071024
449327UK00029B/2545

9 780367 570958